Water Industry Systems:
modelling and optimization applications

VOLUME 2

Edited by

Dragan Savic

Director of Centre for Water Systems, University of Exeter, UK

and

Godfrey Walters

Director of Centre for Water Systems, University of Exeter, UK

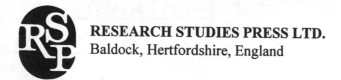

RESEARCH STUDIES PRESS LTD.
Baldock, Hertfordshire, England

RESEARCH STUDIES PRESS LTD.
15/16 Coach House Cloisters, 10 Hitchin Street, Baldock, Hertfordshire, England, SG7 6AE

and

325 Chestnut Street, Philadelphia, PA 19106

Marketing:

Research Studies Press Ltd.
15/16 Coach House Cloisters, 10 Hitchin Street, Baldock, Hertfordshire, England, SG7 6AE

Distribution:

NORTH AMERICA
Taylor & Francis Inc.
47 Runway Road, Suite G, Levittown, PA 19057 - 4700, USA

ASIA-PACIFIC
Hemisphere Publication Services
Golden Wheel Building, 41 Kallang Pudding Road #04-03, Singapore

EUROPE & REST OF THE WORLD
John Wiley & Sons Ltd.
Shripney Road, Bognor Regis, West Sussex, England, PO22 9SA

Library of Congress Cataloging-in-Publication Data

Available

British Library Cataloguing in Publication Data
A catalogue record for this book is available from the British Library.

ISBN 0 86380 249 4

Printed in Great Britain by SRP Ltd., Exeter

Water Industry Systems:
modelling and optimization applications

VOLUME 2

WATER ENGINEERING AND MANAGEMENT SERIES

Series Editors: **Professor Bryan Coulbeck**
                    Professor of Water Control Systems
                    De Montfort University, Leicester, UK

and

Dr Bogumil Ulanicki
Research Director of Water Software Systems
De Montfort University, Leicester, UK

Editorial Preface

This series of books is intended to present reviews of recent developments and associated research, and the application of that work, to engineering, operation and management of water systems.

The basic methods of engineering, operating and managing water systems are well established and well documented. However, the increasing availability and application of computer based methods has had a revolutionary effect on these traditional methods. Significant advances have been made, and are continuing, in such broad areas as: Information technologies, Operational modelling, and Optimal decision strategies.

Very efficient and effective SCADA and GIS systems are now available; these can collect, store and organize large quantities of data. Water companies have taken advantage of these systems to achieve comprehensive information coverage and integration. In turn, this has enabled the development and application of more sophisticated engineering software. The most recent computational modules are based on some new and interesting mathematical and heuristic techniques.

These two books are representative of such advances, and contain contributions from leading experts. The material is expected to be of significant interest to practitioners and researchers in the subject area.

Volume 1 is the third book in the series. This is mainly concerned with analysis and modelling of water systems and includes such topics as:
- Analysis, simulation and design
- Integration and emerging technologies
- Model calibration
- Monitoring and control
- Reservoir modelling and control
- Simulation and modelling
- Transient analysis and design

Volume 2 is the fourth book in the series. This is mainly concerned with operation and management of water systems and includes such topics as:
- Data acquisition and information
- Decision support systems
- Genetic algorithm development and applications
- Geographic information systems

- Operational optimisation
- Optimisation applications
- Quality management and control
- Reliability and failures

The two companion books are based on the proceedings of the Exeter University International Conference on Computing and Control for the Water Industry, 1999. This event is a continuation of the series of conferences which started at De Montfort University in 1981.

Bryan Coulbeck
Bogumil Ulanicki
August 1999

Contributing Authors

Argelaguet R., *Department of Automatic Control, Universitat Politecnica de Catalunya (UPC), Spain*

Barth T., *Dept. of Information Systems, University of Siegen, Germany*

Barton D., *Thames Water, U.K.*

Blakey G.M., *Thames Water Research & Development, Reading, U.K.*

Bowmer R.A., *R.B. Associates, U.K.*

Bremond B., *Cemagref, France*

Burnell D., *Thames Water, U.K.*

Burrows R., *Department of Civil Engineering, University of Liverpool, U.K.*

Carmi N., *Ministry of Planning and International co-operation, Palestinian National Authority, East Jerusalem*

Cembrano G., *Industrial Robotics Institute, UPC-CSIC, Spain*

Colprim J., *Lab. d'Enginyeria Quimica i Ambiental, LEQUIA, Universitat de Girona, Spain*

Cook S.C., *Tynemarch Systems Engineering Ltd, U.K.*

Cooper N.R., *Thames Water Research & Development, Reading, U.K.*

Crerar A., *Thames Water, U.K.*

da Conceicao Cunha M., *Instituto Superior de Engenharia de Coimbra, Portugal*

Dandy G.C., *Dept of Civil & Environmental Engineering, University of Adelaide, Australia*

Davidson J.W., *Centre for Water Systems, University of Exeter, U.K.*

Davis R.B., *Mid Kent Water plc., Snodland, U.K.*

Eisenbeis P., *Cemagref, Bordeaux, France*

Elseid T.E.M, *Faculty of Technological and Developmental Studies, The University of Khartoum, Sudan*

Engelhardt M.O., *Dept of Civil & Environmental Engineering, University of Adelaide, Australia*

Fanner P.V., *Bristol Water plc, U.K.*

Farmani R., *Dept of Civil and Environmental Engineering, University of Bradford, U.K.*

Fernandes C., *Dept. of Civil Engineering, University of Toronto, Canada*

Feron P., *Lyonnaise des Eaux, France*

Feuardent J-P., *Lyonnaise des Eaux, France*

Fotoohi F., *Lyonnaise des Eaux, France*

Foulsham M., *SAUR Services Ltd, U.K.*

Fowler M.R., *Tynemarch Systems Engineering Ltd, U.K.*
Garrard T.P., *Strumap, Redditch, U.K.*
Giel M., *SAUR Services Ltd, U.K.*
Gilbert D., *ENGEES (Ecole Nationale du Génie del'Eau et de l'Environnement de Strasbourg), France*
Goldman F.E., *Department of Civil and Environmental Engineering, Arizona State University, U.S.A.*
Grauer M., *Dept. of Information Systems, University of Siegen, Germany*
Grimshaw D., *Ewan Associates Ltd, U.K.*
Guhl F., *Cemagref, France*
Haarhoff J., *Rand Afrikaans University, South Africa*
Halhal D., *Water and Electricity Distribution Company (RAID), Tangier, Morocco*
Jamieson D.G., *Environmental Software & Services Ltd, U.K.*
Javadi A.A., *Dept of Civil and Environmental Engineering, University of Bradford, U.K.*
Johnson E.H., *Stewart Scott Incorporated, South Africa*
Jolley T.J., *Scottish Environment Protection Agency, U.K.*
Kaden S., *WASY Institute for Water Resources Planning and Systems Research Ltd, Berlin, Germany*
Karney B.W., *Dept. of Civil Engineering, University of Toronto, Canada*
Keane D.S., *Built Environs, Adelaide, Australia*
Köngeter J., *Institute of Hydraulic Engineering and Water Resources Management, Aachen University of Technology, Germany*
Kora R., *Lyonnaise des Eaux, France*
Kozlowski A., *Institute of Geophysics, Polish Academy of Sciences, Poland*
Le Gat Y., *Cemagref, Bordeaux, France*
Lumbers J.P, *Tynemarch Systems Engineering Ltd, U.K.*
Macleod I.M., *Department Electrical Engineering, University of the Witwatersrand, South Africa*
Maksimovic C., *Department of Civil and Environmental Engineering, Imperial College of Science and Technology, U.K.*
Matthew R.G.S, *Dept of Civil and Environmental Engineering, University of Bradford, U.K.*
Mays L.W., *Department of Civil and Environmental Engineering, Arizona State University, U.S.A.*
Michels I., *WASY Institute for Water Resources Planning and Systems Research Ltd, Berlin, Germany*
Montiel F., *S.A.G.E.P. (Société Anonyme de Gestion des Eaux de Paris), France*
Napiórkowski J.J., *Institute of Geophysics, Polish Academy of Sciences, Poland*
Nguyen B., *S.A.G.E.P. (Société Anonyme de Gestion des Eaux de Paris), France*
Odeh K., *Lyonnaise des Eaux, France*
Ouazar D., *Ecole Mohammadia d'Ingenieurs (EMI), Rabat, Morocco*
Perez R., *Department of Automatic Control, Universitat Politecnica de Catalunya (UPC), Spain*

Poch M., *Lab. d'Enginyeria Quimica i Ambiental, LEQUIA, Universitat de Girona, Spain*

Quevedo J., *Department of Automatic Control, Universitat Politecnica de Catalunya (UPC), Spain*

Race J., *Thames Water, U.K.*

Rigola M., *Lab. d'Enginyeria Quimica i Ambiental, LEQUIA, Universitat de Girona, Spain*

Sa Marques A., *DEC-Universidade de Coimbra, Portugal*

Saffi M., *Ecole Supérieure de Technologie, Morocco*

Sakarya A.B., *Department of Civil and Environmental Engineering, Arizona State University, U.S.A.*

Savic D.A., *Centre for Water Systems, University of Exeter, U.K.*

Scheffer E., *Department Electrical Engineering, University of the Witwatersrand, South Africa*

Schindler D.F., *Strumap, Redditch, U.K.*

Schlaeger F., *Institute of Hydraulic Engineering and Water Resources Management, Aachen University of Technology, Germany*

See H.J., *Dept of Chemical Engineering, University of Cambridge, U.K.*

Sheriff S.J., *Connell Wagner Consulting Engineers, Darwin, Australia*

Sherwin C., *Thames Water Research & Development, Reading, U.K.*

Simpson A.R., *University of Adelaide, Australia*

Solomatine D.P., *International Institute for Infrastructural, Hydraulic and Environmental Engineering (IHE), Netherlands*

Sousa J., *Instituto Superior de Engenharia de Coimbra, Portugal*

Sutton D.C., *GHD Consulting Engineers, Adelaide, Australia*

Ta T., *Thames Water Research & Development, Reading, U.K.*

Tarasevich V.V., *Novosibirsk State University of Architecture and Civil Engineering (NGASU), Russia*

Templeman A.B., *Department of Civil Engineering, University of Liverpool, U.K.*

Terlikowski T., *Institute of Geophysics, Polish Academy of Sciences, Poland*

Thomas N.S., *Department of Civil Engineering, University of Liverpool, U.K.*

Uber J.G., *University of Cincinnati, Department of Civil & Environmental Engineering, U.S.A*

van der Walt J., *Magalies Water, South Africa*

van Zyl J.E., *Rand Afrikaans University, South Africa (currently with University of Exeter)*

Vassiliadis V.S., *Dept of Chemical Engineering, University of Cambridge, U.K.*

Walters G.A., *Centre for Water Systems, University of Exeter, U.K.*

Wang C.G., *Institute of Hydrology, U.K.*

Watkins D.C., *Camborne School of Mines, University of Exeter, U.K.*

Wells G., *Industrial Robotics Institute. UPC-CSIC, Spain*

Whiter J.T., *Thames Water Research & Development, Reading, U.K.*

Williams R., *South East Water plc, U.K.*

Wilson D.I., *Dept of Chemical Engineering, University of Cambridge, U.K.*

Woodward C.A., *Thames Water Research & Development, Reading, U.K.*
Wright C.L.M, *Scottish Environment Protection Agency, U.K.*
Zagorulko G.B., *Russian Research Institute of Artificial Intelligence (RRIAI), Russia*

Contents

Introduction

Water Industry Systems: Modelling and Optimization Applications is based on the proceedings of the Fourth International Conference on Computing and Control for the Water Industry (CCWI'99) hosted by the Centre for Water Systems at Exeter University in September 1999. Eighty papers were presented by industry practitioners and leading academics from over twenty countries, providing an informed international view of leading edge computer technology applied to the water supply and waste water disposal industry.

The aims of the conference were:

- To facilitate continued co-operation between academic institutions and industry.
- To examine the current state-of-the-art in computing and control techniques as applied to the water industry.
- To provide a forum for discussion and the dissemination of ideas on applied computing and control for the water industry, with particular emphasis on:
  - Provider's perspective: recent developments in research.
  - User's perspective: industry's experience of the latest techniques.
  - Future needs: current and future planning and operational requirements.

The first volume of the book starts with an industry overview, in the form of five invited papers from internationally recognised authorities. The volume then concentrates on the general areas of hydraulic analysis and control. The second volume covers hydroinformatics, optimisation and quality management. The balance of papers between the two volumes reflects the steady growth in the use of computers, not just for analysis and modelling, important though these applications are, but also for information handling and informed decision making.

In the UK there have been immense changes to the water industry over the last 25 years, with the formation and subsequent privatisation of regional water utilities. In parallel, there has been a workplace revolution led by the availability of cheap and powerful desk-top computers, providing both analytical power and technical communication at the push of a button. Yet the basic hydraulic behaviour of water and most of the engineering design principles remain the same, technical practice now being based on many decades of research, development and experience. We have successfully transformed many of the simulation and analysis processes into computer software applications, improving efficiency, accuracy, and, hopefully, understanding of water system analysis and design.

However, computers now offer a far wider range of opportunities for industry – not just as tools for performing established processes more efficiently, but as ways of exploring and exploiting previously intractable areas such as:

- On-line optimal control of complex systems
- On-line state estimation
- Optimal design of large systems
- Data-mining for information retrieval
- Intelligent decision support systems

As we know, information gained and decisions made are, at best, only as good as the models used and the data available. Hence there is still a need for fundamental research to improve our understanding of the processes involved, particularly in areas such as water quality, otherwise flawed models will lead to unsatisfactory and ultimately expensive solutions, with rejection by industry of the software that delivered them. It is equally important that software developers and users have a clear understanding of the models incorporated in design and analysis packages, particularly concerning limitations and validity, so that models are not called on to simulate situations for which they were not intended.

We hope that these two volumes will go some way not only in keeping industry practitioners in touch with the latest research developments, but also in keeping academics in touch with the practical requirements of a rapidly changing industry.

Finally the editors would like to thank the authors for their excellent papers, the technical committee for their efforts in reviewing the work, and the invited speakers for their inspiring addresses. We look forward to a future water industry in which all engineering and management decisions are supported by well-developed and properly applied computer technology, to the benefit of consumers and industry alike.

Dr Godfrey Walters
Dr Dragan Savic

Centre for Water Systems
University of Exeter
United Kingdom

September 1999

PART I

DECISION SUPPORT SYSTEMS

Water Balance Study - Microscope on Leakage and Demand*

A. Crerar, J. Race, D. Burnell *and* D. Barton

ABSTRACT

Following the Government's "Water Summit" in May 1997, Thames Water Utilities launched their Engineering Leakage Initiative. This paper describes results from one part of this Initiative: a "microscope" on a District Meter Area to improve understanding of the components of the Water Balance. Investigations into leakage and demands are described. The analysis presented here was carried out using a computer model called DISTMET, which was written by Thames Water Corporate Modelling, and is designed to make intelligent use of demand and DMA data.

1 PROJECT BACKGROUND

Following the Government's "Water Summit" in May 1997, Thames Water Utilities launched their Engineering Leakage Initiative. This paper describes results from one part of this Initiative: a "microscope" on a District Meter Area to improve understanding of the components of the Water Balance. Two investigations are described:

- **Leakage** - multiple sweeps were carried out to gain insights into different active leakage control approaches, extent of background losses and the size and growth patterns of leaks.
- **Demand analysis** - pioneering methods to characterise the District Meter Area (DMA) flows into different components (domestic and commercial legitimate use, and leakage which depends on pressure).

Another initiative is covered in a companion paper in section iv of this volume (Burnell and Race, 1999), which describes a leakage framework developed to answer strategic leakage questions.

2 DISTMET COMPUTER MODEL

The analysis presented here was carried out using a computer model called DISTMET. DISTMET was written by Thames Water Corporate Modelling, and is designed to make intelligent use of demand and DMA data. It includes:

- Methods for deriving DMA flow and pressure data from multiple meter feeds
- Methods for data screening e.g. applying pattern recognition methods to highlight unusual behaviour.

---

* © Thames Water

3

- Leakage analysis methods e.g. identifying areas for priority detection effort by looking at the level and growth rate in leakage.
- Demand analysis methods e.g. impact of weather on demand

This paper describes some of these investigations. The results presented here are illustrative only and are not to be taken as representative of Thames Water as a whole.

3   SETTING UP THE STUDY DMA

A district metering area in South London was chosen. The focus has been to concentrate on gaining an in-depth understanding of a particular area. The study DMA was chosen because it has a range of property types, and does not have large scale commercial activity. However, it is not directly scaleable up to company, as it is not practicable to find a single DMA which spans the range of customers and lifestyles found in Thames Water.

The Study DMA has over 3000 properties and has been split into Leakage Control Areas (LCAs). The work was carried out by the Engineering Department of Thames Water as part of the wide-scale DMA Programme. Engineering designed the LCAs, closed them in and installed the meters and telemetry. Most of the areas are homogeneous and thus essentially form "pipework cul-de-sacs" of different house types. The study uses electromagnetic flow meters which can measure pressures as well as flows; usually data is logged at 15 minute intervals. We use telemetry to collect the data on a daily basis. There are 32 meters in total, covering 23 different areas, of which 5 meters monitor the whole DMA. An automatic weather station has also been installed at a local pumping station to allow analysis to be carried out on weather-related demand. We have also collated data on the properties and customers in this DMA. Information has been collected from various sources including our Geographical information System (GIS), our Customer Information System (CIS) and marketing data.

4   LEAKAGE INVESTIGATIONS

4.1   Impact of multiple leakage sweeps

Four separate active leakage sweeps were systematically organised by local leakage teams, with the resulting jobs being repaired shortly after each sweep was complete.

Table 1

Sweep & Method	Leakage Impact	Comments
1. Listening sticks sounding mains, valves and services, then standard correlators	little effect	no mains jobs repaired; one leak on a busy road was known about but not pinpointed exactly
2. Method as 1	significant reduction	mains jobs fixed
3. Method as 1	little effect	no burst mains identified
4. Aqualoggers (60) deployed then standard correlators to pinpoint leaks	significant reduction	mains jobs fixed, many of which were new bursts. Aqualoggers analysed frequencies of sounds to identify leaks. If results were positive, normal correlation was carried out to pinpoint the leak

Other findings were:
- Average loss rates vary considerably from standard loss rate assumptions
- The majority of leaks fixed (80%) showed minimal leakage savings. The best way to significantly reduce leakage is to find and fix leaks on mains. 72% of the leakage savings were made from fixing leaks on mains.
- Only the mains jobs leakage savings could be seen in the DMA-level flow data
- Aqualoggers were successful in finding some mains leaks which were not found by listening sticks
- No evidence found to support a theory that smaller noisier leaks mask the larger quieter leaks. Sweep 1 did not fix any leaks in the vicinity of the elusive leak which was not pinpointed exactly until sweep 2. It may be that leak detection is an inherently variable process, rather than one type of leak masking another.

4.2   Background losses

We assessed the point at which the nightline levelled out which, after subtraction of legitimate night use, gives us an approximate estimate of background losses for this DMA. By examining the night flows in different LCAs we can investigate the spread of background losses.

Here we compare the unadjusted nightlines for areas of different types of houses. The areas chosen are fairly homogeneous in terms of house types and contain no 'non household' demand (i.e. no shops, schools etc).

6

Table 2

LCA	DISTMET code	Description
8	FU0108	Mostly terraced, some flats
13	FU0113	Modern 4 storey flats (marketing information indicates majority of people living here are retired)
18	FU0118	Mixed sized semis

The following figure compares Minimum Night Flows per property.

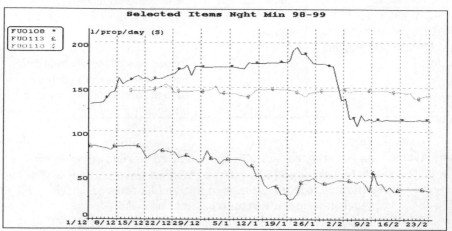

Figure 1 Comparison of night flows for LCAs 8,13 and 18

A sizeable service repair was carried out in LCA8 in late January, which explains the drop in night flow. In early February, after all repairs had been carried out, we recorded a range of night flows per property. The modern retirement flats area (LCA 13 FU0113) showed a consistently lower night flow.

The variation in night flows per property cannot be explained fully by variations in legitimate night use, leading us to conclude that background losses are not evenly spread within the DMA.

4.3 Estimating Recurrence

Around 2000 litres/hour was saved from the repairs carried out in the first sweep. This was calculated from looking at the changes in each of the individual LCA nightlines and summing them. However, the nightline for the study DMA remained broadly level. This suggests that the level of recurrence at the beginning of the year had counteracted the repairs carried out. The sweep 1 repairs were carried out over a 20 day period, which gives an estimate of recurrence for this study area at 100 litres/hr extra leakage per day.

4.4 Life-cycle of Bursts

Some of the LCAs had burst incidences which clearly show up in the nightlines. Examples of these now follow. The units in the graphs are litres/second. These results provide insights into how bursts start and grow and are potentially very powerful in adding to our understanding of leakage recurrence.

LCA 21 Burst Main

The following graph shows how a burst main started and when it was fixed. It appears to have started on the 22nd January 1999, which was after sweep 3. It rose dramatically from 0.02 litres/sec to 0.8 litres/sec in the space of 2 days. It then rose steadily again to reach 1.35 litres/sec over almost 2 weeks. The short downward trend was a meter blip. The burst rose to 1.45 litres/sec (5000 litres/hour) at its peak until it was fixed on 17th February 1999.

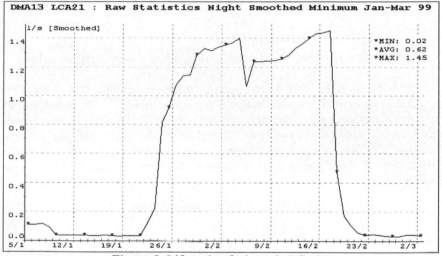

Figure 2 Lifecycle of a burst in LCA21

8

Figure 3 illustrates the weather for London at this time.

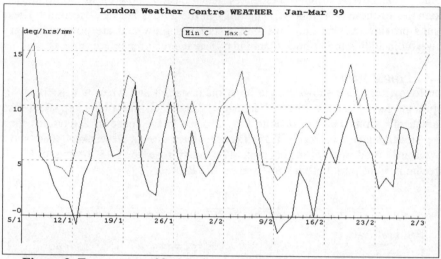

Figure 3 Temperatures °C Maximum (upper line), Minimum(lower line)

The burst occurred after a cold period, as the temperature was rising. The flow and pressure profiles were analysed for this area and it was found that the flow was *directly*-related to the pressure, because of the extent of leakage in the area. The following picture shows the LCA flow profiles just before and just after the repair. The flow profile changes from one dominated by pressure (higher night time flows) to one dominated by legitimate usage (night time flows are lower than day time).

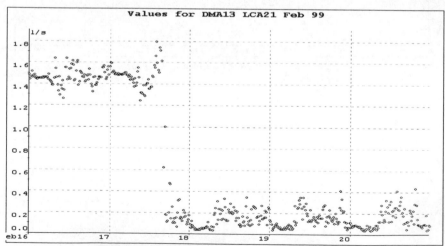

Figure 4 Change in flow diurnal profile as burst fixed

LCA 9 Burst Main

The following graph shows the history of a burst 3 inch main in LCA 9. The LCA had shown a gradual rise in nightline since November, but no leaks were found during the first three sweeps. The burst main was eventually found during sweep 4 and fixed on 11th March 1999, at which point the nightline reverted to a lower level which suggests that the rise in leakage was mainly due to the growth of this running burst main.

The sudden jump in the nightline suggests that there was a change in the state of the pipe. Could it have started out as a gradually widening crack but then suffered a catastrophic failure? This burst main was not found in any of the first 3 sweeps, but was in its gradual growth stage which may have made detection more difficult.

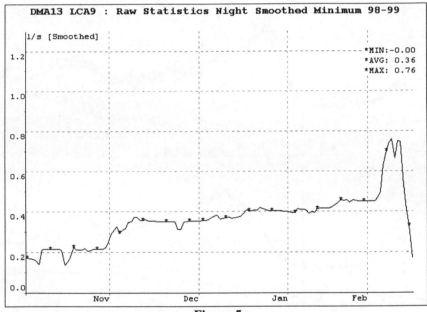

Figure 5

5   DEMAND ANALYSIS

5.1   Choosing analysis period

The flow meters on the individual LCAs will be recording the sum of:

- genuine demand in the households
- wastage and/or losses within the houses
- supply pipe leakage in customer's underground pipes
- leakage in company assets (service pipes, joints, mains, valves etc)

Thus to focus on water being used in houses (genuine demand + wastage) we need to analyse a time period when leakage in underground assets (customers +

company) is at is lowest. Hence this demand analysis was carried out for the period immediately following the intensive leakage detection and repair activity. This was a period in early 1999, during which time any remaining leaks can be considered as 'background losses' since they have eluded 4 leakage sweeps.

5.2 Diurnal Variations

DISTMET can examine the diurnal profile for different days of the week. This allows differences between weekdays and weekends to be studied. In the following figures we can clearly see the effect of the weekend lie-in because the morning rise in demand is shallower and flows peak at around 10-11 am instead of 8am for weekdays.

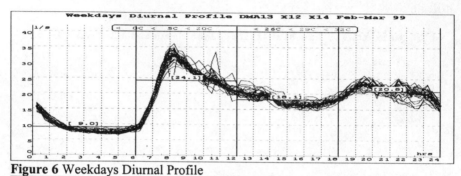

Figure 6 Weekdays Diurnal Profile

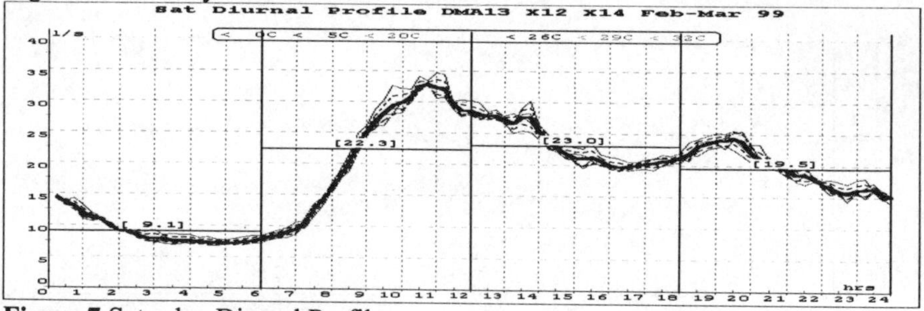

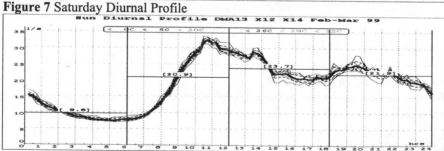

Figure 7 Saturday Diurnal Profile

Figure 8 Sunday Diurnal Profile

5.3   Developing efficient ways of characterising DMAs

The study DMA is being used to develop methods for characterising DMAs which can be applied more generally, without the need for a large amount of data gathering. The plan is to verify the methods against the more detailed information available for the study DMA. An example of this approach is described below, which estimates the leakage and demand elements of DMA flows.

Traditional methods of estimating leakage in an area have focused on analysis of night use e.g.

- establish the night use for the area
- research the numbers of properties in the area, for domestic and commercial customers
- using standard assumptions about night use per property, subtract the estimated genuine usage from the observed night use

However, this method requires reliable data on the numbers of properties in the area, and puts a great reliance on the legitimate night use per property estimate. In a highly populated area such as London, it is difficult to disentangle the exact number of flats, as larger houses are split into converted flats. Another complication is the number of multiple occupied houses.

DISTMET contains a statistical method, which we have termed *component analysis*, which can be used to estimate the proportion of flows split between domestic demand, commercial demand and leakage. The next sections describe component analysis and show initial results of applying it to the study DMA.

5.4   Description of component analysis

DISTMET component analysis splits a given weekly flow profile into:

- domestic demand
- commercial demand
- residual leakage, which can be related to pressure

It considers one week profiles at a time because the pattern of weekdays is different from weekends. The steps in the component analysis are :

- feed in a typical shape for domestic demand and commercial demand
- calculate the weekly profile for the area under study
- apply statistical techniques to find the mix of the domestic and commercial and leakage pressure profiles which best explains the area's weekly profile

This technique makes use of the entire profile, rather than just the night use. It therefore should be a more robust estimate. The method requires only the shapes of the domestic and commercial profiles, it does not require property counts, volumes or any other information which would require special collection. If the period under study contains hot/dry weather then DISTMET will augment the domestic demand profiles with the additional weather related demand.

5.5   Profiles for use in Component Analysis

Thames Water possess a sizeable Domestic Water Use Study, whereby 15 minute data is collected for nearly two thousand households. The DWUS panel is managed by Thames Water's Water Resources team. DWUS provides information

12

for Thames' per capita consumption figure in the Company Water Balance. The detailed 15 minute data, from a range of house types across the Thames area, is a valuable source for demand studies. Two legitimate use domestic profiles (flats and houses) were produced, both derived from DISTMET analysis of DWUS data. This is because flats and houses have different diurnal patterns and weekday/weekend behaviour.

The commercial demand shape was derived from Thames Water's study of commercial demand which has logged a range of commercial premises (different sizes and industry categories).

Fifteen minute pressure data is available alongside the flows. In this example leakage was assumed to be directly proportional to pressure, however, the method can incorporate different power laws relating pressure to leakage.

5.6   Applying component analysis to DMA13

The illustrative results shown below are for the study DMA in early 1999, which is just after the intensive leakage work. The figure below shows the comparison of the actual points with the prediction line.

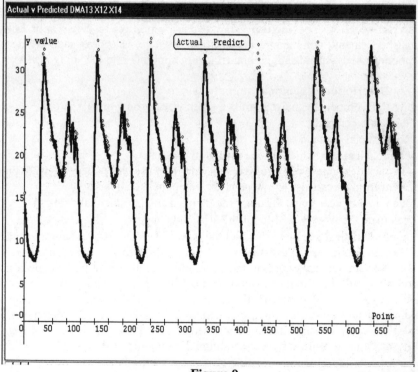

Figure 9

The mix of components explains 95 % of the variance, and captures the weekday/weekend differences. Experimentation with the model showed that the commercial profile is needed to explain the daytime demand, and pressure related leakage is required to explain the night flows.

The confidence range around the flows ascribed to the domestic demand (flats and houses) gives a number of properties within the range of the actual detailed data gathered for this DMA. The model was also applied to the DMA flows before the intensive leakage detection activity; the results gave comparable property estimates and (as expected) showed a higher leakage element. The figure below shows the make-up of the prediction.

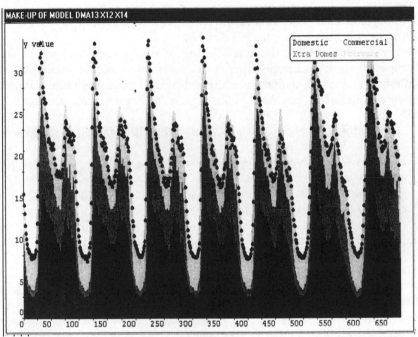

Figure 10

The make-up shows the contribution of the different components to the DMA flow. At night about half of the flows are ascribed to pressure related leakage. This gives an indication of potential leakage savings from pressure reduction.

6   CONCLUSIONS

This paper has described leakage and demand work in an intensively metered study DMA. A general purpose computer model DISTMET has been developed to perform DMA analysis.

Results from the leakage sweeps show:

- Multiple sweeps were effective in reducing the night flows in the study DMA
- There is a wide variation in loss rates, whereby many leakage repairs had an indiscernible impact on the night flow in the LCA. Gains from these repairs

must be masked by ongoing recurrence. Mains repairs showed significant savings in leakage.

- Tracking the life-cycle of some mains bursts suggests a gradual growth stage, followed by catastrophic failure. Some bursts are associated with cold weather. Investigation of bursts against weather will continue by longer term observations in the study DMA. The LEAKPLAN model, described in the companion paper (Burnell and Race 1999) relates leakage recurrence to cold weather, based on analysis of several years repair data for London.

- After the intensive leakage sweeps and repairs, the remaining leakage represents background losses for this study area. Examining the residual flows in different LCAs suggests that background losses are not evenly spread throughout the DMA. This is an area which requires further study to try to identify possible reasons for this variation.

The demand analysis illustrates the use of Component Analysis which is a statistical method for decomposing DMA flows into constituent components of domestic, commercial and pressure related leakage, which gave results which are within confidence ranges when checked against the detailed data gathered for this study DMA.

The contribution of the DMA flow from the different components gives the following important insights:

- the behaviour of leakage components can be used to prioritise leakage reduction efforts

- the mix of houses against flats and extent of weather related demand can guide customer demand management and metering strategies

These methods will be developed further as we progress through the summer period, for example examining the influence of weather on demand.

ACKNOWLEDGEMENTS

We would like to acknowledge the hard work and enthusiasm from colleagues in Thames Water who have worked on this project, including the local leakage teams, Engineering, Customer Field Services and R&D. This was a multi-disciplinary project, where our role was in data analysis and modelling. Our Engineering colleagues worked hard to manage the project, design the scheme, install meters and setup telemetry. Local leakage teams responded very well to the challenge of intensive leak detection and repairs. We are also grateful for the support and encouragement of Thames Water management. The views expressed are those of the Authors.

REFERENCES

Burnell, D. and Race, J. 1999 Network Asset Strategy Modelling, *Water Industry Systems,* Vol. 2, Research Studies Press Ltd, UK.

Water Information Systems in Developing Countries: Community-based Analysis and Design

T. E. Mohamed Elseid

ABSTRACT

The representation of stakeholders and the empowerment of the whole community are the current front-line challenges facing water managers in developing countries. The problem domain and solution space is becoming sufficiently complex to warrant a shift in the methodologies used for the analysis and design of water information systems to incorporate such complexity. Approaching water information systems with community orientations and structural linkage improves the possibilities of "making a difference" in the water environment. This paper sheds light on community-based analysis and design of water information systems. It is based on the research initiation on *"leveraging water-related DSS in developing countries"* targeting the effectiveness of water decision-support systems (DSS). It also builds on the paper on improved learning in the water environment in developing countries (Elseid 1999a).

1 INTRODUCTION

The field of information systems has witnessed tremendous developments and innovations that resulted in improved performance and increasingly sophisticated enterprise-wide information platforms. Developments range from technological settings (data capturing, processing and communication technologies) to methodologies and techniques of information analysis and representation. However, the polarising effects of such revolution continued to shift the swing of the pendulum of control over time. In the digital economy and information age, some organisations are "bedazzled" while others are "frightened". In real terms, such advances should be means for improved interaction and co-operation in a wider organisational context rather than an end in themselves. It is information utilisation, not information supply, which leads to managerial advantages and situational merits.

Over the last two decades, information systems (particularly decision support systems) have been widely used for the management of water resources. Water allocation, water quality investigation, hydropower scheduling, reservoir management, and the preservation of the aquatic ecology are, among others, areas

15

of applications of information systems in the field of water resource management. A wide range of techniques is being used for data structuring and management, modelling, and simulation.

The effectiveness of such systems and the satisfaction of their users are situation-based, despite the common functionality. However, organisational factors as well as system-specific features usually shape the framework of success of such systems. In many developing countries, the models used do not reflect the actual representation as well as the interplay among the decision variables. As a result, many users (at different organisational landscapes) are unable to make use of such systems or interpret their results (Elseid, 1999b). Moreover, the inflexible organisational forms, the changing portfolio of decision partners and the complex hierarchical setting of decision making endanger the performance of information systems. Despite the technological advances and the increasingly enhanced system functionality, many systems failed to improve the quality of decision making and create responsive water organisations.

The way water information systems are approached, analysed and designed is a major determinant of success. Within the technological context, wide attention is being paid to "systems infrastructure" with little focus on the concepts and methodologies of systems analysis and design. Many conventional methodologies are being used for the analysis and design of water information systems. Such methodologies are based on the framework developed by Anthony and Scott Morton (Gorry & Morton, 1971, 1989). The majority of these methodologies combine Anthony's taxonomy of managerial functions (strategic, tactical and operational planning) with Simon's concepts of "decision programming" and "decision phases" (Simon, 1947, 1967).

In many situations, the emerging water information systems are being characterised by the following features (Elseid, 1998):
1) The focus on structured quantitative settings with special emphasis on highlighting information variations across the managerial scope.
2) The use of "defensive" patterns as a means for scanning the agents for change and avoiding uncertainties in the decision environment. The central concern is to protect the decision-making "portfolio" from the fierce attacks of risk. Targeting and influencing the environment through initiation is seen as an unwarranted activity that "awakens the sleeping dogs". Therefore, long-standing mechanisms are in action to "pull the fat of the organisation from the fires of uncertainty". Despite the "declared" perspectives of openness and comprehensiveness, environmental diagnosis reflects internal orientations in many systems.
3) The intensive use of digital mapping and interfacing techniques that disregard the information processing and decision support scope and become, in many circumstances, serious limiting factors.

Within the context of global development and change, the field of water resources management cannot remain an isolated island in an integrated and sophisticated archipelago of disciplines. Over recent years, the control and management of water resources has become increasingly complex. The tremendous environmental shifts, the escalating costs of water development projects and the considerable expansion of water demand are the main change agents. The objective function is no longer a set of structured variables with predictable coefficients. The continuous economic shifts, the diversified natural ecology, the changing demographic settings and the comprehensive social reforms are becoming active vectors in the matrix. Particularly in developing countries, the representation of stakeholders and the involvement of the whole community are front-line challenges that are currently looming very big.

Meeting these challenges in the new millennium is not possible through the articulation of situation-specific information and the provision of the necessary decision aids. Under all circumstances, partial solutions and remedial actions can not ensure the optimisation of the utility matrix in the water system. The decision-based transformation warrants a shift in the way water information systems are analysed and designed. The naivety characterising the analysis of water systems (Schultz, 1989) and the analytical pre-occupations (Elseid, 1999b) that rendered many information systems to be "theoretical" and "documentary" need to be "reinvestigated". We are looking for effective water information systems that support decision-making and make a real "difference" by involving and empowering the whole community. The efforts of system analysis and design should go beyond the organisational boundaries and favourably influence the environment. Keeping the community informed and involved builds up a lasting spirit of reconciliation and compromise and facilitates the adjustment to discontinuity and uncertainty.

Some experiments are being conducted to analyse and design water information systems with efforts being directed toward providing stakeholders with direct access to models rather than through an intermediate group of consultants (Reitsma et al, 1996). Model access is expected to help in informing the participants about the constraints of the decision-making problem and the alternatives and in setting the agenda and framework for conflicts (Kraemer, 1985). However, in addition to the laboratory and experimental domains that endanger the managerial and learning potentials of such an approach, the following prerequisites are essential to maximise the chances of success:

1) Sophisticated technological platforms that support, among others, advanced and comprehensive model-calibration, multitasking, and information communication.

2) Stand-by control mechanisms that ensure the security, priority handling, equilibrium-assurance, privacy and integrity across the various nodes of the whole system.

3) Reliable system management capabilities. In the absence of reliable planning and monitoring, the cost of data acquisition and processing will be high. This will discourage organisations and users from not only using the information system but also from maintaining its database (Elseid, 1998).

Given the situational challenges and associated change agents, many developing countries are not in a position to meet such requirements. The unfavourable ramifications of the new economic order have moved many water organisations into serious liquidity traps. At the abstract level, there is a considerable lack of effective organisations capable of managing technology-intensive projects and ensuring a situational "match" (Elseid, 1999a). Approaching "ineffective" information systems through model access is expected to add to the existing data redundancy (the overabundance of irrelevant information that many nodes may bring in) and the lack of relevant information (i.e. of weak databases).

2 COMMUNITY-BASED ANALYSIS AND DESIGN OF WATER INFORMATION SYSTEMS

The use of model-accessibility domains for the analysis and design of water information systems, particularly in developing countries, is not expected to stand in the face of the front-line challenges. Community-based analysis and design is a view targeting the involvement of the whole community in the process of water conservation through information communication. The articulation, acquisition and communication of information among the nodes making up the system's environment are regarded as activities connecting partners in an interactive, on going and comprehensive learning and information partnership process. The analytical and design efforts are directed towards "discharging" relevant information into the whole community and "monitoring" responses through intensive open communication. For methodological concerns, the flow of information and response tracking is viewed in terms of the organisational sets (Evan, 1967) making up a learning society in pursuit of shared visions, flexible negotiations, and adaptation to organisational defensive mechanisms (Argyris, 1990).

At the highest levels of abstraction, water-related organisations, educational institutions, and the community organs are regarded as the principal organisational sets in the water system (Elseid, 1999). Each organisational set is assumed to have its own learning capacity that contributes to the overall learning curve. Water organisations in many developing countries are public authorities with limited and constrained involvement of private organisations called for by the efforts of privatisation. Despite the loose structure of the term, "community organs" is used to denote the formal and informal organisations and groupings sharing an interest in the conservation process. They represent the main actors in the water system. Moreover, higher education and research institutions are playing central roles in the process of knowledge diffusion and the empowerment of the water community.

Figure 1 provides a framework for the interaction among the proposed organisational sets. The two-direction arrows indicate joint communication of information or responses that can be valid inputs in the respective databases or pre-processed inputs. In all circumstances, the media used for such communication, the formats of information communication and the interval domains are expected to vary across and within sets. Each set has it own linkages and information sources other than the sets depicted in Figure 1. However, our focus tends to be on the communication and interaction that contribute in one way or another to water information or improved learning in the water environment to facilitate water resources management.

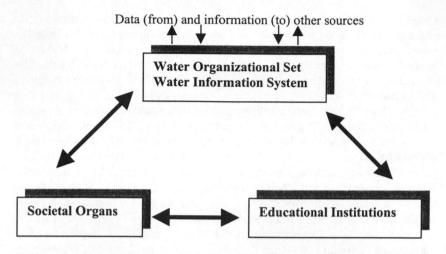

Figure 1 Community-based Analysis and Design

Selection of the three organisational sets is being motivated by the need for building sound foundations for sustainable development through:

1. Creating and maintaining the learning curve of the community through improved learning, communication and interaction among the main sets.
2. Avoiding sub-optimisation by creating a mandatory network of "monitors" including the main "interest-bearers".
3. Fostering information partnership and inter-organisational linkage.

Based on the proposed framework, information acquisition and communication in the water management system takes the following directions:

1. Intensive communication initiated by the water organisational set and directed to the community: depending on the nature of "what to be communicated and the format to be used", continuous discharge of "procedural" and "awareness" information can improve learning and ease involvement, negotiation and representation. Procedural information include information about the factors interacting in the physical system such as the available and forecasted water supplies, the current and projected water demand, the consequences of water usage, spatial variability indicators, financial parameters, and hydro-meteorological information. Awareness information, on the other hand, encourages the change of behaviour in response to spatial and situational transformations. It usually stems from the internal processing of the water organisational set and the external interactions with the other two sets. Improving communication with the community at large can facilitate the resolution of conflicts that pump out under conditions of water shortage and the suspension of water rights. Moreover, it also ensures the appropriate involvement that maintains the financial sustainability of the whole system. However, the process of information flow is not that automatic. Water organisations must employ effective mechanisms to acquire feed back, monitor the resulting change of behaviour and develop plans accordingly.

Based on such interaction, the database of the system will be maintained (one of the problems facing many DSSs in developing countries) (Elseid, 1999b).

2. Interaction between water organisations and educational institutions: it cultivates mutual relations and supports cooperation between both sets. The technological and scientific access provided by the educational set result in new or modified axioms for managing water resources. The resulting research creates a basis for problem handling and societal empowerment. Information about the educational programs represents a part of the training and development database of the water information system. On the other hand, educational institutions will gain insights into water-related problems and phenomena that enrich their learning capacities. The financial as well as non-financial inputs support their contribution to the conservation process. Improved awareness is expected to move "water-based" specialisation many steps up the "professionalisation" ladder (Elseid, 1999a). Educational institutions are also contributing to the whole process of water resources management through knowledge diffusion to the community. Such interaction strengthens the level of awareness and facilitates transformation and adaptation.

3 CONCLUSIONS

Information systems have the potential to act as decision catalysts in increasingly complex decision environments. Water information systems analysed and designed with community-orientations, rather than on the basis of mere decision analysis and conceptualisations are expected to have greater chances of "making a difference" particularly in developing countries. The comprehensive framework of analysis incorporates the necessary decision supports and enhances the ability of water organisations to act beyond their organisational boundaries. Instead of being frightened by the change agents in the environment, initiation and responsiveness will ensure appropriate control throughout the water system. The emphasis on linkage among the key partners in conservation and learning provides a foundation for data acquisition, requirements analysis and information communication. Moreover, the continuous interaction facilitates information engineering and management by focusing on "what to communicate, to whom and in what format". However, focusing on empowering and involving the whole society through information communication is an important, yet complex assignment. It requires not only organisational transformation but also a dramatic shift in the approach and methods. In real terms, the articulation of the appropriate organisational sets, the "conceptualisation" of the appropriate communication patterns and the reliable "tracking" of the interplay among the change agents are some expected problems. However, intensive research that supports enhanced placements of water information systems into "human", "managerial" and "societal" contexts is our key to the new millennium.

REFERENCES:

Argyris, R. Organizational learning Patterns, in R.Anderson: *The Process of Learning in complex systems*, KU Press, 1990

Elseid, T. *"Leveraging Water DSS in developing countries"*, The Methodological Report, Phase One, Limited Distribution, The Netherlands, 1998.

Elseid, T. Towards improved learning in the water environment in developing countries, forthcoming, the *proceedings of the international symposium on the learning society and the water environment,* UNESCO, Paris, June 1999a.

Elseid, T. Water DSS in developing countries: Operationalization, orchestration and intelligence building, a paper for *the 2nd International conference on Water Resources and Environmental Research,* Australia, July 1999b.

Evan, W.M. "The Organization Set: toward a theory of International Relations", in J.D. Thompson (ed.): *Approaches to Organizational Design*, University of Pittsburgh Press, 1967.

Gorry. A and M.S. Morton, A framework for Management Information Systems, *Sloan Management Review*, 13 (1) (1971), 55-70.

Gorry. A and M.S. Morton, Retrospective Commentary on the Gorry and Scott Morton Framework, *Sloan Management Review*, spring 1989, 58-60.

Kraemer, K.L. "Modeling as negotiating: the political dynamics of computer models in policy making." *Adv. in Information processing in Organizations*. 2, 275-307.

Reitsma, R, Zigurs. I; Levis. C; Wilson. V. and Sloane. A. Experiment with Simulation Models in Water Resource Negotiations, *Journal of Water Resource Planning and Management*, January-February 1996, 64-70.

Schultz. G.A. "Ivory tower versus Ghosts: The interdependency between systems analysts and real-world decision-makers in water management", in: Loucks. D.P: *Closing the gap between theory and Practice*, proceedings of the Baltimore Symposium, IAHS publication no. 180, 1989, 23-32.

Simon. H, *Administrative Behavior*, Macmillan, New York, 1947.

Simon. H, Information can be managed, *Think33* (3) (1967), 9-12.

Decision Support and Distributed Computing in Groundwater Management

M. Grauer, T. Barth, S. Kaden *and* I. Michels

ABSTRACT
This contribution is intended to present the solution of three typical classes of optimization and control problems for decision support in groundwater management: 1. Facility optimization for steady-state groundwater problems, 2. Optimal control problems for a given time period, and 3. Feedback optimization problems for operational management. The solution of these kinds of optimization problems needs coupling of optimization with simulation, managing the involved difficulties concerning the interface and synchronization between simulation and optimization. Due to the large computational effort in solving simulation-based problems, the computation time for optimization must be reduced significantly. This can be done by introducing different kinds of parallelism to the solution process. For the solution of these optimization problems a network of workstations is proposed as the platform of the software system. A software architecture is presented and a prototypical implementation of a coupled simulation (FEFLOW) and optimization system (OpTiX) is discussed. With this prototype, groundwater management problems of the aforementioned problem classes are solved.

1   INTRODUCTION

The main part in decision support for groundwater management is the solution of mathematical optimization and control problems. For the solution of these problems the following three classes will be analyzed: 1. Facility optimization for steady-state groundwater problems, 2. Optimal control problems for a given time period, and 3. Feedback optimization problems for operational management.

In the case of optimization and control problems from groundwater management practice, objective and constraint functions are mostly nonlinear and describe complex physical processes, e.g. subsurface flow or transport processes. The well known "curse of dimensionality" of this class of problems is shown in Fig. 1. by the (exponential) growth of solution time with increasing dimension of the objective function [BGG91]. Additionally, in groundwater management an analytical formulation of the non-linear optimization problem is mostly unavailable and the computation of a single objective function value and/or values for constraints implies numerically expensive simulation. Therefore, the solution of

24

these simulation-based optimization problems requires excessive computation time which leads to the use of nonsequential solution concepts, e.g. mathematical decomposition methods (s. e.g. [Lasd72]), to reduce the total computation time.

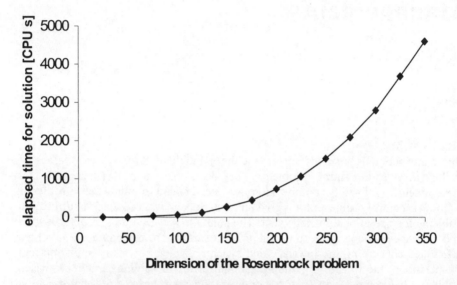

Figure 1 Exponential growth of computation time for the solution of the Rosenbrock problem (s. e.g. [BGG91], [GB97]) by changing problem dimension

The objective and constraint functions of the optimization problem are in general non-smooth, non-differentiable and non-convex. Thus, there is no "traditional" numerical optimization algorithm guaranteeing global and local convergence. One approach to overcome these problems is the use of hybrid methods ([BG95], [BGG91]). The basic idea behind hybrid methods is the potential benefit of combining globally and locally convergent algorithms to find the global solution. The solution of this kind of optimization problem poses several problems from different scientific domains: the development of adequate optimization algorithms and the design and implementation of software environments supporting the distributed solution of these problems.

In this paper approaches to the solution of different problems raised by the coupling of simulation and optimization software are presented. In the first section three problem classes and their appearance in water engineering are introduced: static optimization problems, optimal control problems and feedback optimization problems in groundwater management. A software system supporting the solution process of these problems has to fulfill certain requirements. These requirements and their software engineering implications are presented in section 3. To show practicability of such a coupled simulation/optimization software architecture, a prototypical implementation is presented. In the prototype the groundwater simulation system FEFLOW® ([Dier96], s. section 4.1) was used for simulation and coupled with the distributed optimization environment OpTiX® ([BG91], [BG95], [Bode96], [Brue97], [GB97]), s. section 4.2). Using this prototype, test problems of the aforementioned problem classes were solved. These results are discussed in section 5. The paper is concluded with a[1] summary and topics for future work.

2   PROBLEM CLASSES IN GROUNDWATER MANAGEMENT

The following subsections specify three general classes of mathematical nonlinear optimization or control problems in decision support for groundwater management.

In the following sections $f(x): R^n \to R$ denotes the objective function of an n-dimensional optimization problem. The m equality constraints are given by $g_j(x) = 0, \forall j = 1, \ldots, m$ and k inequality constraints by $h_p(x) \le 0, \forall p = 1, \ldots, k$.

2.1   The class of optimal design or facility optimization problems

The Problem of Optimal Design (POD) can be stated as follows:

$$U_{POD} = \left\{ x \in R^n \mid g(x) = 0, h(x) \le 0 \right\}$$

defines the set of feasible points and the optimization problem is specified by

$$\left\{ \min f(x) \mid x \in U_{POD} \right\}.$$

This is an important problem in the water industry for instance in the optimization of water distribution pipe networks [DSM96], the problem of optimal aquifer remediation design ([ML95],[ZW99]), optimization of artificial recharge-pumping systems [JZNM97], optimization problems in groundwater quality management [TKP98] or finding optimal locations for groundwater monitoring wells [CER95]. A problem of this class is solved and the results are analyzed in section 5.1.

2.2   The class of optimal control problem

In this section the Problem of Optimal Control (POC) is defined. The vector of time-dependent input variables of the controlled system consists of z(t) of input variables and u(t) of q decision variables. The output vector is denoted by y(t) and the state equation is given by s(t). Dynamic behaviour is described by a system of differential equations $\dot{x} = \varphi(z(t), s(t), y(t), u(t), t)$ with start conditions $x(t_0) = x_0$. The control problem can be solved using the same algorithms as for POC if it is transformed to a discrete time problem [Veli97]. Therefore, we consider that any appropriate time discretization leads to a sufficient solution of the original problem. The discrete problem (POCD) is defined over a time period

---

® FEFLOW is a trademark of WASY Ltd., Berlin, Germany
® OpTiX is a trademark of Co.Com Ltd., Berlin, Germany

M=[t_0,t_e] with start condition x(t_0)=x_0. The set U_{POCD} of feasible discrete controls is defined as:

$$U_{POCD} = \left\{ u \in R^{q \cdot l} \middle| \begin{array}{l} g(z(t), s(t), y(t), u(t)) = 0, h(z(t), s(t), y(t), u(t)) \leq 0, \\ x = \varphi(z(t), y(t), u(t), t), x(t_0) = x_0, t \in M \end{array} \right\}.$$

The optimal control problem is therefore:

$$\left\{ \min \int_{t_0}^{t_e} \phi(z(t), s(t), y(t), u, t) dt \,\middle|\, u \in U_{POCD} \right\}.$$

The time-discretization of the decision variables transforms problem POC to the solution of a problem of the class of static nonlinear optimization problems.

Problems of this type in the water industry are e.g. the optimal management of hydroelectric generation and water supply [NM95], pump-and-treat optimization problems [HM97] or optimal operation of a river/reservoir system [KFW97]. The solution of a problem belonging to this class is discussed in section 5.2.

2.3 The class of feedback optimization

In contrast to the optimal control problem POC, the decision variables of a problem of feedback optimization (POF) do not depend directly on time. The time dependency is introduced via a feedback of time-dependent output variables y(t). Thus, the decision variables can be defined as u(y) and the set of feasible solutions is

$$U_{POF} = \left\{ u(y(t) \in R^q \middle| \begin{array}{l} g(z(t), s(t), y(t), u(t)) = 0, h(z(t), s(t), y(t), u(t)) \leq 0, \\ \dot{x} = \varphi(z(t), y(t), u(t), t), x(t_0) = x_0, t \in M \end{array} \right\}.$$

The optimization problem in this case is

$$\left\{ \min \int_{t_0}^{t_e} \phi(z(t), s(t), y(t), u, t) dt \,\middle|\, u \in U_{POF} \right\}.$$

With the decision variables u being not directly depending on time, this problem class can be solved with the same algorithms as problems POD and POCD.

The solution of a problem of this type is analysed in section 5.3.

3 SOFTWARE ARCHITECTURE FOR SIMULATION-BASED OPTIMIZATION

From the analysis of the presented problem classes and their solution process arise different requirements for an integrated software environment supporting the solution of simulation-based optimization problems in groundwater management. Partly, they arise from the computationally expensive numerical solution process, partly from the necessity to integrate simulation code with optimization code.

The following software engineering subjects were identified:

- **Wrapping of legacy code**

Mostly, numerical simulation software packages are written in programming languages like FORTRAN or C. Since modern software design follows the object-oriented paradigm, it is useful to integrate these software systems by wrapping them, i.e. to design and implement classes which transform a call to an interface method of the class to a call to the legacy software. If there is no application-level interface available, this "call" may be identical to the start of the program with appropriate parameters e.g. in the form of an input file for the program.

- **Interface management**

The optimization software must provide an interface to the simulation software which allows control of the simulation and retrieval data from the simulation model. For instance, the optimization software must be able to set parameters of the simulation model according to certain values of decision variables of the optimization problem. Similarly, after a simulation run is completed, the optimization must obtain values for constraints from the simulation model. At present, most simulation systems, e.g. for structural mechanics, aerodynamics or aeroelastics ([MSC89], [SKH96]), provide only a file-based interface. Access to data of the simulation model via a programming interface is usually not possible.

- **Synchronization**

If the optimization algorithm has requested the evaluation of a solution vector from the simulation system, the algorithm has to wait for the completion of the simulation. Vice versa, the simulation software has to wait for the next request after finishing the current one. Thus, between these two components at least two processes have to be synchronized, and the data exchange has to be co-ordinated.

- **Distributed computation**

For the distributed computation of the optimization and simulation tasks, networked high-performance workstations are a preferable platform ([Bode96], [GB97]). The coupled optimization/simulation system should be shielded from platform-specific hard- and software matters, such as inter-process/object communication, by an intermediate software layer (middleware).

A system architecture for coupling optimization and simulation software systems according to the above requirements is shown in Fig. 2.

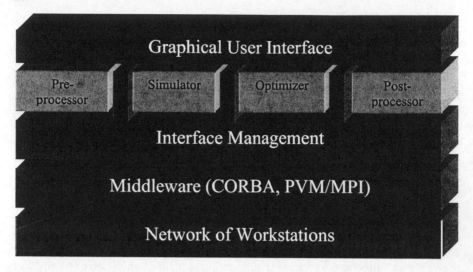

Figure 2: Software architecture for distributed coupled
simulation/optimization.

The two lower layers provide (platform-independent) functionality to start the individual components of the system and to provide the communication between them. Implementing the middleware can be realized using an object-oriented approach like the Common Request Broker Architecture (CORBA, [OMG98]) but alternatively, an implementation of the Message Passing Interface standard (MPI) [MPI94], or the Parallel Virtual Machine (PVM)[PVM94], can be used. Besides this functionality, the two lower layers also offer functionality for load distribution to improve performance of distributed applications on a network of workstations [AFKT98]. The layer above implements the interface management functions to provide the basis for application-level communication, i.e. exchange of values for decision variables and constraints between the components. This layer encapsulates the application-specific interface, whether it is file-based or a programming interface, and makes a common interface for data exchange available, e.g. by creation and/or transformation of files. Furthermore, synchronization between the components will be handled in this layer. Components like the pre- and postprocessor have their own (graphical) user interfaces. The topmost layer has to provide a user interface for the convenient formulation of the optimization problem (e.g. by allowing nodes of a finite element model, representing constraints or decision variables, to be selected graphically, or by enabling the user-friendly specification of the objective function), and probably additional visualization techniques for the results of the optimization. This layer should present the components of a coupled optimization and simulation system consistently, and initiate and control the data flow between distributed components: from model generation in the pre-processing stage and simulation/optimization to the visualization of the optimization results in the post-processing stage.

4 PROTOTYPICAL IMPLEMENTATION

The following sections are intended to give a brief overview of the components used for the prototype implementation. The simulator (see Fig. 2) is in this case FEFLOW comprising pre- and post-processing functionality. The optimizer is included in the software package OpTiX. Interface management for communication and synchronization is a built-in feature of FEFLOW. The distribution is implemented based on PVM utilizing the resource management (load distribution) capabilities of WINNER [AFKT98].

4.1 The groundwater simulation system FEFLOW

FEFLOW (Finite Element subsurface FLOW system) is an interactive, graphics-based groundwater modeling system for three-dimensional (3D) groundwater flow and mass transport. It is a computer aided engineering (CAE)-oriented software package for modern UNIX engineering graphics workstations and Windows 95/NT PC's. Application examples are

- to describe the spatial and temporal distribution of groundwater heads and/or contaminants,
- analyze moisture dynamics and seepage,
- estimate the duration and travel times of a pollutant in aquifers,
- plan and design remediation strategies and interception techniques,
- assist in designing alternatives and effective monitoring schemes.

FEFLOW is an interactive, fully graphics-oriented and menu-driven, hierarchically structured software system that contains graphical editors and mesh generators for the geometric design and discretization of possibly complex study areas as well as for the problem attribute specifications. It relies on general computational techniques to solve a wide class of subsurface flow and mass transport problems characterized by flexibility and robustness, and many additional graphical tools to manage the entire solution process and model data.

Databases in line-, point-, polygon- and annotation-related formats, such as the ESRI Shape file format and the ASCII (Generate) format widely used in the **Geographical Information Systems** (GIS) ARC/INFO and ArcView as well as Raster Databases (TIFF) and Hewlett-Packard Graphics Language (HPGL)-formatted data, can be imported and applied for purposes of data regionalization and visualization.

FEFLOW employs a fully three- and a two-dimensional finite element method (FEM) to solve the governing partial differential equations that describe amongst others the following interdependent processes:

- groundwater flow dynamics which can be fluid density-dependent and can involve free surfaces (perched water table),
- variably saturated flow and transport (3D and 2D),
- convective and dispersive contaminant transport, in which the chemical species may be subjected to: adsorption, hydrodynamic dispersion and first-order chemical reaction.

The prescription of the initial and boundary conditions can be relatively general so that formulations on arbitrary geometries with different types are possible. Accordingly, it also allows the handling of mixed conditions (e.g., surface water interactions or pumping and injecting well functions) as well as specific mass flux-occupied boundaries (e.g. leaching of substances, diffuse intake of chemicals from landfills, etc.) based on an alternative formulation of the basic equations, so-called divergence form transport formulations.

The simulator FEFLOW is completely written in ANSI C and C++ so that dynamic memory allocation resulting from the interactive capabilities of the system can easily be handled. Thus, no real software limits exist. The problem measures are only restricted by the computer storage capacity available for the user. Dynamic memory monitoring allows efficient use of the editors, the mesh generators, the simulator kernel routines, and postprocessing modules.

FEFLOW runs under Windows NT/95 using an X-Server and under UNIX and the X Windows and OSF/Motif graphics standard, so that it is available for a wide range of today's engineering graphics workstations. The present FEFLOW version is based on the recent releases of the X Window system (version 11.4 or later) and the OSF/Motif (version 1.2 or later).

The **Interface Manager** controls the configuration of FEFLOW's **Programming Interface**. It allows the linking of FEFLOW with third party software or self-created codes. Data can be exchanged for preprocessing, during the simulation run and for postprocessing. The interface allows creation of additional submenus and menu entries in the FEFLOW GUI, too. The IFM provides support in all phases of building and maintaining modules. An assistant guides the initial creation process, the project management rescues from editing *Makefiles* and others. A callback editor generates the source code for each event handled. Further tools and editors complete the module development environment.

For the purpose of coupling optimization and simulation, this interface manager is very important. Since nodes of this finite-element model represent locations of decision variables and constraints, the optimization system must have access to these nodes, arcs or elements to set and get values. Besides the traditional interface based on files, the Interface Manager (IFM) of FEFLOW enables the loading of libraries (shared objects, dynamic link libraries) at runtime. The IFM provides a bi-directional interface: callbacks transfer control to external code at certain points in the FEFLOW-internal cycle (e.g. before or after every time step in the simulation). This allows control of the simulation, for instance to wait for a request using synchronous communication. Likewise, the interface provides access to the finite element model with functions e.g. to retrieve the hydraulic head in a certain node of the model.

4.2   The distributed optimization environment OpTiX

OpTiX is an optimization environment supporting the entire process from problem formulation until distributed solution of a problem on a network of workstations. The optimization problem can be formulated in a formal mathematical language, specifying decision variables and their upper and lower bounds, an objective function and constraints.

The parallel solution of optimization problems is supported by the design of so called "visual optimization schemes", i.e. manager/worker schemes for the parallel solution of a decomposed problem formulation or a hybrid approach. The object-oriented design of OpTiX offers the following benefits:

- flexibility concerning the problem formulation,
- integration "wrapping" of non-object-oriented implementations of optimization algorithms,
- transparent management of the distributed computations across a network of workstations,
- a graphical user interface for the formulation and the control of the solution of an optimization problem.

The first two aspects imply the existence of an abstract layer between the optimization algorithms and the problem formulation. This interface reduces algorithms to their essential attributes (stopping criteria, maximal iteration number etc.) and optimization problems to the objective function, constraints etc. These abstractions enable the combination of any problem with any algorithm (if appropriate) for its solution. Instances of algorithms can be implemented in C, C++ or FORTRAN, problem formulations can be analytical or simulation-based. The distributed computation of an optimization problem must be transparent to the user. Therefore, the user must have the opportunity to graphically design an optimization strategy (parallel or sequential) which is then employed on the network of workstations without further interaction with the user.

4.3   The coupled simulation/optimization using FEFLOW and OpTiX

For the implementation of a prototype according to the presented architecture, the simulation system FEFLOW was integrated with the optimization software OpTiX. For this purpose, OpTiX was used as library without graphical user interface functionality. The static structure of the prototype is shown in Fig. 3 as a component diagram in the notation of the Unified Modelling Language (UML, [BRJ99]).

Simulation and optimization are basically two different software packages communicating with each other via some kind of inter-process communication facility. In the case of this prototype all communication is based on message passing (package "PVM" in Fig. 3). To improve performance of distributed applications on networks of workstations, resource management is provided by WINNER. WINNER realizes load distribution by selecting the currently most appropriate host (i.e. the host with the least overall load) for starting a new process. Therefore, WINNER replaces the PVM built-in process placement strategy and a component which starts a process on a remote host automatically utilizes the capabilities of WINNER.

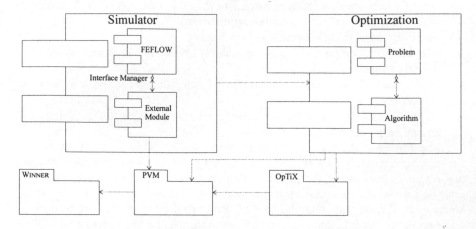

Figure 3 Unified Modeling Language (UML) component diagram of the developed prototype implementation

5   DISCUSSION OF RESULTS

5.1   An example problem for facility optimization

Optimization problems in (ground-)water engineering typically deal with the effects of human interference with nature, e.g. when building facilities for water supply or hydroelectric generation without harming the environment by an excessive rise or fall of the groundwater level. As an example for this class of optimization problems we discuss a minimization problem with limits on the acceptable rise of groundwater level. The problem can be stated as follows. The process of building a sluice in a channel in Berlin causes an increasing infiltration of surface water and therefore a rise of the groundwater level. To protect the tree population in a nearby park, this rise of the groundwater level has been restricted to 0.1 meters in each of five observation points. The objective function of the minimization problem is the sum of the quantity of water – as a measure of the operational costs – to be extracted by four wells in order to lower the groundwater level. The optimization problem has four decision variables (the individual quantities of extracted water per day of four wells), five (implicit) constraints (upper bounds of groundwater level in five observation points) and four (explicit) constraints (technical restrictions of the pumps).

All the presented problems were solved using an implementation of the proposed software architecture with the simulation system FEFLOW and an improved variant of the Complex Box optimization algorithm [BGG91]. In Table 1, the optimal solution of different optimization runs is compared with the reference solution. The reduction of the objective function value is about 23%.

Table 1 Comparison of the reference solution and the optimal solution to the example control problem.

	Reference solution [m^3]	Optimal solution [m^3]
Well #1	400	193.5
Well #2	400	480.0
Well #3	400	191.9
Well #4	400	368.5
Total	1600	123.9
	100%	77%

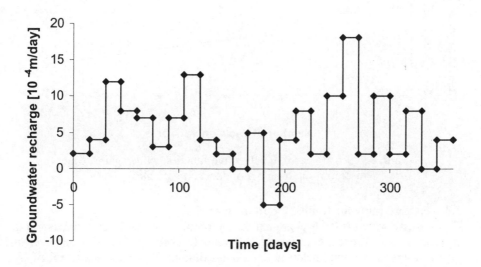

Figure 4 Groundwater recharge rate as input variable for the optimal control problem

5.2 An example problem for optimal groundwater management

In this section optimal groundwater management as an example for optimal control problems in the water industry is described. The problem is to find the optimal control strategy over one year for one well with a maximal groundwater level in one observation point about 500 m away from the well location. In the observation point, the groundwater must be below a level of −3m. As a boundary condition a groundwater recharge rate is given (see Fig. 4). The reference solution used five intervals; in three intervals no water was extracted. In the remaining two intervals a constant quantity of water was extracted yielding a total quantity of 2.45 mil. m^3/year.

To estimate the sensitivity of the objective function value against different time discretizations, the problem was solved using two, six and twelve decision variables at fixed points in time. In Fig. 5 the objective function values yielded by the different discretizations are given. The improvement of the solution using twelve decision variables compared to the reference solution is about 35%.

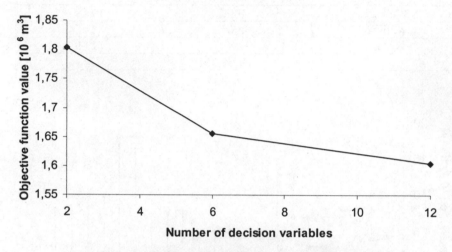

Figure 5 Sensitivity of the objective function value against time discretization (number of decision variables)

5.3 An example for feedback optimization

The scenario which is the background for the problem of feedback optimization is as follows: a well extracts a quantity of water to keep the groundwater level in given observation points below a certain maximum. This well is controlled by another observation well in such a way that this observation well switches between different water levels (quantities) depending on the measured groundwater level. The groundwater recharge rate is again given as a boundary condition.

The optimization problem has five decision variables (levels where the well switches between quantities) for the five different quantities: 0, 1036.8, 2073.6, 3110.4 and 4147.2 m³/day. In Fig. 6 the trajectory of the groundwater level in the observation point (using the optimal strategy) is shown. The pumping rates from the well using the optimized levels for switching are shown in Fig. 7. The optimized levels improved the reference solution by almost 13%.

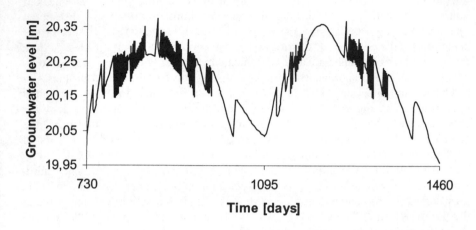

Figure 6 Groundwater level in the observation point of the optimal control

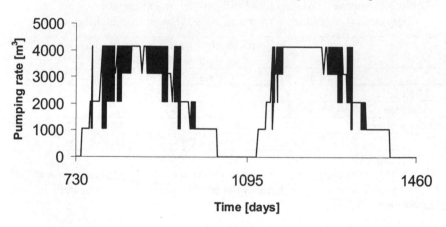

Figure 7 Pumping rates from the extraction well for optimal control

6 CONCLUSIONS AND FUTURE WORK

In the paper typical problem classes in groundwater management (facility optimization, optimal groundwater management, and feedback optimization) and their numerical characteristics and solution methods were analyzed. From this analysis arise several requirements for a software environment: enabling reuse of legacy code, interface management and synchronization between simulation and optimization, and the implementation of distributed solution concepts. Based on the prototypical implementation several groundwater management problems were solved and proved the feasibility and performance of the proposed concepts.

36

As a first step the results presented in the previous sections were computed on a single host for optimization and simulation using sequential algorithms. Focus of future work will be put on the design and implementation of distributed optimization algorithms for the presented problem classes. Earlier work demonstrates the potential of distributed optimization by reducing the elapsed time for the solution of mathematical test problems by one to two orders of magnitude [BG95]. Distributed approaches were also used to solve optimization problems in mechanical engineering ([GB93], [EG99]). As a first step towards distributed optimization the algorithm for nonlinear constrained optimization used in the presented examples was enhanced by a parallel initialization phase. In Table 2 the comparison of the computation times of the sequential and the parallel implementation are shown, demonstrating the potential of distributed computing. The parallel implementation reduces the computation time to one third. Additionally, the benefit of adaptive load distribution is also demonstrated in Table 2 by comparing the computation times of a parallel initialization with and without the use of WINNER. The adaptive load distribution strategies of WINNER improve the performance of the parallel code by 45%

Table 2 Comparison of sequential and parallel implementation phases of the Complex Box algorithm with and without WINNER load distribution

	Computation time of initialization phase for eight solutions [s]		
	Sequential	Parallel (PVM)	Parallel (WINNER)
Maximal	714	441	172
Minimal	696	118	86
Average time	703	232	124
	100%	33%	18%

Therefore, inherently parallel algorithms for the presented classes of optimization and control problems will be designed and integrated in the implementation.

ACKNOWLEDGEMENTS
This work is partially funded by the German Federal Ministry of Science and Technology (FUEGO 0033701E8).

REFERENCES
[AFKT98] Arndt, O., Freisleben, B., Kielmann, T., Thilo, F., Dynamic load distribution with the Winner system, in: Proc. Workshop ALV'98, Munich 1998.
[BGG91] Boden, H., Gehne, R., Grauer, M., Parallel nonlinear optimization on a multiprocessor system with distributed memory, in: Grauer, M., Pressmar, D., (eds.) Parallel Computing and mathematical optimization, Springer 1991.
[BG91] Brüggemann, F., Grauer, M., VOpTiX – an objet-oriented environment for parallel optimization, in: Grauer, M., Pressmar, D. (eds.), Parallel computing and mathematical optimization, Springer 1991.

[BG95] Boden, H., Grauer, M., OpTiX-II: a software environment for the parallel solution of nonlinear optimization problems, Annals of OR, J.C. Baltzer Science Publisher, 1995.

[Bode96] Boden, H., Multidisciplinary optimization and cluster computing, (in german), Springer/Physica, Heidelberg 1996.

[BRJ99] Booch, G., Rumbaugh, J., Jacobson, I., The Unified Modeling Language User Guide, Addison-Wesley, 1999.

[Brue97] Brüggemann, F., Object-oriented and distributed solution of optimization problems, (in german), Springer, 1997.

[CER95] Cieniawski, S., Eheart, J., Ranjithan, S. Using genetic algorithms to solve a multiobjective groundwater monitoring problem. Water Resources Research, Vol. 31, No.2, 1995.

[Dier96] Diersch, H.-J., Finite element analysis of three-dimensional transient free convection processes in porous media, In: Computation of Three-Dimensional Complex Flows,)Proc. IMACS-COST Conf. Lausanne, Sept. 13-15, Switzerland), ed. by M. Deville et al., Notes on Numerical Fluid Mechanics Vol. 53, Vieweg & Sohn, Braunschweig/Wiesbaden, 1996.

[DSM96] Dandy, G., Simpson, A., Murphy, L., An improved genetic algorithm for pipe network optimization, Water Resources Research, Vol.32, No.2, 1996.

[EG99] Eschenauer, H., Grauer, M., Decomposition and parallelization strategies for solving large-scale MDO problems, Design Optimization – Int. Journal for Product & Process Improvement, Vol. 1, No. 1, MCB University Press 1999.

[GB93] Grauer, M., Boden, H., On the solution of nonlinear engineering optimization problems on parallel computers, in: Proc. of ASME Design Technical Conference, Albuquerque 1993.

[GB97] Grauer, M., Barth, T., Multidisciplinary optimization and cluster computing using the OpTiX-workbench, Proc. Conf. On Opt. In Industry, Palm Coast, Florida 1997.

[HM97] Huang, C., Mayer, A. S., Pump-and-treat optimization using well locations and pumping rates as decision variables, Water Resources Research, Vol. 33, No. 5, 1997.

[JZNM97] Jonoski, A., Zhou, Y., Nonner, J., Meijer, S., Model-aided design and optimization of artificial recharge-pumping systems, Hydrogeologistical Science Journal, 42(6), 1997.

[KFW97] King, J. P., Fahmy, H. S., Wentzel, M. W., A genetic algorithm approach for river management, Dasgupta, D., Michalewicz, Z. (Eds.), Evolutionary algorithms in engineering applications, Springer, 1997.

[Lasd72] Lasdon, L., Optimization theory for large systems, Macmillan, London 1972.

[ML95] McKinney, D., Lin, M., Approximate mixed-integer nonlinear programming methods for optimal aquifer remediation design, Water Resources Research, Vol.31, No.3, 1995.

[MPI94] Gropp, W., Lusk, E., Skjellum, A., Using MPI: Portable Parallel Programming with the Message--Passing Interface, MIT Press, 1994.

[MSC89] Cifuentes, A., Using MSC/NASTRAN: Statics and Dynamics, Springer 1989.

[NM95] Nardini, A., Montoya, D., Remarks on a min-max optimization technique for the management of a single multiannual reservoir aimed at hydroelectric generation and water supply, Water Resources Research, Vol.31, No.1, 1995.

38

[OMG98] The Common Object Request Broker: Architecture and Specification - Revision 2.2, Object Management Group, (ftp://ftp.omg.org/pub/docs/formal/98-07-01.ps), 1998.

[PVM94] Geist, A., PVM: Parallel Virtual Machine – A Users Guide and Tutorial for Network Parallel Computing, MIT Press, 1994.

[SKH96] Schweiger, J., Krammer, J., Hörnlein, H., Development and application of the integrated structural design tool LAGRANGE, AIAA 96-4169, 1996.

[TKP98] Tucciarelli, T., Karatzas, G., Pinder, G., A primal method for the solution of the groundwater quality management problem, Operations Research, Vol. 46, No. 4, 1998.

[Veli97] Veliov, V., On the time-discretization of control systems, SIAM Journal on Contr. And Opt., Vol.35, No.5, 1997.

[ZW99] Zheng, C., Wang, P., An integrated global and local optimization approach for remediation system design, Water Resources Research, Vol.35, No.1, 1999.

Influence of Inflow Prediction on Performance of Water Reservoir System

J.J. Napiórkowski, A. Kozlowski *and* T. Terlikowski

ABSTRACT
The two-layer hierarchical technique with three different prediction methods was applied to a part of the Wupper Reservoir System. The reservoir system consists of two reservoirs in series with additional inflow to the lower reservoir. The tasks of these reservoirs are flood control, recreation, hydropower and low flow augmentation with the aim of water quality improvement. It is shown that the introduced optimisation concept improves considerably the system performance in comparison with the Standard Operation Rule.

1   INTRODUCTORY COMMENTS

A method for determining the yield of a multireservoir water supply system has been applied to a part of the Wupper Reservoir System in Germany. The major objectives of this particular system are flood control, recreation, hydropower and low flow augmentation. The proposed technique may be reduced to the following associated parts: the optimisation of a simplified quantitative model of the actual system and the multiobjective verification and/or comparison through simulation. The first part consists of constructing a relatively wide class of control schemes based on a two-level optimisation technique method. We focus our attention on the implementation of a number of prediction techniques of the system inflow (ARIMA, Deterministic Chaos, Artificial Neural Networks) that result in different operation rules. The second part is based on simulation performed for historical data over a long time horizon (39 years). This simulation consists of testing the control rules for chosen scalar objectives. The diagrams of frequency (reliability) criteria, calculated on the basis of simulation for a number of scalar criteria are analysed to obtain the final comparison results.

Several control schemes corresponding to the prediction models considered have been proposed in the form of computer programs. The simulations have been performed for a large number of years and for many objectives. To present advantages of the control schemes corresponding to the prediction systems, they are compared with so-called Standard Decision Rule (SDR) and Stochastic Dynamic Programming (SDP).

39

2   DESCRIPTIONS OF THE CASE SYSTEM MODEL

The catchment of river Wupper is located in the southern part of North Rhine Westfalia. The hydrological features of this catchment are characterised by a massive rocky underground covered only by a small layer of soil and an average yearly precipitation of about 1300 mm per year. The absence of underground water storage leads to dangerous floods as well as to extreme droughts. To accommodate this problem several reservoirs were built. Here we are just interested in the management of the two reservoirs governing the discharges in the city of Wuppertal, which lies about 20 km downstream of reservoir No. 2. Figure 1 shows the simplified Wupper Reservoir System.

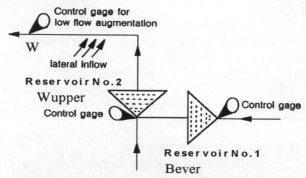

Fig. 1. Basic structure of reservoir system.

It contains two reservoirs located in series, the control centre at reservoir No. 2 and several runoff and rainfall gages. The release of the reservoirs depends mainly on the runoff at the control gage in Wuppertal. A runoff of 5 m³/s at this gage is sufficient for the required water quality, runoff less than 3.75 m³/s should be avoided and runoff less than 1 m³/s has to be regarded as ecologically disastrous. The basic hydrological and reservoir characteristics are given in Table 1.

The purpose of the model is to describe relationships between flow rates in the rivers over a long time horizon (one year) with a discretization period of 10 days. Therefore, only the dynamics of the storage reservoir are considered, while effects of flow dynamics in the river channels are neglected.

For brevity, the following notation is used: j- number of 10-day intervals, V^j - state of the reservoir, d^j - natural inflow, u^j - flow in a given cross-section, z^j - water demand, m^j - outflow from the reservoir, $1,2$ - denote the Bever and Wupper reservoirs, 3 - denotes the lateral inflow, W- cross-section at Wuppertal.

Table 1. The basic characteristics of the Wupper Reservoir System.

Reservoir	Bever	Wupper
total storage V_{max} (mln m^3)	23.70	25.90
dead storage V_{min} (mln m^3)	0.70	2.10
max. outflow (m^3/s)	17.00	180.00
min. outflow (m^3/s)	0.10	1.00
annual average flow (m^3/s)	0.94	3.51
catchment area km^2	25.7	212.00

According to the introduced notation, the state equations for the reservoir system and flow balance equation for the selected cross-section W are:

$$V^{j+1} = V^j - B * m^j + C * d^j \tag{1}$$

$$V = [V_1, V_2]; \; m = [m_1, m_2]; \; d = [d_1, d_2] \tag{2}$$

$$B = \begin{bmatrix} 1 & 0 \\ 0 & 1 \end{bmatrix}; \; C = \begin{bmatrix} 1 & 0 \\ -1 & 1 \end{bmatrix} \tag{3}$$

$$u_W^j = m_2^j + d_3^j \tag{4}$$

3 THE OPTIMISATION PROBLEM

The objective function of the optimisation problem under consideration for any time instant k (for any 10-day period) and for annual time horizon T can be written in the form of a penalty function:

$$Q(m,V) = \sum_{j=k}^{k+T} [\, a_1^{+j} (m_1^j - z_1^j)^2 + a_2^{+j} (m_2^j + d_3^j - z_W^j)^2 \tag{5}$$

$$+ b_1^j (V_1^j - V_1^{*j})^2 + b_2^j (V_2^j - V_2^{*j})^2 \,]$$

In equation (5), symbols a and b with respective subscripts denote weighting coefficients. The performance index $\mathbf{Q}$ is expressed explicitly on controls m^j and the state trajectory V^j (reservoir contents) as follows:

$$\mathbf{Q}(m,V) = \sum_{j=k}^{k+T} Q(m^j, V^j) \qquad (6)$$

The objective function during each 10-day period is subject to the constraints on the state of the system, controls and flows in given profiles:

$$V_{min}^j \leq V^j \leq V_{max}^j$$
$$m_{min}^j \leq m^j \leq m_{max}^j \qquad (7)$$

3.1 Required retention trajectory V*ʲ

It is assumed that the operation of the reservoir system is carried out on an annual basis in the following way:

- by late December, the reservoirs are normally returned to low level to prepare the system for the next flood season, completing the annual cycle.
- the storage reservation for flood control on January 1 was determined for controlling the maximum probable flood. During the normal filling period, January-April, the reservoirs should be filled up completely.
- during the May-August period the first reservoir should be filled up to meet recreational requirements.
- during the May-November period the water stored in and released from the reservoirs is used for low-flow augmentation and hydropower.

3.2 Weighting coefficients aʲ⁺ and bʲ

According to the general objective of the control problem, which is aimed at the rational protection against water deficits and at reaching the desired state at the end of April, the following values of weighting coefficients in the optimisation problem are used: $a^{j+}=1$ if demands are greater than supply and $a^{j+}=0.01$ otherwise, for k=[1,36]. As far as the second coefficient is concerned, in order to avoid a good performance in one year followed by a very poor performance in the next year, $b^j =0.01$ for j=[1,12] (May-August), $b^j =0.001$ for j=[1,30] (September-February), $b^j=0.004$ for j=[31,33] (March) and $b^j=0.01$ in April, for j=[34,36].

4 TWO-LEVEL OPTIMISATION TECHNIQUE

To solve the aforementioned problem we adjoin the equality constraints (1) with the Lagrange multiplier sequence λ (prices). The Lagrangian function has the form:

$$L(m,V,\lambda) = \sum_{j=k}^{k+T} [Q(m^j,V^j) + \lambda^j(V^{j+1} - V^j + B*m^j - C*d^j)] \quad (8)$$

To include the state-variable and outflow constraints the above problem is solved by means of the two-level optimisation method in a decentralised (co-ordinated) fashion. At this stage we make use of the additivity of the Lagrangian function (9) and the possible separation of the decision variables.

The Lagrangian function has a saddle point which can be assigned by minimising $L(\lambda, V, m)$ with respect to V and m, and then maximising with respect to λ. Finally, the optimisation problem can be expressed in the form:

$$\max_{\lambda} [\min_{V,m} L(\lambda,V,m)] \quad (9)$$

with inequality constraints on state and control and no constraints on Lagrange multipliers. Figure 2 illustrates how the two-layer optimal control method works.

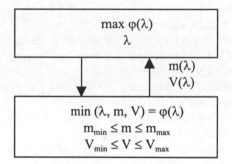

Fig. 2 Two-level optimisation method.

At the *lower level* for given values of the Lagrange multipliers we look for the minimum of the Lagrange function. The required condition is the zero value of the gradient with respect to m and V. The task of the upper level is to adjust the prices, λ, in such a way that the direct control of the reservoir, affected by λ, results in the desired balance of the system (the mass balance equation (1) is fulfilled satisfactorily). In the upper layer, in the maximisation of the Lagrange function with respect to λ, the standard conjugate gradient technique is used.

44

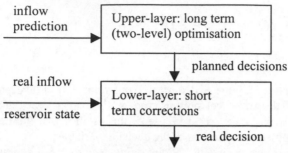

Fig.3 Two-layer control method.

In the applied Two-Layer optimisation control method (TLM) illustrated in Fig.3 the solution of the two-level optimisation problem (9) is the essential "upper layer part".

5   INFLOW PREDICTION MODELS

Three inflow prediction techniques that were used for inflows predictions in the two-layer control method are briefly presented below.

5.1   Box-Jenkins ARMA model

A classic multiplicative decomposition was applied to deseasonalise the observed data and then the ARMA (Auto Regressive Moving Average) model was used in the prediction of inflows to the reservoir system. In the practical calculations the set of relevant procedures from Microsoft IMSL Library of Professional Edition of Microsoft Fortran Power Station v. 4.0. was adopted.

These procedures enable the computation of estimates of auto-regressive and moving average parameters of the ARMA(p,q) model and then the calculation of values of inflow estimates for specified number of points to be included in the forecast of a fitted model.

Calculations showed that the most effective model was ARMA(2,1). It enables better forecasts to be made than using a model in the form of average historical values for 20-days time horizon.

5.2   Artificial Neural Network model

Inflow predictions based on the neural network simulation were generated with the help of the NeuroSolutions software package (NeuroDimension, Inc., 1997).

NeuroSolutions adheres to the so-called local additive model. A processing element (PE) simply multiplies an input by a set of weights and nonlinearly transforms the result into an output value. The principles of computation at the PE level are deceptively simple. The power of neural computation comes from the massive interconnection among the PEs which share the load of the overall processing task and from the adaptive nature of the parameters (weights) that interconnect the PEs. Under this model, each component can activate and learn using only its own weights and activations and those of its neighbours. The neural network architecture used is the multilayer perceptron (MLP) (Lippman, 1987).

The performance of an MLP is measured in terms of a desired signal and an error criterion. The output of the network is compared with a desired response to produce an error. NeuroSolutions uses an algorithm called back-propagation (Rumelhart et al., 1986). The network is trained by repeating this process many times. The goal of training is to reach an optimal solution based on the performance measurement.

The simulation results obtained justified 3 points as the maximum that can be included in the forecast with the following parameters of applied MLP:

- Hidden Layers = 1
- PEs = 8
- Transfer function = TanhAxon (hyperbolic tangent -1/+1)
- Learning Rule = Momentum (gradient and weight change, momentum = 0.7)
- Transfer function specified for output layer = LinearTanhAxon (piecewise linear -1/+1).

It should be noted that the ANN model gives the best predictions of inflows to the system.

5.3   Model Based On Deterministic Chaos Concept

The real process in its whole complexity is defined by a given generalized-state evolution mapping: $t \mapsto X_t = (\underline{X}_t, \ldots)$, where X_t is a generalised state value, often of unknown character and dimension. If the dimension of X_t is infinite, we deal with a case of *chaos*. However, according to the concept of *deterministic* chaos, a distinguished "sub-vector" $\underline{X}_t$ of X_t may satisfy a functional relationship between precedent and the next state value. $\underline{X}_t$ is assumed to evolve on the so-called *attractor*. The phenomenon is of deterministic nature: there exists a functional relationship: $\underline{X}_{t+T} = F(\underline{X}_t)$ strongly non-linear and unstable. $\underline{X}_t$ evolves on an *attractor* $M \subseteq X$, in *the* space $X = R^n$ (Takens, 1981), where M is a smooth, compact manifold of specific properties, see also (Soukhodolow et al., 1996).

The dimension of attractor M, e.g. the so-called Hausdorff dimension, is relatively small in practical applications. It is usually not a topological dimension (it is less than n) and may be not an integer (*strange attractor*).

The fundamental task is to determine the proper prediction model, that is, to construct a good approximation of function F. Practically, we are faced with measurement and approximation errors that grow exponentially in time, due to process unstability. Hence, it is possible to have a prediction of a given accuracy only for a short time horizon. The second difficulty results from the limited observability of the process: one disposes of a sequence $\{x_i\}$ of measured scalar values only (time series) and not of a whole vector $\underline{X}_i \in R^n$. Therefore, while constructing a prediction model from a given inflow time series, we apply the so-called *embedding approach*. The idea of this concept (Takens, 1981) consists of determining the relationship between the value of state X_{j+T} at time instant j+T and a finite sequence y^j_m of its m past values.

$$x_{i+T} = f_T(y^i_m) = f_T(x_i, x_{i-1}, \ldots, x_{i-(m-1)}) \tag{10}$$

According to the embedding theorem of Takens, it is possible to adjust the value m, called *embedding dimension*, such that that the resulting evolution (10) of x_i reconstructs the topological properties of the original attractor M. The value m is closely related to the attractor Hausdorff dimension. To find it, we use the so-called correlation integral concept, introduced and developed in (Grassberger and Procaccia, 1983; Packard et al., 1980). In Takens, the embedding dimension m can be defined as: $m=2m^*(\upsilon) + 1$, where $m^*(\upsilon)$ is the minimum integer number greater than attractor dimension υ.

When constructing an approximation of function f_T in relation (10), almost all authors (e.g Porporato and Ridolfi, 1997; Casdagli, 1989), propose an approximation of f_T by polynomials of a given order (*local model* concept): for a given "embedding point" y^i_m the set $Y^i = \{y^j_m : j \in K_i\}$ is determined, as the set of K nearest (in the norm $\| \bullet \|$) *neighbours* to y^i_m. Then, function f_T is adjusted, to obtain: $\min \Sigma \mid x_{j+T} - f_T(y^j_m) \mid^2$, over $j \in K_i$, where x_{j+T} is the first component of vector y^{j+T}_m.

Function f_T is searched as a linear function (2nd order approximation), or as a polynomial of 2nd degree (3rd order approximation). This then gives respectively:

$$f_T(x_i, x_{i-1}, ..., x_{i-(m-1)}) = < a, y^i_m > +b \qquad (11)$$

$$f_T(x_i, x_{i-1}, ..., x_{i-(m-1)}) = (y^i_m)^T C(y^i_m) + < a, y^i_m > +b \qquad (12)$$

where 'a' is a vector, 'b' is a constant, C is $m \times m$ matrix, $< >$ is inner product.

Numerous computations for inflows in the Wupper Reservoir System were performed, in order to verify the hypothesis of deterministic chaos and to find the embedding dimension m. It has been shown that the data represent the chaotic dynamics of dimension $m = 7$. Two prediction models (11) and (12) were then built. The best quality of forecast (the minimum error between forecasts and the original data) was obtained with $m = 4$ and $m = 5$; thus, less than $m=7$. The quadratic approximation model (with prediction horizon $T=3$) showed better results than the linear model and the model in the form of average historical values.

6 COMPARISON OF CONTROL METHODS BY SIMULATION

The simulation of some of the chosen control methods was carried out over the long time horizon of 39 years, with the real, historical data of natural inflows to the system. The methods under investigation have been partially discussed in the previous sections. Let us mention here once again those which - after an initial stage of synthesis consisting of adjusting their parameter values - have been thoroughly compared by simulation.

1) TLM - Two-layer optimisation method with:

a) complex, long-term planning aiming at the optimisation of all the particular goals in a compromising manner.

b) realisation of the planned decisions (water supplies and discharges) in the real, current conditions.

2) SDR -The standard decision rule was developed by simulation techniques on the basis of a historical record of 39 years and ten synthetic records of 50 years (Schultz and Harboe, 1989).

3) SDP - Sequential Stochastic Dynamic Programming (Napiórkowski et al., 1997).

In the first method, requiring solution of the optimisation problem (9), the long-term prediction of inflows (for 36 10-day periods) consists of two parts. For 10-day periods $j = [1,3]$ the results of one of the discussed inflow prediction models were used and for $j = [4,36]$ the average values of historical data were applied. Furthermore, to compare and investigate the 'power' of optimising methods, the variants denoted OPT and AVR have been considered, which differs from the optimising methods only in the fact that real/average values of inflows are put in place of predicted values.

In order to compare in a clear, well-ordered manner the results of different controls and the results of the other control techniques, we introduce the following scalar criteria goals (Napiórkowski and Terlikowski, 1996):

-global deficit time TD:

$$TD = Card(\{ j : u_W^j < z_W^j \})\qquad(13)$$

- average relative deficit AvD:

$$AvD = \sum_{j=1}^{36} \frac{(z_W^j - u_W^j)_+}{z_W^j}\, \frac{1}{36}\qquad(14)$$

- maximum relative deficit MxD;

$$MxD = \max(\{\frac{(z_W^j - m_W^j)_+}{z_W^j} : j = 1,...,36\})\qquad(15)$$

- average losses in recreation area in the summer period for Bever Reservoir:

$$RE_B\,Av = \sum_{j=1}^{12} \frac{REmax - f_V(V_B^j)}{REmax}\, \frac{1}{12}\qquad(16)$$

where REmax corresponds to maximum possible water area.

As a result, we obtain a sequence of 4 numbers, characterising system performance in a synthetic way. This could be sufficient to evaluate and compare the different functions for one year, e.g. with the aid of any multiobjective optimisation method. However, it is more complicated, because we have to

48

compare the control effects not for a particular year, but for a long historical record.

To solve such a problem it is necessary to use a specific approach, which is arbitrary to some extent and makes use of intuition. To obtain the final comparison results we analyse the diagrams of s.c. frequency (reliability) criteria calculated on the basis of simulation for 4 scalar criteria (13-16).

Those frequency criteria are also functions, but defined over the set of values of respective scalar criteria. Their values represent the number of years for which the respective scalar criterion has its values in a given range. Formally, e.g. for MxD we have:

$$F_{MxD}(x) = Card(\{ I : MxD^I \le x \}) \tag{17}$$

where MxD^I denotes the value of criterion MxD for the year I. As it is seen, F corresponds to the notion of cumulative distribution function of the "random variable" MxD^I, when I is treated as representing the elementary events.

7   RESULTS AND CONCLUSIONS

Some of the simulation results for the control methods considered, namely SDR, TLM, SDP and OPT are presented below by means of the reliability criterion F. Fig.(4-7) show the diagrams of distribution F corresponding to the criteria (13-16).

Fig.4 Maximum relative deficit at W cross-section.

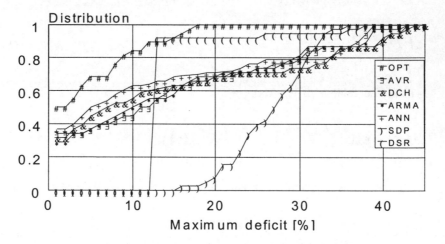

The advantage of TLM, for all considered inflow prediction models (AVR, DCH, ARMA, ANN), but especially for ANN (the best forecast) and DCH, is evident in the sense of MxD criterion (Fig.4). It stems from the fact that TLM takes into account the co-operation of the whole system and better co-ordinates the partial decisions when compared with the other methods discussed.

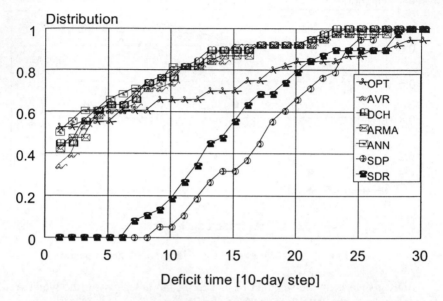

Fig.5　　　Global deficit time at the W cross-section.

For TD criterion (Fig.5) the plot of OPT is below the plots of ANN, DCH, ARMA and AVR models. It reflects the fact that the "system" prefers longer and small deficits rather than short and deep ones and, of course, the knowledge of future inflows guarantees the lowest maximum deficit.

For the criterion AvD (Fig.6) the differences between diagrams corresponding to 4 prediction models are smaller, but the method TLM is still shown to be better than SDP. Moreover, these diagrams are then closer to the "optimal" ones (those for OPT).

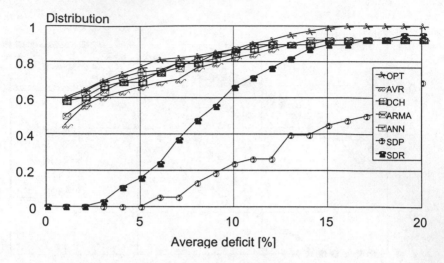

Fig.6   Average relative deficit at W cross-section.

SDR gives results between TLM and SDP; the latter giving the worst results for all but the recreational loss criterion (fig.7).

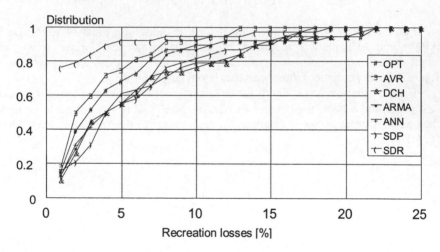

Fig.7   Losses in recreation area for Bever reservoir.

SDP gives the worst results for all criteria. This results from the character of this technique. SDP requires the discretization of both inflows and storages and due to "curse of dimensionality" that discretization cannot be too dense.

To recap, the method called TLM proved to be the best for reservoir system simulation with short time prediction obtained by means of ANN.

ACKNOWLEDGEMENTS
The authors would like to thank Mr. H. Kissler, Wuppertal Water Authority, for 39 historical scenarios.

REFERENCES
Casdagli M., 1989. Nonlinear prediction of chaotic time series, Physica D 35, 335-356.

Grassberger P. and Procaccia I., 1983. Characterization of strange attractors, Physical Review Letters, Vol. 50, No 5, 346-349.

Lippman R. 1987. An introduction to computing with neural nets. IEEE Trans., ASSP Magazine 4, 4-22,.

Napiórkowski, J.J. & T.S. Terlikowski 1996. Operational control for a multireservoir system - multiobjective approach. In P. Zannetti & C.A. Brebbia (eds.) *Development and Application of Computer Techniques to Environmental Studies VI*, Computational Mechanics Publications, Southampton, Boston.

Napiórkowski J.J., Terlikowski T.S., Wolbring F., 1997. Multiobjective Approach to the operational control synthesis - The Wupper Reservoir System Case Study. In Operational Water Management, Eds. Refsgaard and Karalis, 165-170, A.A.Balkema/Rotterdam/Brookfield.

NeuroDimension, Inc., 1997. The Neural Network Simulation Environment, Manual Version 3.

Packard N.H., Crutchfield J.P., Farmer J.D. and R.S. Shaw., 1980. Geometry from a time series, Phys. Rev. Lett., 45, 712-716.

Porporato A. and Ridolfi L., 1997. Nonlinear analysis of river flow time sequences, Water Resources Research, Vol. 33, No 6, 1353-1367.

Rumelhart D., Hinton G. and Williams R. "Learning internal representations by error propagation." In Parallel Distributed Processing, (eds. Rumelhart and McClelland), MIT Press, 1986.

Schultz, G.A. and R. Harboe 1989. Development of reservoir operation rules for Wupper Dam during floods and low flow periods. Wasserwirtschaft 79, 1-4.

Sukhodolov A., Napiórkowski J.J., Rowinski P., 1996. Applicability of Flow Prediction Based on Attractors. Acta Geophysica Polonica, XLIV, No.3, 277-286.

Takens F., 1981, Detecting strange attractors in turbulence. In: D.A. Rang and L.S. Young (eds.), "Topics in Dynamical Systems and Turbulence", 366-381, Springer-Verlag, Warwick.

Computation of Flow Distribution in a Pipe Network with the Help of NeMo+ System

V.V. Tarasevich *and* G.B. Zagorulko

ABSTRACT

The paper is devoted to the development of new approaches based on intelligent technologies, for the solution of pipe network problems in water distribution and other similar systems. The problem of calculating flow distribution in the pipe network is considered as an example. The formulation of the problem is given in a general form. The difficulties arising in the application of traditional methods, such as the rigidity of the algorithm, uncontrollability of data, complexity in accounting for constraints, etc are characterised. New opportunities are described based on knowledge-based systems, constraint programming, subdefinite computational models, etc. An approach based on integrated intelligent systems is offered, uniting the opportunities and advantages of many of the above-mentioned technologies. This approach uses the interpretation of a simulated system as a semantic network. The application of the subdefinite computation technique requires changing the very concept of "solution". The process of solving the problem is regarded as a process that reduces the degree of subdefinity and uncertainty of a problem. Thus, obtaining the solution and specification of the initial data are not divided as consecutive stages of the solution process, but represent an integrated information process. The description of the NeMo+ system and the example of calculating a specific problem of flow distribution through this system are given. The technology described is quite general to cover many problems of water distribution and other water systems.

1   INTRODUCTION

By a pipe network (pipe system) we mean a set of inter-connected pressurised pipelines. The pipe junctions and various devices (pumps, valves, regulators, filters, etc) are also parts of the system, which will be designated as system nodes hereinafter. Besides hydrodynamic processes in pipes, the modes of operation of such systems can include mechanical, thermal, electrodynamic and other processes [1–3].

Complex pipe systems (CPS) are important and widespread in practice. Pipe systems can serve as water distribution systems (WDS) or be their essential part (combined pressure -and - free-flow system). Although the concepts 'pipe system' and 'water distribution system' in many respects overlap, they are not identical,

53

since pipe systems cover a wider class of objects. The systems of water supply, thermal networks, gas and oil pipelines, hydrodrives of machines and mechanisms are examples of such systems.

On the other hand, many problems in the analysis of CPS and WDS are identical or quite similar. These are the problems of calculating flow parameters in a system, control, optimization of systems, etc. Therefore, the technique offered in this paper does not make the special distinction between CPS and WDS.

The problems of calculating parameters of steady and unsteady flows in systems of the kind considered belong to a category of important and widespread problems arising in practice. Such problems can emerge both at a design stage of the systems, and at a stage of their operation, including also control process and management.

The structure of such systems can be represented as a directed graph, where the vertices correspond to nodes of the system and arcs correspond to the pipelines. The orientation of arcs specifies a positive direction of flow. The subsystem can be represented as a node of the system. Such problems belong to a category of problems of flows in graphs and networks [1–4].

The parameters of the problem considered should satisfy constraints. These constraints are due to physical and technological opportunities, requirements of optimality and standards, etc. Moreover such constraints can be mutually opposite, for instance, requirement of minimal cost and requirement of maximum reliability. The positioning of networks in the ground should satisfy constraints on depth, requirements for non-crossing with other networks, convenience of construction, bypass of buildings and structures, provide necessary modes of operation, and be optimal from the point of view of construction and operational expenses.

Large dimensions, the complexity of a system and the presence of many constraints create significant difficulties in solving such problems by traditional methods. For example, the systems of technological pipelines of nuclear power plants contain more than 2000 pipes and more than 2000 nodes. In this case the usual methods of solving become ineffective and special methods are required which take into account the topological structure and individual features of the network considered.

The application of intelligent technologies can considerably facilitate this process and help the transition to a new and qualitatively higher level of design and control of systems of the kind considered.

2   GENERAL FORMULATION OF PROBLEM

Let Γ be a direct graph of the system, where the vertices correspond to nodes of the system, and arcs correspond to pipes. Let us designate an arc through index k (subscript) and a vertex through index j.

We shall designate through $\vec{\Lambda}$ a vector of eigen parameters of an element of the pipe system. Let vector $\vec{u}_k$ be a vector of parameters of a flow in k-th pipe. Let's designate through $\vec{u}_k^j$ a vector of parameters of a flow on the extremity of

k-th pipe adjoining node j. Then $\vec{u}^{\,j}$ will designate a vector made of all components of vectors $\vec{u}_k^{\,j}$.

The flow parameters in each pipe of system should satisfy the equation

$$H_k\left(\vec{u}_k,\vec{\Lambda}_k,t\right)=0 \tag{1}$$

describing the current state of a liquid in this pipe, where t is time.

The functioning of nodes of the system is described by equations of the kind:

$$S^j\left(\vec{u}^{\,j},\vec{\Lambda}^j,t\right)=0 \tag{2}$$

It is necessary also to specify the relationships, describing conjunction of flows in pipes with nodes:

$$g_k^j\left(\vec{u}_k^{\,j},\vec{\Lambda}^j,\vec{\Lambda}_k,t\right)=0 \tag{3}$$

Depending on particular specifics of a problem, the operators H, S and g can be of various kinds (systems of differential or algebraic equations, or their mix, etc.).

We deal with a steady-state problem if any parameter of a problem does not depend on time; the problem considered is unsteady if there is a dependence on time. In the latter case it is necessary alongside relationships (1)–(3), to specify also the parameters of an initial state of the system for unique characterisation of the system's evolution over time. If, on the contrary, the parameters of a final state or other additional conditions are specified, and the parameters of an initial state are unknown values, the inverse problem can be considered as a problem of identification, recognition, control, etc.

The problem in a general case is formulated in the following way: some values (components) of parameters $\vec{u}$ and $\vec{\Lambda}$ are known; it is necessary to determine the unknown values, i.e. to solve the problem (1)–(3) whilst satisfying all constraints. The solutions will be functions of time in the case of an unsteady problem.

2.1 The Problem of Flow Distribution

The problem of finding the parameters of steady isothermal pressurised flows in pipe networks is a special case of the above problem. This rather simple problem is nevertheless quite typical and is frequently met in practice.

Parameters describing the pipeline are length L, diameter d, pipe roughness Δ, etc., i.e. $\vec{\Lambda}_k=\left(L_k,d_k,\Delta_k,\ldots\right)$. A pressure h and discharge Q can serve as flow parameters, i.e. $\vec{u}_k=\left(h_k,Q_k\right)$. Then $\vec{u}_k^{\,j}=\left(h_{0,k},Q_k\right)$, if k-th pipe is outlet for node j, and $\vec{u}_k^{\,j}=\left(h_{L,k},Q_k\right)$, if k-th pipe is input into node j, where $h_{0,i}$ is pressure at the upstream end k-th pipe, and $h_{L,i}$ is pressure at the downstream end of k-th pipe.

The constraint equation (1) will be the well-known Darcy-Weisbach law:

$$h_{0,k}-h_{L,k}=R_k\left|Q_k\right|Q_k \tag{4}$$

where R_k is the hydraulic resistance of k-th pipe [1 – 4].

A "pipe branching" node will be described by the system of equations

$$\sum_{in} Q_k^j = \sum_{out} Q_k^j + q^j \tag{5}$$

$$h_k^j = h^j \tag{6}$$

where a pressure at node h^j and outflow (loss of discharge) from node q^j will act as parameters of the node, i.e. $\vec{\Lambda} = \left(h^j, q^j \right)$.

The relationship (5) describes the balance of the discharges, where the first term is the sum with respect to all input pipes into node j and the second term is the sum with respect to all output pipes from node j. $q^j = 0$ for pipe branching without outflow. This equality considered together with (5) is represented by (2).

The relationship (6) represents equality of pressures on the pipe extremities attached to node j and serves as a condition of conjunction (3).

Special cases of equations (5) – (6) are the boundary conditions for hanging vertices: the assignment of a pressure h^j in a node and the assignment of the discharge q^j from a node.

3   DIFFICULTIES WITH TRADITIONAL COMPUTATION METHODS

Numerous publications (for example, [1,2,4]) are devoted to solving the problem formulated above. The range of techniques used is quite wide, including iterations, sweep methods, special algorithms, relaxation methods, least-squares method, etc. However, as a rule, these methods (which we shall name "traditional" ones) do not give a completely satisfactory solutions to problems. The main defects of traditional methods are:

- Rigidity and stickiness of the algorithm, its non-malleability to specific features of a problem (impossibility of data driven process);
- Fatal character of errors; impossibility to correct the problem solving during execution;
- Sensitivity to incompleteness of the initial data;
- Sensitivity to inexact and contradictory initial data; impossibility to operate in an "uncertain" situation;
- Absence of a universal mechanism taking the constraints into account.

Rigidity of traditional methods appears in the fact that the effective algorithms are specialised for a narrow class of problems as a rule. For example, the design of an effective quasi-direct algorithm using an elimination approach is possible [2] for the tree graph. Other effective techniques require the preliminary processing of the initial data by the user (this processing takes account of individual features of the network considered). The universal methods, (for example, relaxation methods, method of minimisation of the residual, etc.) suitable for a wide class of problems, have a rather low rate of convergence.

The data preparation is separated from the computation under the traditional process of solving. The first step is undertaken by the user, and the second one is executed by computer. The intervention of the user is admissible only at a preliminary stage, but not during the solution process. The algorithm can reach only two possible states: "the problem is solved" or "the problem is not solved".

The constraints are usually perceived as a vexatious hindrance in the traditional scheme of calculations. Some additions to the calculation algorithms are designed to take account of constraints, generally conforming to the specific problem.

4   USE OF NEW INFORMATION TECHNOLOGIES

A spectrum of information technologies under the umbrella of "artificial intelligence" is actively being developed. These technologies (e.g. knowledge-based systems, constraint programming, subdefinite computational models, etc. [5,6]) give the user new opportunities and approaches to solving the modelling problems. The most recently developed integrated intelligent systems (such as NeMo+ [7] or Semp-TAO [8]) are especially attractive because they incorporate the means and advantages of many of the above-specified information techniques.

The approach presented in this paper uses the interpretation of the simulated system as a semantic network on a directed graph. The objects are represented as vertices of this graph and the binary relations serve as arcs. An object (relation) of such a network can be any entity of the data domain allocated by the user (Fig.1). Each object is characterised by its name and values of attributes (slots).

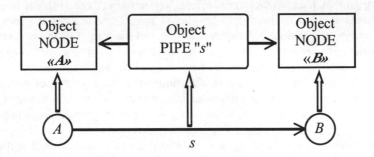

Fig. 1 Interpretation of pipe network as semantic network.

The constraints are defined as values of object slots. The above-mentioned computational models are activated and provide recomputation (more precise definition) of appropriate object slots when the semantic network is created and also whenever there is a change.

4.1   New Opportunities of Intelligent Technologies

In view of modern requirements the process of problem solving should allow:

1) Intervention throughout the solution process at intermediate stages. Note that this can be realised purely by program facilities. However, the point is that the

problem formulation and computational technique should support opportunities for an interactive interface, i.e. interaction "user - computer";

2) In the case of incompleteness of problem solution, to analyse the solution process and to reveal the reasons for incompleteness. To achieve these goals, it is necessary to define an intermediate state between concepts "the problem is solved" and "the problem is not solved";

3) To use specifics of a particular problem for acceleration of the computation, i.e. to have at one's disposal a flexible set of computation methods, which can be varied depending on individual problem features and solution step. (Frequently it is possible in practice to find a "fast" way of solving a problem using individual features of the specific system. The problem consists of creating the computing techniques, which can flexibly adapt to specific problems and their individual features.) The other purpose is "to learn" which algorithm to operate, when it is impossible to find the solution in a traditional sense.

4.2 Towards the Concept of "Solution"

The concept of "solution" should be redefined for achieving the above-mentioned objectives. Namely, the solution process should be considered as a process of reducing the subdefinity of a problem, resulting in a decrease in the level of uncertainty.

The initial parameters of the system have the greatest degree of uncertainty. During the solution process there is improvement and re-improvement according to the scheme described above. The process comes to an end when further improvement of parameters is impossible. The concept "the problem is solved" corresponds to exact solutions in a traditional sense (intervals are contracted to a point), and to a solution with interval values, whose range does not exceed an admissible error. All other states can be considered as fitting the concept "the problem is not solved". However, the new concept of "solution" has not so fatal a nature. The user can look through the under-specified values of parameters, correct them and/or enter the additional information, whereupon the appropriate computational models will be activated and a new improvement of parameters will result.

The initial data processing and solution process are realised within the framework of a united process under such an approach. This supports an opportunity for intervention of the user in the solution process at any stage. It is possible, working in a dialogue mode, to correct errors, to adjust the data, to find "bottlenecks" of a problem containing incomplete or inexact data.

Moreover, the process of "solving" is concurrent with the process of "designing" a system, since all the system parameters are specified with each addition of a new node or new pipe. This gives to the user new opportunities, absent in the traditional approach, to find a rough "outline" solution at first, designing the basic "skeleton" of a system, and then gradually to make it more exact, adding new more detailed particulars. Thus, even the solution of a steady problem can be executed by a dynamically increasing semantic network.

An unsatisfactory result (poorly specified values or obvious contradiction) also grows out of computations and provides additional information about insufficiency of initial data or about errors in formulation of a problem.

5   APPLICATIONS OF INTEGRATED INTELLIGENT ENVIRONMENTS

5.1   Toolkit: NeMo+ System

The NeMo+ system [7] is intended for solving problems by means of subdefinite models. The concept of subdefinity provides certain advantages in comparison with other methods for processing partially known information (fuzzy sets, interval mathematics, constraint programming).

The basic means of description of the problem in the NeMo+ system is the language for model specification. It is possible to note the following features:

- Object orientation. Users have an opportunity to design their own classes and to derive them from already existing ones.
- Declarativeness. All operators and functions of language are constraints, which compute not only a value of the result, but also values of the arguments. For classes of NeMo+ it is possible to specify a set of properties as constraints above slots.
- Modularity (separate compilation, creation/use of libraries, etc.).
- Access to a base level. The user can copy and expand a set of base data types and relations in C++ language.

The user describes the problem as a set of constraints, which the object values in the model should satisfy. The objects of the user's model can be numbers, strings, sets, arrays, and also concepts of a higher level (e.g. pipeline, node of a pipe network, time interval). The constraints on object values are specified as a set of algebraic equations and inequalities of any complexity, which form a mathematical model of the researched problem.

The user can, alongside with exact values, specify intervals of allowable values which will be defined more precisely during computation. The problem can have objects without precisely specified values. If a number of initial constraints are insufficient for obtaining an exact solution to the specified problem, it is possible to take advantage of special functions for finding them.

The system NeMo+ is supplied with the friendly multiwindow graphic interface including a menu system and the help that essentially facilitates the work of the user.

5.2   Description of an Application Domain

The specified technology was applied to the solution of a problem arising frequently in practice, about computation of liquid (or gas) flows in a pipe system under steady-state conditions. Some problems of the above-described kind were solved with the help of the NeMo+ system.

As an example of calculation illustrating the used approach, we shall consider an internal water-supply system for an industrial factory. The graph of the system is shown in Fig.2.

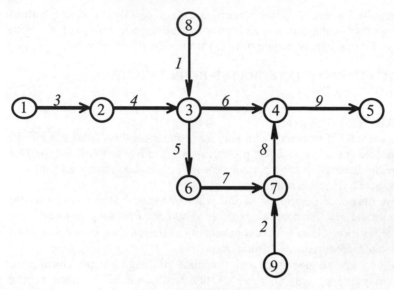

Fig. 2 The graph of a water-supply system.

Here arrows denote system pipes (arcs of the graph), and circles denote its nodes (vertices of the graph). All pipes and nodes are numbered. Pressures in nodes 1,8,9 are known, the consumed demand is given at node 5. The nodes 2 - 4, 6, 7 represent pipe junctions without leaks.

Constraints are imposed on the discharges and pressures in pipes:

$$h_{\min} \leq h_k \leq h_{\max}, \quad |Q_k| \leq Q_{\max}, \quad R_{\min} \leq R_k \leq R_{\max}, \tag{7}$$

where h_{min}, h_{max}, R_{min}, R_{max} and Q_{max} are minimum and maximum allowable values of parameters, arising from technological requirements.

It is required to find the distribution of the discharges and pressures in all pipes of the system, satisfying constraints (7), under the known pipe resistances.

5.3 Specification of the Problem in Language NeMo+

Let's introduce the classes PIPE and NODE, appropriate to concepts «pipeline» and «node» of system.

Their specification in language NeMo+ will assume the form:

```
uses "opr";
class PIPE;                          // Class PIPE describes pipeline.
   slots( h0, hL  : real;            // Slots h0, hL are the pressures
                                     // at the extremities of pipe

          Q      : real;             // Discharge of liquid in pipe.
          R      : real );           // Pipe resistance
   h0 = [100.0, 150.0];              // the constraints (7) on values of slots
   hL = [100.0, 150.0];              // are specified as a priori intervals.
```

```
Q  = [-3.0, 3.0];              // Here hₘᵢₙ = 100 ,
R  = [0.001, 1.0];             // hₘₐₓ = 150  Qₘₐₓ = 3.0³/.
                               // Rₘᵢₙ = 0.001; Rₘₐₓ = 1.0.
h0 - hL = R*Q*abs(Q);          // The relationship (4),
                               // connecting values of slots
end;


class NODE                     // This class is node
slots(  pin     : list of PIPE;     // The list of incoming pipes into node
        pout    : list of PIPE;     // The list of outgoing pipes out of node
        press   : real;             // The pressure hʲ in node.
    q : real;                       // The liquid leak qʲ in node.
    qin : real ;                    // The first sum in (5) , i.e. the total
                                    // influx in node
    qout  : real );                 // The second sum in (5) , i.e. the total
                                    // outflow from node
    press = [100.0,  150.0];        // The constraints (7) on the parameters
    qout  = [-3.0, 3.0];            // of node, similarly the "PIPE" class
    qin   = [-3.0, 3.0];
    qin =  +pin.Q;                  // The discharges of outgoing pipes
                                    // are summarised
    qout = +pout.Q;                 // The discharges of incoming pipes
                                    // are summarised
    qin – qout = q;                 // The discharge balance (5)
                                    // at the node
    pin.hL  = press;                // The realization (6)
                                    // for all incoming pipes
    pout.h0 = press;                // similarly for all outgoing pipes
end;

inf priority=10 "=" ( x, y : PIPE ) :   // The notion "equality" for "PIPE"
bool;                                   // objects is described
    x.h0 = y.h0;   x.hL = y.hL;
    x.Q  = y.Q;    x.R  = y.R;
end;
```

After describing a data domain it is possible to proceed to the specification of subdefinite model of the specific problem considered:

```
main pipe_test;
p1, p2, p3, p4, p5, p6, p7, p8, p9 : PIPE;   // The objects are introduced
                                             // for 9 pipes
v1, v2, v3, v4, v5, v6, v7, v8, v9: NODE;    // The objects are introduced
                                             // for 9 nodes
v1.pout[1] = p3;                             // The constraints on its slots are
v2.pin[1] = p3; v2.pout[1] = p4;             // introduced, which specify
v3.pin[1] = p1;v3.pin[2] = p4;               // the circuit diagram
```

```
v3.pout[1] = p5; v3.pout[2] = p6;
v4.pin[1] = p6; v4.pin[2] = p8;
v5.pin[1] = p9;
v6.pin[1] = p5; v6.pout[1] = p7;
v7.pin[1] = p2; v7.pin[2] = p7;
v7.pout[1] = p8;
v8.pout[1] = p1;
v9.pout[1] = p2;
  p1.R = 0.353;                              // The resistance values are
  p2.R = 0.266;                              // specified for each of 9 pipes
  p3.R = 0.337;
  p4.R = 0.34;
  p5.R = 0.29;
  p6.R = 0.694;
  p7.R = 0.007;
  p8.R = 0.45;
  p9.R = 0.012;
  v1.press = 144.48;                         // The pressures are specified
                                             // equal to 144.48m at node v1,
  v8.press = 143.69;                         // equal to 143.69 m at node v8,
  v9.press = 144.46;                         // equal to 144.46 m at node v9.
  v5.q  = 2.62;                              // The discharge on an outflow
                                             // (node v5) is specified
                                             // equal to 2.63m³/s
  v2.q = 0.0;  v3.q = 0.0;  v4.q = 0.0;      // The leak of liquid from internal
  v6.q = 0.0;  v7.q = 0.0;                   // nodes is absent (equal 0).
  p1.Q+p2.Q+p3.Q = p9.Q                      // Superfluous constraint (*)
exact(p1.Q, p3.Q, p3.h0, p3.hL, p5.h0,       // The search of the exact solution
p5.hL,);                                     // is specified
end;
```

We shall find the following values of the discharges and pressures in pipes (unique solution) as a result of calculations by the NeMo+ system.

```
p1.Q = -0.1340, p1.h0 = 143.690; p1.hL = 143.696;
p2.Q =  1.6781, p2.h0 = 144.460; p2.hL = 143.711;
p3.Q =  1.0759, p3.h0 = 144.480; p3.hL = 144.090;
p4.Q =  1.0759, p4.h0 = 144.090; p4.hL = 143.696;
p5.Q = -0.2218, p5.h0 = 143.696; p4.hL = 143.711;
p6.Q =  1.1637, p6.h0 = 143.696; p4.hL = 142.757;
p7.Q = -0.2218, p7.h0 = 143.711; p4.hL = 143.711;
p8.Q =  1.4563, p8.h0 = 143.711; p4.hL = 142.757;
p9.Q =  2.620,  p9.h0 = 142.757; p4.hL = 142.674;
```

Various kinds of constraints, especially non-linear ones, can be modelled more effectively by the NeMo+ system when compared to traditional methods.

One more feature of the calculations by the NeMo+ system is the opportunity to solve the over-determined system of equations. The over-determination is caused by additional constraint (*). The existence of over-determination gives additional information, which only accelerates the solution process.

6 CONCLUSIONS

The representation of objects of a pipe network by the objects of the appropriate classes "PIPE" and "NODE" seems quite natural. The pipes and nodes of real systems are simulated by objects of those classes, and the parameters of elements of the system correspond to slots of those classes. The known parameters are described as exact values, and the unknown quantities are described by subdefinite values. The system constraints are represented as constraints on the above objects. The activation and execution of the model results in a reduction of the subdefinity of unknown quantities of objects down to «exact» values (solution in the classical sense).

We shall note that all possible constraints on the flow parameters and of the pipe network are specified in a natural manner within the framework of the NeMo+ system instead of being represented as an inconvenient addition to the algorithm, as in the case of application of classical methods.

The principle of data driven computations used in NeMo+ automatically allows the finding of weaknesses in the problem, where the revision of subdefinite values is possible, i.e. it can flexibly adapt to the individual features of a specific problem, that raises the efficiency of computation.

In the case of the absence of a solution in the classical sense, the NeMo+ system allows the user to obtain useful information. First, the reduction in the degree of subdefinity of a problem takes place that narrows the range of possible solutions. Second, the analysis of "zones of subdefinity" suggests to the user which initial data requires updating or additions in order to obtain an acceptable solution.

The approach described can be easily spread to cover other systems defined on graphs: biological systems, natural systems (for example, system of river channels), etc. For example, for simulation of water quality, the slots of objects should describe water quality features, together with hydrodynamic parameters, and the set of constraints should include ecological requirements. An example of use of the Semp-TAO environment is described in [3].

Thus, the approach presented for the solution of a problem of calculating the flow parameters in pipe networks gives wide additional opportunities for development of effective procedures for solving complex problems.

REFERENCES
1. Voevodin A.F., Shugrin S.M. Methods of the solution of one-dimensional evolutionary systems. – Novosibirsk: Nauka, 1993. – 368 pp. (in Russian).
2. Tarasevich V.V. The solution of problems of one-dimensional pressure currents in hydraulic networks. – in "Methods and programs of the solutions of optimization problems on the graphs and networks". – Novosibirsk, Computer Centre SB AN USSR, 1980, 186-188 pp. (in Russian).

3. Tarasevich V.V. The Simulation and Mathematical Modelling of the Complex Pipe Systems. 5-th (IMACS) World Congress on Scientific Computation, Modelling and Applied Mathematics. Berlin, August 1997. Proceedings, vol. 3., Computational Physics, Chemistry and Biology. p. 115-120. Editor by Achim Sydow.

4. Merenkov A.P., Hasiliev V.Ya. The theory of hydraulic circuits. – M.: Nauka, 1985. (in Russian).

5. Mayoh B. Constraint Programming and Artificial Intelligence. In: Constraint Programming. Springer-Verlag, 1993. NATO ASI Series F: Computer and Systems Sciences, Vol.131, pp.17-50.

6. Narin'yani A.S. Subdefinite Models: A Big Jump in Knowledge Processing Technology // Proceedings of East-West Conference on AI: from theory to practice, EWAIC'93, – Moscow, September 7-9, 1993, p. 227-231.

7. Telerman V.V., Sidorov V.A., Ushakov D.M. Problem Solving in the Object-Oriented Technological Environment NeMo+ // Lecture Notes in Computer Science, 1181, Springer, 1996. – p. 91 - 100.

8. Zagorulko Yu.A., Popov I.G. Object-Oriented Language for Knowledge Representation Using Dynamic Set of Constraints. // Knowledge-Based Software Engineering, P.Navrat, H.Ueno (eds). – (Proc. 3rd Joint Conf., Smolenice, Slovakia). -Amsterdam: IOSPress, 1998. – p.124-131.

PART II

GEOGRAPHICAL INFORMATION SYSTEMS

Extending the Intranet to Mobile Working

P.V. Fanner

ABSTRACT
Utilities are increasingly considering the implementation of mobile working systems to improve the efficiency and productivity of field staff, improve customer service and improve the capture of accurate work and asset data. Mobile computer systems used for these applications take a variety of forms from text-only handheld computers to graphic computers operated by pen systems.

The mobile computing solutions currently on the market which support GIS are mainly based on Windows computers and are generally based on a fully functional GIS application, with a simplified user interface. These systems are expensive to implement, have high maintenance costs and have complicated data management processes. For these, and other reasons, many utilities have yet to implement mobile field systems.

In order to overcome these problems, and considerably reduce the cost of implementing a mobile field system, Bristol Water has developed a novel mobile field system, based on internet technologies; called Cartesia. This paper describes the Cartesia system, discusses the advantages of this system over conventional solutions and outlines the business benefits provided by the system at Bristol Water.

1   INTRODUCTION

Utilities are increasingly considering the implementation of mobile working systems to improve the efficiency and productivity of field staff, improve customer service and improve the capture of accurate work and asset data. Mobile computer systems used for these applications take a variety of forms from text-only handheld computers to graphic computers operated by pen systems. Applications involving the field use of GIS are the most demanding systems to implement on a mobile computer and are normally implemented on a graphic computer operated by a pen system.

The mobile computing solutions currently on the market which support GIS are mainly based on Windows computers and are generally based on a fully functional GIS application, with a simplified user interface. These systems are expensive to implement, have high maintenance costs and have complicated data management processes. For these and other reasons, many utilities have yet to implement mobile field systems.

In order to overcome these problems, and considerably reduce the cost of implementing a mobile field system, Bristol Water has developed a novel mobile field system, based on internet technologies. The system, called Cartesia, has been developed by Tadpole Technology plc, based on Java software to provide a fully functional information and data capture system. It runs on a ruggedised pen-operated Java mobile computer. The system effectively provides mobile access to the corporate intranet, with persistent local storage of downloaded data. All software is server-based, simplifying system maintenance and updating. The system can communicate with the server using PSTN, GSM or ethernet links. The GMS communications effectively provide real time two-way data transmission with field staff.

This paper describes the Cartesia system, discusses the advantages of this system over conventional solutions and outlines the business benefits provided by the system at Bristol Water.

2   BUSINESS DRIVERS

Utility customers and the infrastructure assets that supply them are distributed throughout a utility's area of supply. The utility staff and contractors who maintain these assets and provide front line customer service are similarly distributed. However, the coverage of the IT systems that control the work of these staff and provide them with the information to do their jobs do not generally extend further than the local depot office. Paper-based systems are usually used to provide information to field staff and to capture data from the field.

There are a number of limitations to paper-based systems:

- They are inefficient and require considerable administrative support
- The data in paper-based systems can be out of date.
- Paper based systems normally provide the minimum information required to undertake planned jobs. The additional information required to respond to unexpected events is not usually available to field staff, often resulting in unplanned visits to the depot.
- There is no immediate feedback to the work management or customer service systems on job progress and status. As a result, managers of field staff cannot easily track job progress and customer service staff are unable to give up to date information to customers.
- Managers cannot easily reassign jobs to other field staff in the event of a job taking longer, or shorter, than planned.
- In most cases, the gang who excavate the hole are the only staff who can provide improved data about the asset they have exposed and the work undertaken. However, if the data provided on paper is inconsistent, it is too late to go back when inconsistencies are identified in the office.
- Field staff must make regular journeys to their depot to return paperwork and pick up new jobs.

In order to overcome these disadvantages, utilities are increasingly considering replacing these paper-based systems with mobile computer-based systems. The main business drivers forcing utilities to look at these systems for managing their front line activities are:

- Competitive and regulatory pressures to improve efficiency and productivity.
- Competitive and regulatory pressures to improve customer service by providing access to corporate data in the field.
- The need to improve the management of field staff and obtain immediate feedback on job status.
- The need for accurate, validated asset and job data to improve asset management planning and reduce unit costs.
- Competitive and regulatory pressures to minimise the impact of incidents on customers by rapid and effective response to emergencies.

3   EXISTING FIELD SYSTEM SOLUTIONS

Mobile computer systems used for field applications take a variety of forms from text-only handheld computers to graphic computers operated by pen systems. Applications involving the field use of GIS are the most demanding systems to implement on a mobile computer and are normally implemented on a graphic computer operated by a pen system. A mobile GIS application is necessary for a utility to realise all the benefits available from computerising field work.

The mobile computing systems currently on the market which support mobile GIS are mainly based on Windows pen computers and the mobile GIS systems are generally based on a fully functional GIS application, with a simplified user interface. There are several difficulties with this approach:

- The mobile computer hardware required to run such systems is expensive. Each mobile unit can cost between £3,000 and £4,000, which makes widespread implementation of these systems an expensive project.
- The volume of GIS data to be held on the mobile system is large. It is usual to hold the complete foreground and background data for the area covered by a depot on the mobile units. The volume of this data can easily run into several Gb's, resulting in long data transfer times.
- It is necessary to track changes on both the corporate GIS and the mobile GIS in order to identify updates to the corporate GIS that need to be transferred to the mobile and to identify data captured on the mobile to be passed back to the corporate system.
- The interface between the corporate and mobile GIS systems is specific to the data models on both systems. Any changes made to the data model on either system will entail changes to the interface between the two systems, imposing a maintenance overhead to development of the corporate system. This is a particular problem where a utility company is still actively developing its corporate GIS system.

- Experience has demonstrated that hardware failure rates are high, usually due to hard disk failure. Hardware and software maintenance costs are also high because of the geographic distribution of the machines.

For the above reasons, and others, many utilities have yet to implement mobile field systems despite the business pressures to improve their field operations.

4   THE MOBILE INTRANET SOLUTION

4.1   Overview

In order to overcome the problems inherent with existing mobile field systems, and reduce the cost of implementing such a system, Bristol Water has investigated the potential for utilising a new type of mobile field system, based on intranet technologies. The system, called Cartesia, has been developed by Tadpole Technology plc, using Java software with software designed to meet our specifications. The system is designed around workflows and the mobile client only holds the information required for each job, but is able to connect to the corporate systems for additional information in the field at any time by communicating over GSM telephones. This communication link is also used to provide immediate feedback of job status information and time critical asset data, such as a change of valve status. All software is server-based, simplifying system maintenance and updating.

Cartesia runs on a ruggedised pen-operated mobile network computer equipped with the minimum components required to support the business requirements. For example, the machine will have no hard disk because it is not necessary to hold much data on the mobile client. This means that data can easily be held in RAM and Flash RAM. The machine has a large colour screen designed for viewing in bright sunlight and a range of communication options including GSM, PSTN and Ethernet links. The system in effect provides mobile access to the corporate intranet, with persistent local storage of downloaded data.

The first phase of system development was concentrated on the development of a mobile GIS system. Mobile work management was excluded from this phase because it was first necessary to make our corporate work management system year 2000 compliant. This work has now been completed and the mobile work management and customer service components are currently under development.

4.2   Cartesia Architecture

The Cartesia system is made up of four main components, a Mobile GIS module, a Mobile Work module, a Dispatcher Module and a QA Module.

Figure 1 illustrates the system operation. Planned jobs from our En Garde work management system are automatically linked to the GIS map base each night. The GIS Operator uses the Dispatcher module to allocate a map extract to each job and allocate the job to the appropriate crew. The job card and customer service data for the job are extracted from EnGarde and our customer service system (CSS) and also held on the Dispatcher.

Whenever a mobile user links to the network by any communication route, any completed jobs and data captured are uploaded to the Dispatcher module and new jobs and maps are downloaded to the mobile. The District Manager checks the data returned for each job before committing it to update the EnGarde and CSS corporate systems. The GIS operator then checks the connectivity of any vector GIS data and digitises any sketch data using the QA module, prior to updating the Smallworld database.

The Dispatcher module is a server based Java application. The Mobile GIS and Mobile Work modules are both Java applications that reside on the mobile client pen computer. The QA module is a Smallworld GIS application written in the Smallworld Magik language that runs on the depot GIS operators' workstation. The functions of each of these modules are detailed in the following sections.

Figure 1 - Cartesia System Operation

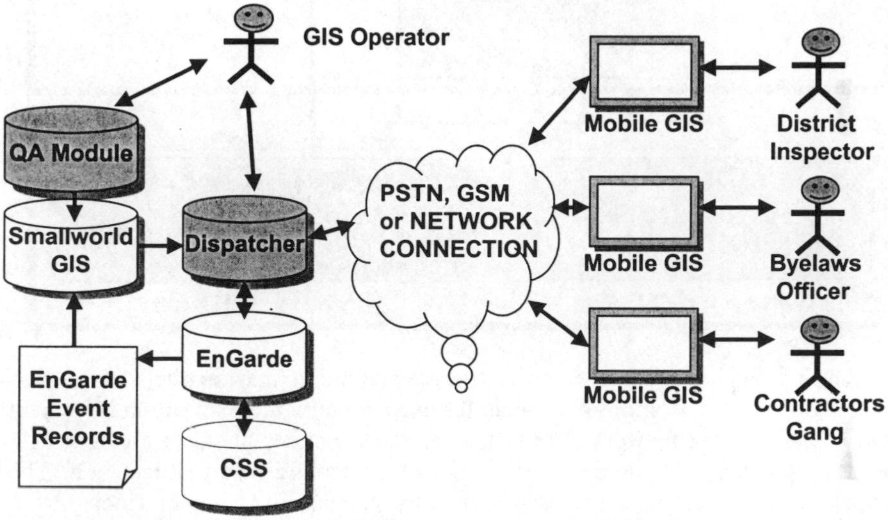

4.3   Dispatcher Module

The Dispatcher module is a three-tier client-server Java component, with the architecture illustrated in Figure 2.

Figure 2 - Dispatcher Architecture

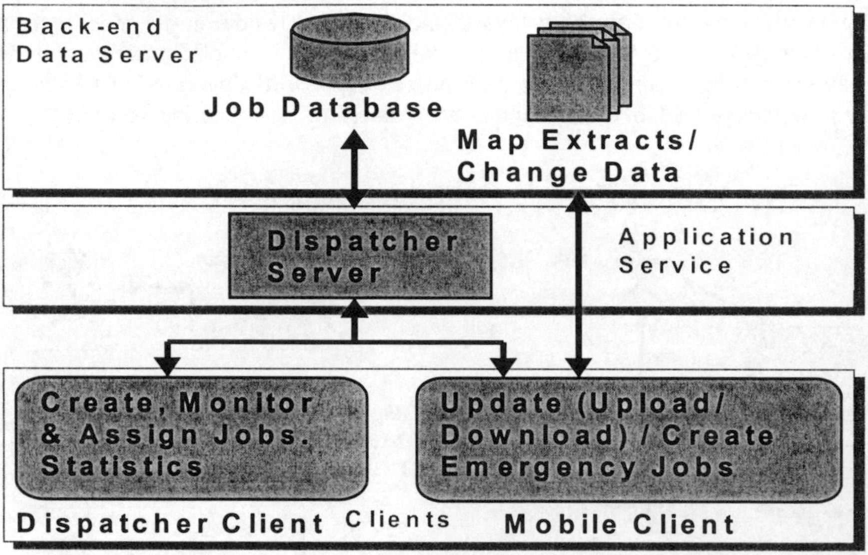

The Dispatcher module controls the generation of map extracts and stores mapping, job and customer service data for each job until the relevant mobile client next connects to the network. The Dispatcher server stores this data together with the job allocation data used to determine which mobile receives each job. The Dispatcher client may be accessed from any connected computer (including a mobile client) and has facilities for the District Manager to monitor actual job progress, re-allocate any job from one mobile to another in response to this feedback and to authorise updates to the corporate systems. The Dispatcher module also manages the communications between the server and the transfer of data to and from the Mobile Work module. The Dispatcher server sits between the back end data servers and the dispatcher and mobile clients.

4.4   Mobile GIS Module

The Mobile GIS module provides comprehensive mobile GIS functionality. Localised vector and raster foreground and background mapping is stored for each planned job, centred on the job. For a job at a point location, the mobile client will normally receive mapping for a 200 m square around the job location. For a linear job such as a main laying job, the mapping will be a 200 m wide strip along the line of the job. The Mobile GIS has a full gazetteer for the whole company area and has pan and zoom, and visibility control functions. The Mobile client also provides real-time access to mapping not held locally via direct browsing of the corporate GIS using GSM data communications. Selected mapping can be extracted from the server and downloaded on line. The Mobile client is also designed to hold all the schematic line diagrams of the network for a depot area as well as other utility digital mapping for each job.

Vector foreground data on the Mobile GIS module holds all the attribute data held on the corporate system and the module has facilities for the user to update any of this data using an editor which has the full data validation used on the corporate GIS. Form-based screens are provided for capture of job, asset and condition data from all jobs, with full data validation on the Mobile client. The Mobile GIS also has freehand red-lining sketch facilities as well as new vector data capture facilities with full data validation and integrated surveying facilities such as bilateration and chain and offset to ensure correct location of vector data.

4.5   Mobile Work Module

The Mobile Work module provides functionality to log on and off the Mobile client and communicate with the Dispatcher module to download any new maps and jobs and upload completed jobs and all data captured. The Mobile Work module is centred around a worklist, which shows the user all jobs on the mobile, with their current job status. Mapping, job card and customer service data are accessed for any job from the worklist and there are facilities for the user to update the job status as the work progresses. The "Completed" status is used by the system to determine that data linked to this job is complete and ready to be uploaded to the Dispatcher module.

4.6   Mobile Hardware

Figure 3 illustrates the mobile client, which is of a rugged magnesium alloy construction with tight fitting rubber covers over all external ports, providing a machine that is suitable for use in the rain. The machine can be equipped with either an internal or external GSM phone. The unit illustrated has an internal phone. All data is held in RAM and is instantly available when required by the user. Similarly, the machine will instantly revert to Suspend mode when the user has temporarily finished using it. These features result in a battery life of some 6 to 8 hours of continuous use, which is more than adequate for most users. All data captured in the field is held in Flash RAM to ensure that it is not lost in the event of total battery failure. The system is powered off the vehicle cigarette lighter when in a vehicle and used in a carrying case that allows it to be hooked over the steering wheel, when the vehicle is parked, for ergonomic operation of the system in the vehicle. The unit cost of the mobile client hardware is considerably lower than the cost of comparable Pen Windows machines.

Figure 3 - Cartesia Mobile Client

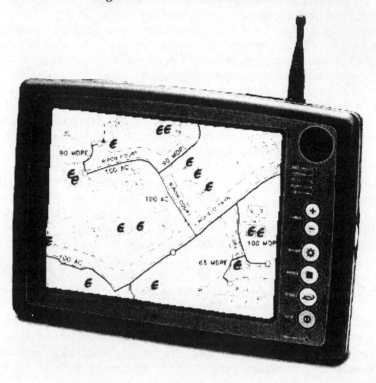

4.7   QA Module

The QA Module is a Smallworld application which sorts the incoming updates into various types and prioritises the data capture workload, provides facilities for checking incoming digital data and access to the normal Smallworld functions for digitising from freehand sketches.

4.8   Business Benefits

Trials of the Cartesia system have been undertaken at our Millbrook depot by representatives from six groups of potential users; district managers, district inspectors, byelaws inspectors, leakage inspectors, trunk mains teams and contractor's gangs. The trials have been designed to evaluate the benefits, productivity improvements and practicalities of using Cartesia with each type of field worker. At the time of writing, the trials have reached an advanced stage, but the analysis of business benefits has yet to be completed.

Our preliminary conclusions are:

- The Cartesia mobile intranet solution offers significant improvements over conventional Pen Windows solutions.

- The trials have confirmed that it is not necessary to hold mapping for the whole depot area on the mobile because the GSM data link provides a perfectly workable means of retrieving mapping data for unplanned and emergency jobs in the field. The speed of accessing this data is adequate now, and it will improve as faster GSM data technology is introduced in the future.
- The ability to update job and valve status data immediately a job is completed will enable us to improve productivity by facilitating work rescheduling to improve the utilisation of field workers as well as improve customer service.
- The availability of mapping and job information in the field for unplanned jobs has considerably reduced the number of return trips to the depot.
- The capture of validated digital data in the field has significantly improved the quality of data returned from the field and reduced the administrative workload in the depot.
- The trial users have adapted to the system well.

5 CONCLUSIONS

The trial work we have undertaken with Cartesia has confirmed that internet technologies can be used to extend access to corporate systems into the field by effectively extending the intranet to field workers. The trials have demonstrated that GSM data communications can be used effectively for this application. The Java based system offers utilities significant capital and operating cost savings over conventional Pen Windows based solutions and has the potential to significantly improve operational productivity and customer service, while also improving the quality of data captured in the field for asset management planning purposes.

The Development of a GIS for Flood Risk Assessments in the West of Scotland

T.J. Jolley *and* C.L.M. Wright

ABSTRACT

The Environment Act (1995) defines the Scottish Environment Protection Agency (SEPA) as a statutory consultee to planning authorities with respect to the flood risk associated with planning applications. The subsequent publication of the *National Planning Policy Guideline 7* (NPPG7) by the Scottish Office (1997) has provided a framework for the consideration of flood issues and identified SEPA as one of the principle authorities with respect to the provision of data and professional advice for its successful implementation.

SEPA recognise that if advice on flood risk is to be useful to planners it must be based on a sound hydrological understanding of the causes of flooding and on reliable data and background information. There can be many different local factors which need to be considered when providing advice. This information comes from a wide variety of sources and takes many different forms. Moreover, experience has shown that flood related data is extremely varied in terms of quality and coverage. Consequently, SEPA has investigated the potential of a Geographic Information System (GIS) as a tool for the collation and presentation of data to facilitate the assessment of flood risk.

This paper discusses the nature of flood related data and describes the development of a GIS facility to aid the assessment of flood risk in the West Region of SEPA. The first section of the paper considers the nature of flood data and in particular addresses the issues of data sources (e.g. flood plain maps, aerial photographs, model studies and anecdotal information), quality, and spatial and temporal coverage. The second section describes the approach taken for the development of a GIS application, focusing on the data model adopted and particular problems associated with flood data (e.g. scale and resolution). The third section considers a number of examples to illustrate how the GIS has been used in practice to bring together data from a variety of sources and to assess flood risk. The final section assesses the success of this approach and considers the future of flood risk assessment.

1 INTRODUCTION

The Environment Act (1995) defines the Scottish Environment Protection Agency (SEPA) as a statutory consultee to planning authorities with respect to flood risk. The subsequent publication of the *National Planning Policy Guideline 7* (NPPG7) by the Scottish Office (1997) provides the framework for addressing the issue of flooding within the planning process and explicitly identifies SEPA as one of the principle authorities with respect to the provision of information and professional advice to planning authorities.

If SEPA's advice on flood risk is to be useful to planners then it must be based on a sound hydrological understanding of the causes of flooding and on reliable data. However, data is typically of variable quality and does not lend itself to traditional hydrological analysis. For example, quantitative information is rarely available at specific locations, flood data is often qualitative information provided by eye-witnesses, and continuous data for single events is rarely available at the regional scale. Hence, SEPA have investigated the potential of a Geographic Information System (GIS) as a tool for the collation and presentation of data to facilitate the assessment of flood risk.

This paper discusses the nature of flood related data and describes the development of a GIS facility to aid the assessment of flood risk in the West Region of SEPA.

2 CHARACTERISTICS OF FLOOD DATA

One of the main problems associated with information on flood risk is that it comes from a diverse range of sources and takes many different forms. For example, SEPA have sourced information from SEPA's own hydrometric database, local councils (and from several different departments within a single council), consultants' reports, Police reports, media reports, members of the public and from SEPA staff in the field. For significant floods such as the extreme flood of December 1994 that affected the Strathclyde region, information will be available from numerous sources, whereas for minor local floods affecting rural areas information may only be available from SEPA staff or local residents. SEPA recognise that all information has a value and that any information (no matter how uncertain) is generally better than no information.

The importance of qualitative information depends on the location of the planning application and the quality of the hydrometric records. For applications close to river gauging stations, records of water levels are available. However, the quality of this data can vary greatly according to the record length and the nature of the gauged cross section. Even at gauged locations qualitative information can be important for quantifying the area of inundation associated with a flood level or for extrapolation in cases where a flood inundated the gauging station.

3 THE APPROACH

In the West of Scotland Region SEPA have taken a twin approach to addressing the problem of providing accurate flood risk information and advice. Firstly, greater resources have been allocated to the collection of flood data from all potential sources. Secondly, a Geographic Information System (GIS) has been introduced as a tool for maximising the use of all available information for any planning application, and to provide a consistent approach to assessing flood risk.

Prior to April 1996 flood data had been collected by SEPA's predecessor body in the West region, the Clyde River Purification Board (CRPB). As the CRPB did not have the same statutory duty as SEPA with respect to the provision of flood risk advice they tended to focus on data for significant flood events only. The Flood event of December 1994 was the first event where the CRPB actively commissioned a flood survey (Babtie Group, 1995). Moreover, this was the first time that a flood survey had been carried out in Scotland with the requirement that the data should be delivered in a digital format appropriate for a GIS (SEPA have adopted Arc/INFO and ArcView).

Since April 1996 SEPA have actively converted the CRPB archive of flood information into a digital format (with associated attribute data) either by digitising mapped flood extents or by recording records as point data. For example, the January flood of the White Cart Water (Babtie Shaw & Morton, 1987) and coastal flooding during January 1991 (Townson and Collar, 1993) were both digitised.

Different data sources have different map scales and extents that make integration extremely difficult without the use of a GIS. For example, the December 1994 event was digitised at a scale of 1:10 000 and covers much of the Strathclyde region and parts of Ayrshire including Kilmarnock and Irvine. Whereas the maps of the 1991 coastal flooding are sketched onto photocopied 1:25 000 maps and only cover relatively small areas of Tarbet, Saltcoats, Rothesay and Dumbarton.

The result of this concerted effort is a detailed digital coverage of flood events, with attribute data attached. This attribute data includes date and scale of capture, period of inundation if known, name of person and organisation that surveyed and digitised the flooding. In order to allow the recording of flood extents for different flood events at the same location, they are stored as Arc/INFO "Regions".

In addition to digitised flood extent data, ad-hoc flood locations are recorded in a spreadsheet. This includes information on flooding where the area affected is not known. It contains flood reports from councils, the Police, members of the public, and SEPA staff, and includes information on the cause and extent of flooding where known. This spreadsheet is periodically converted into a format which can be used with the GIS.

In addition to the main flooding data sets described above, there is also data on known flood prevention works, SEPA gauging station locations, and rain gauge locations.

All the flood related data is made available to SEPA's Hydrology and Planning staff through the use of the ArcView desktop GIS package. The data has been added into a standard "view", along with some background Ordnance Survey data. Each data set is set up to only display at appropriate viewing scales to avoid being used inappropriately. Viewing all the information together in this way allows planners to get an instant view of all information available in an area. This package also allows the addition of more specialised functions as and when they are required.

4   EXAMPLE APPLICATIONS

4.1   The River Clyde at Cambuslang in Glasgow

Figure 1 shows the extent of inundation during the December 1994 flood event and illustrates the uncertainty introduced by overlaying the digitised 1:10,000 scale data onto a small scale OS Strategi map (1:250 000) and the dependency on the detail of the original ground survey. Great care has to be taken when identifying specific sites especially if they lie close to the flooded extent.

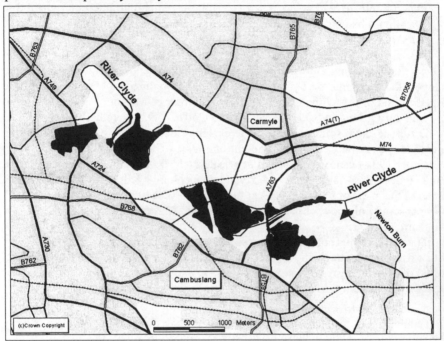

Fig. 1. Flooding in the Cambuslang area of Glasgow

4.2   The White Cart Water at Spean Street, Shawlands

The White Cart Water has been known to flood frequently at this location. Figure 2 illustrates the extent of the January 1984 flood and the estimated extent of the 120 year flood as predicted by modelling. This information is supplemented by more recent information recorded as point data. In residential areas such as this, uncertainties caused by a change of map scale or by a coarse survey can mean the difference between a high and a low flood risk for particular streets. Flood risk advice has to be robust to these causes of uncertainty such that the advice given is not highly sensitive to small changes in uncertain data.

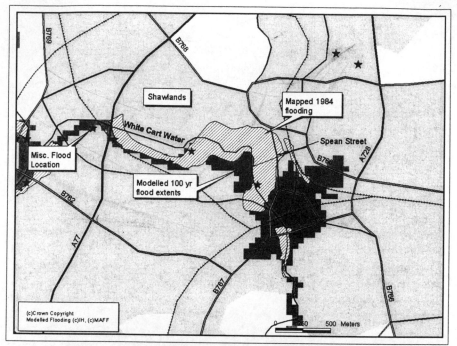

Fig. 2. White Cart Water at Spean Street, Shawlands, Glasgow.

4.3   The River Gryffe at Crosslee Park, Houston

This site flooded shortly after a new housing estate was constructed. As a consequence Renfrewshire Council are proposing to build flood banks for protection. Figure 3 illustrates that without complete information (including flood alleviation measures) flood risk advice will be of poor quality.

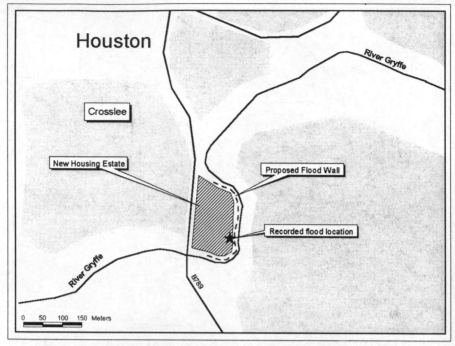

Fig. 3. Illustration showing the River Gryffe at Crosslee Park, Houston

4.4   Ferguslie Park, Paisley

The need for complete information is emphasised again in Figure 4 where flooding was caused by a surcharged culvert. Information on culverts can only be obtained through a site visit or through information from a local source. This example is another example of a relatively new housing estate being inundated shortly after construction. A large number of flood incidents in the Glasgow region are caused by surcharged or blocked culverts which can only be identified through site visits and are not generally easily identified from areal surveys of flooded extent.

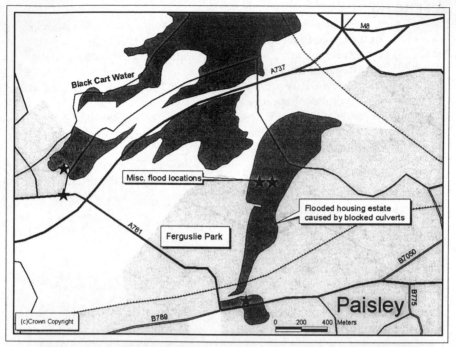

Fig. 4. Ferguslie Park, Paisley. Illustrating the effects of blocked culverts.

5   CONCLUDING REMARKS

This paper has demonstrated the potential of a GIS as a tool for assessing flood risk. However, it has also emphasised through example that the quality of a flood risk assessment ultimately depends on the quality of the data made available to the GIS. Equal effort has to be applied to the collection of data as to the development of the GIS.

This paper also demonstrates that for the majority of locations there is little flood information. Hence, it is rarely possible to quantify the risk of flooding based on observed data. To assess risk the Hydrologist either has to revert to modelling (e.g. Babtie, Shaw & Morton, 1987) or to refer to national flood risk maps such as the ones developed for England and Wales (Morris and Flavin, 1996) and the ones currently being finalised for Scotland. In general it is only possible to assess whether the site is known to have been inundated in the past and whether the flood risk is either high or low. In the case of high risk sites SEPA will advise that the applicant address the issue of flood risk through a detailed study.

84

REFERENCES

Babtie Group, 1995. *10-12 December 1994 Flood Event - Floodplain Mapping* (A report for the Clyde River Purification Board), May 1995.

Babtie, Shaw & Morton, 1987. *Flooding from the White Cart Water* (A report for Strathclyde Regional Council), June 1987.

Morris D.G., and Flavin R.W., 1996. *Flood Risk Map for England and Wales*. Institute of Hydrology Report No. 130.

Townson, J.M., and Collar, R.H.F., 1993. *Assessment of the Potential of a Coastal Flood Warning Scheme for the Firth of Clyde,* Report to the Clyde River Purification Board, May 1993.

The Scottish Office, 1995. *Planning and Flooding*, National Planning Policy Guideline (NPPG7), The Scottish Office Development Department Planning Series.

HMSO, 1995. *Environment Act 1995 - Chapter 25*

GIS Supported Analysis of Pressure Dependant Vulnerability of Distribution Networks to Leakage

C. Maksimovic *and* N. Carmi

ABSTRACT

The paper presents an innovative methodology of assessment of water distribution systems' pressure dependent vulnerability to leakage. It is based on the assumption that the vulnerability can be assessed on two major groups of factors: statistics of diameter and material dependent pipes' damage and on the integral of pressure. These two groups of factors are processed and presented in GIS form for an easier application of Boolean algebra in quantifying, scaling and mapping the spatial distribution of vulnerability. A test case of the Srbac distribution network has been used for testing the details of the methodology. More reliable data are needed for further development and confirmation of the methodology.

1 INTRODUCTION

On the threshold of the 21st century, existing water supply systems are expected to meet the increased demand at reduced costs and a higher efficiency. One of the ways to achieve better levels of service and to reduce the withdrawal from almost exhausting natural resources is to reduce the losses in existing systems. Leakage, being the most dominant loss component, is the first to be targeted. Many attempts have been made to reduce leakage losses in water supply systems, which gave rise to the different leakage control policies adopted today. Present methods of leakage reduction consist mainly of repairing bursts that are either visible or detected by direct methods such as the use of correlators and other leakage detection devices or indirectly by analysing the abrupt changes in the night flow pattern. The methods of leakage assessment based on assumed night flow "allowance" have been subject to criticism. One of the problems, which cause high uncertainty in assessment of leakage in UK, is a low percentage of consumption being metered. In the analysis of water balance it is thus not easy to distinguish between the real consumption and leakage. The project recently launched by four UK universities (Imperial College, Brunel, Bradford and East Anglia), funded by EPSRC GR/M 17808 and co-sponsored by UKWIR, introduces several innovative methods in leakage detection and quantification, reduction of uncertainty and improvement of operational management of water supply.

One of the leakage control methods which has attracted most resources in the water industry is pressure reduction since the effect of pressure on leakage has long been recognised. However, the users of this method frequently report the problems of fluid transient caused by pressure reduction devices operation. The introduction of network analysis models 20-25 years ago has enabled the simulation of the behaviour of water supply systems (WSS). Several approaches have been adopted in the simulation of leakage losses; however, few attempts have been made to incorporate the pressure-leakage relationship into the simulation of leakage losses. Furthermore, the effect of pipe material and diameter has not been appropriately incorporated into the simulation of leakage. This seems a bit non-intuitive since obviously different pipes behave differently. Obviously, what is needed is an approach in simulation and assessment of leakage which takes into account the effect of both pressure and pipe characteristics on leakage.

The present paper is based on the MSc research performed by the second author at Imperial College in 1997 under the supervision of the first author. It introduces a new approach in the matching of the hydrodynamic simulation of losses with the vulnerability maps created from pipe characteristics. The approach introduced uses the analytical power of both network's hydrodynamic analysis and Geographical Information Systems (GIS).

2   THEORY OF MAPPING THE VULNERABILITY TO LEAKAGE

2.1   General aspects
There are different approaches adopted in the simulation of leakage losses that vary in their level of complexity. Network hydrodynamic simulation of leakage losses by taking into account spatial and temporal distribution of leakage has not been broadly applied. Models which take leakage into account tend to assume that leakage is evenly distributed throughout the network following the prescribed daily profile. The assumption that losses are evenly distributed is not realistic.

The approach, implemented in this paper, distributes the leakage losses, separate from consumption, unevenly over the water distribution system and proportional to the local vulnerability. It is an iterative process that takes into account the combination of the pipe's vulnerability and pressure dependent leakage.

The methodology of the vulnerability assessment and mapping presented here is based on the following assumptions:

- each pipe (or group of pipes with the same characteristics) in the network can be quantified for its vulnerability to leakage based on its age, type of material, diameter, type of joints,
- for each of the above characteristics (type of material for example), different vulnerabilities can be assigned based on the statistics of the bursts in the past,
- vulnerability to burst and thus to leakage depends on the integral of the pressure to which each pipe is exposed during its regular daily variations,
- all these variables can be either strictly quantified (pressure integral), assessed from the past statistics or by other means,

- significance of each of the factors can be presented in its spatial distribution (GIS layers) and "weighted" by assigning certain weight to each of them,
- linear or other combination between layers results in the spatial distribution of vulnerability to leakage.

Spatial distribution of vulnerability is assessed by creating vulnerability maps in GIS software (in this case Idrisi for Windows). Various sources of data are organised in the form of GIS layers, enabling the GIS functions to be applied in the creation of intermediate results in the form of additional GIS layers. When assigning the weighting factors to the selected layers and after performing Boolean operations with the layers, the resulting spatial distribution of the network's vulnerability is obtained.

2.2 Leakage as a function of the integral of pressure

The discharge rate through an orifice of fixed dimension (representing a crack or hole in the pipe) can be obtained by an elementary hydraulics (energy equation) written for the flow through an orifice:

$$\frac{V_i^2}{2g} + \frac{p_i}{2g} + z_i = \frac{V_0^2}{2g} + \frac{p_0}{2g} + z_0 + \frac{\xi V_o^2}{2g} \tag{1}$$

where p is the pressure, V is the velocity, z is the elevation above zero datum and ξ is minor energy loss factor. The subscripts i and 0 correspond to the cross section inside the pipe and outside the respectively.

If p_0 is assumed to be zero, i.e. in the case of a free jet, and the difference between the elevations of the jet and the pipe's axis can be neglected, equation (2) yields:

$$V_0 = \frac{1}{\sqrt{1+\xi}} \sqrt{2g \frac{p_i}{\rho g}} = C_1 \sqrt{p_i} \tag{2}$$

where C_1 is constant for a given geometry of the crack or hole. The discharge through this opening is equal to:

$$Q_0 = A_0 V_0 \tag{3}$$

where Q_0 is the rate of leakage, A_0 is the cross section area of the jet. Combining Eqs. (2) and. (3) one obtains:

$$Q_0 = A_0 C_1 \sqrt{p_i} \tag{4}$$

However, the area of the jet is smaller than the area of the crack. To take this into consideration, a coefficient (C_a) is usually included to compensate for the contraction (Daughtery et.al, 1989), combined with the area of the crack ($A_0 = A_c C_0$) and by introduction of discharge coefficient $C_d = C_0 C_1$ one finally obtains :

$$Q_0 = C_d A_c \sqrt{p_1} \tag{5}$$

where C_d is the overall coefficient of discharge (ratio of actual to ideal discharge) and A_0 is the area of the crack (Chadwick & Morfett, 1993). Thus, should the area of the crack remain constant the instantaneous leakage in the given point would be proportional to the square root of the local pressure. It is thus seen that leakage is spatially and temporally variable and this paper presents the means of assessing this variability in combination with the spatial variability of the pipes' vulnerability.

The relationship of rate of leakage and pressure is depicted as in Fig. 1.b. for two different pipe materials. For the rigid pipe the assumption is that the area of the crack is constant, i.e. independent of the pressure. However, in practice, the pipes used in the water supply system will exhibit some kind of elasticity and some of them may deform under pressure. As a result, the area of the crack will be altered. Fig. 1.a. depicts the area-pressure relationship. This problem had been investigated by the Water Research Centre, which carried out a series of experiments (Goodwin, 1980).

Hence the leakage rate is proportional to some exponent (a) of the pressure

$$Q_0 \sim p^a$$

The total (daily) volume of water leaking form one single crack equals:

$$V_{leakage} = \int_0^{24} Q_0 dt = \sum_{i=1}^{N} Q_{0,i} \Delta t_i = \sum_{i=1}^{N} (cp^a)_i \Delta t_i \qquad (6)$$

where N equals the number of time intervals Δt used in daily simulation.

The relationship between the pressure and the area of the crack has been investigated by May (1994). He introduced a classification for the leakage paths in a network. Paths could be considered either 'fixed' or 'expanding' with pressure/flow relationships to the power 0.5 and 1.5 respectively. In any distribution system, leakage can be assessed as leakage from bursts where the leakage paths do not expand with pressure and leakage from joints where the leakage areas are totally dependent on pressure. In this study, the effect of pressure on the area of the crack is analysed under 3 different assumptions for the values of the exponent factor a (0.5, 1.0 and 1.5). Similar considerations apply to a single pipe along which there is one or more cracks and to the whole network.

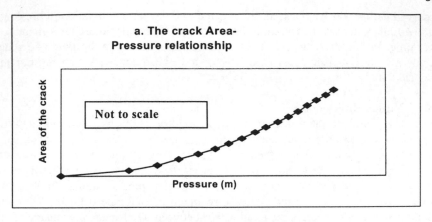

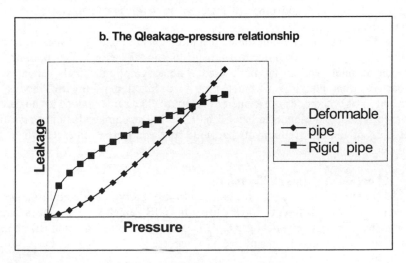

Fig. 1. Two types of pressure – leakage relationships

2.3 The effect of pipe characteristics on vulnerability

A rating survey was given to managers of water utilities where pipe materials were rated according to their overall performance in the reduction or prevention of water losses. The survey shows that asbestos cement (AC) contributes least to leakage reduction (Male et al., 1985).

In terms of the effect of pipe diameter on rate of failure, a series of papers were published with the increased interest in maintenance, replacement and upgrading of WSSs. In general, an increased rate of failure was observed for smaller- diameter pipes: see Morris (1967), Sullivan (1982), Kettler and Goulter (1983). All studies reviewed identified a relationship between failure rate and pipe diameter. In 1985, Kettler and Goulter further investigated the rates of pipe breakage with increasing diameter and time. In cast-iron, the same relationship observed earlier was emphasised. In AC pipes, no significant variation in breakage rate with diameter was observed. In both types of pipes, the failure rates increased over time. In this study the results of the statistics obtained in the Belgrade water supply system (Bobusic and Krstic (1995)) have been used for carrying our the pilot case study (Fig. 2)..

2.4   Vulnerability maps

In order to build up the vulnerability map based on pipe characteristics, data on damage percentage per pipe characteristic was needed. In the city of Srbac, such data was not available. Hence, the statistics of a study area in Belgrade were used. Since Belgrade and Srbac share similar levels of WSS maintenance and type of water, then the assumption that the behaviour of a pipe is similar in Srbac to Belgrade is acceptable. Observations were made in Belgrade between 1980 and 1988 in an attempt to define the relationship between the damage percentage along the pipes of the WSS and pipes of different material and diameter (Bobusic & Krstic, 1995).

The pipe factor expressing the contribution of pipe characteristics to leakage varies along the pipe. In order to estimate the nodal factor values, a simplified approach of interpolation has been used. The assumption, which validates this approach, is that the flow within a pipe is a function of the two nodes; the contribution of both nodes is assumed to be equal. The same applies to the damage factor; each node contributes to half of that factor. The damage factor at each node is the sum of the contribution of that node to the different pipes to which it is connected.

2.5   Creation of vulnerability maps

The derivation of the factor losses (describing partial vulnerability), can be achieved in two different ways: use of GIS algebraic operators and/or spreadsheets. The mathematical principle is the same. Using equation (6), the total volume of leakage can be quantified. To acquire the new factor, the following relationship was used:

$$f_p = \frac{\sum_{i=1}^{N} p_i \Delta t_i}{\sum_{j=1}^{M} \sum_{i=1}^{N} p_i \Delta t_i} \tag{7}$$

where N is equal to the number of time intervals used in daily simulation and M equals the number of nodes in the network to which leakage is assigned.

Using the GIS approach, the pressure values have been input as separate data files into Idrisi using edit. Using the SCALAR module, the sum of the integral pressure was calculated and the factors were derived using the above relationships.

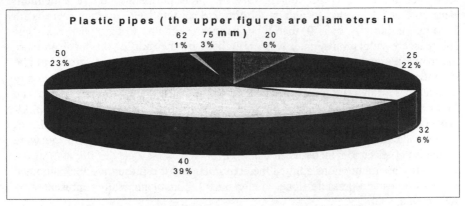

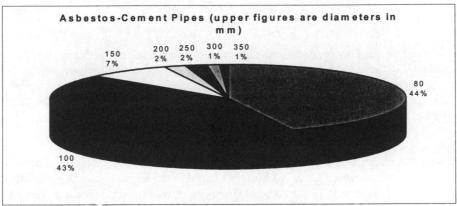

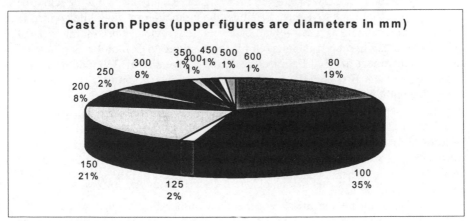

Fig 2. Diameter dependent pipe damage statistics for various materials

In order to be able to overlay the two vulnerability maps (pipe material vulnerability and pressure dependent leakage), both maps should have the same scale. Therefore, the pipe vulnerability map was normalised by its maximum value so that it is scaled to max = 1, min=0. This implies that when there is no damage, i.e. pipe factor is 0, then there is no leakage due to pipe characteristics. The pressure vulnerability map was normalised by the ratio of the local pressure and maximum pressure, and scaled between 1 (point with maximum pressure) and the ratio between minimum and maximum pressure is obtained during daily simulation. Normalisation was achieved by using the SCALAR module: one of the three Idrisi modules which perform mathematical operations. SCALAR mathematically changes every pixel in an image by a constant. Using SCALAR, the pipe factors representing the effect of pipe characteristics on leakage were normalised (i.e. factors lie between 0 and 1).

Under the assumptions applied the combined factor representing the combined effect of pipes and pressure (local vulnerability) is mathematically represented by the formula:

$$F = f_p * f_{pc} \qquad (8)$$

where F if the combined vulnerability factor and f_p and $f_{p,c}$ are the partial vulnerabilities due to pressure and pipe material respectively. In Idrisi, the two vulnerability factor maps were multiplied using the OVERLAY function.

Using EXTRACT module, the loss factors at the different nodes were acquired from the combined dependent vulnerability map created by overlaying the two separate vulnerability maps.

3   CASE STUDY

It was difficult to find a suitable case in which the methodology described above would be thoroughly tested. This will hopefully be subject to specially contemplated field tests in the future. However, it was possible to use a sample case of the Srbac (IRTCUW, 1996a & 1996b) water supply system to develop the necessary procedure and to demonstrate how the methodology would be applied in the case where reliable data sets would be available. Previous work on the WSS of the city of Srbac had been undertaken by the Institute of Hydraulic Engineering and the Department of Civil Engineering of the University of Belgrade under the supervision of the first author. Based on measurements of single reservoir level drop (late night consumption assumed to be zero), combined daily leakage losses (scaled to average daily pressure variations) in the WSS of Srbac have been estimated at 20.5 l/s. ($\approx$ 50 % of total water supplied). The extracted loss factors were used to redistribute the 20.5 l/s among the 41 nodes. The new loss values were used to re-run the simulation. Once again, the results of the minimum pressure at the required nodes were recorded as snapshots at 6-hour intervals, peak demand and low demand.

In order to create the pressure dependent vulnerability map, results of the earlier hydrodynamic simulation based on the even distribution of losses were used as an initial iteration. The following procedure has been applied:

a. in the initial simulation, leakage losses assessed from the level drop in the reservoir were evenly distributed,
b. results of the pipe damage statistics shown in Fig 2. (although obtained in different WS system) have been hypothetically used for creation of the partial vulnerability due to pipe failure,
c. results of the pressures obtained in the first iteration have been used for creation of initial partial vulnerability due to pressure,
d. spatial leakage distribution has been created taking into account the initial vulnerability map,
e. hydrodynamic simulation has been iteratively repeated until and acceptable agreement of measured (in several nodes) and simulated pressure has been obtained,
f. final vulnerability map has been crated (Figure 3),
g. various tests have been performed in examining the effect of the value of the exponent a (Equation 6).

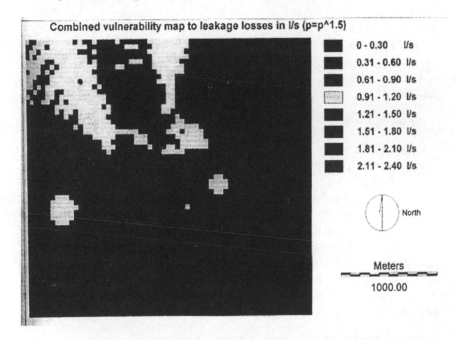

Fig. 3. Total (combined) vulnerability of the consumption area.

The results obtained with and without the effect of pipe characteristics are compared in the Figure 4.

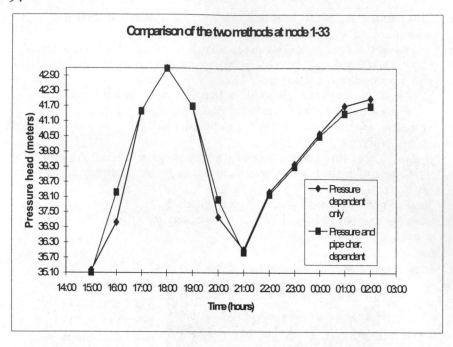

Fig. 4. Effect of the neglect of one of the partial vulnerabilities

It is argued here that, providing that reliable field measurements were available, the above procedure could be used routinely in water companies for creation of leakage vulnerability maps. These could potentially be used for concentrating the conventional leakage combating efforts to the areas of higher vulnerability. This could result in significant savings on leakage combating costs.

It is expected that with the introduction of more reliable measurements, telemetric data transmission, automatic meter readings and higher level of informatic support in the UK water industry the appropriate data will be acquired to enable more rigorous verification of the hypothesis presented here. Future computer software should incorporate complex equations, which include pressure dependent leakage.

4   CONCLUSIONS AND RECOMMENDATIONS

This study had one primary objective: the development of the methodology for vulnerability analysis and vulnerability to leakage map creation.   Different approaches in assessing vulnerability of water supply systems to leakage losses have been tested and the GIS supported method has been applied and tested in a case study.   Based on the analysis undertaken, the following conclusions have been drawn:

1. leakage losses in a water supply system are significant. Henceforth, not only should they be taken into account when setting up water management

strategies, but also it should be acknowledged that the reduction of leakage is a prerequisite in the development of a sustainable WSS,

2. leakage should be incorporated into a pipe network analysis as a function of nodal pressure and pipe characteristics. To achieve this goal, it is suggested that, the network vulnerability to leakage has to be analysed as a function of pressure, pipe characteristics and other factors

3. the choice of the approach in the simulation of leakage losses depends on data availability,

4. it is difficult to arrive at a firm opinion as to the hypothesis made in this study. The study area is too small, the number of nodes experiencing losses is too small, the quality of data not extremely high. However, the approach does indicate some improvements in simulation,

5. since, in the UK, individual water consumption has been traditionally poorly measured, the introduction of new water meters which could be read automatically and remotely dramatically increases the possibilities of applying the approach introduced in this study,

6. the problem is handled with more objectivity using the GIS approach than with traditional approaches since GIS offers the visual representation,

7. there is a need for specialists who understand the problem and have the knowledge and willingness to deal with it. It is people who usually impede adoption of new ideas and technology.

5   RECOMMENDATIONS

1. To obtain more representative results, the methodology hypothesised here should be implemented on an area, which has a reliable data storage, processing and data management system.

2. for collecting data and building up a proper database, pressure dependent modelling can be applied either in a non-iterative or in an iterative fashion similar to the one adopted in this project. If the second approach is applied, the mean values of leakage should also be corrected.

3. the relationship of pressure and crack area should be looked into in more detail.

REFERENCES

Bobusic, M. and Krstic, M. (1995) Statistics of damage in water distribution systems. In proceedings of the symposium "Man made reservoirs as sources of drinking water" Leskovac, 13-15 December 1995, Published by YU Association for Water Technology and Sanitary Engineering. (private communication).

Chadwick, A. and Morfett, J. (1995) Hydraulics in Civil and Environmental Engineering, E & FN SPON, Chapman & Hall, London.

Daughtery, R.L., Franzini, J.B. and Finnemore, E.J. (1989) Fluid Mechanics with Engineering Applications, McGraw-Hill Inc. Singapore.

EPSRC GR/M 17808 (1998) project Management of Uncertainty and data Accuracy for Monitoring and Burst Detection in Water Networks, Imperial College, Brunel University

Goodwin, S.J. (1980) The results of the experimental program on leakage and leakage control, Tech. Rep. TR 154, Water Research Centre, 52

IRTCUW (1996a) SRBAC Community Water Supply System Rehabilitation, Document 1: Urgent Rehabilitation and Repairs, Belgrade and Srbac.

IRTCUW (1996b) SRBAC Community Water Supply System Rehabilitation, Document 3: Project of Rehabilitation and Upgrade, Belgrade and Srbac.

Kettler, A.J. and Goulter, I.C. (1985) An analysis of pipe breakage in urban water distribution networks. Can.J.Civ.Eng.,12, 286-293

Male, J.W., Noss, R.R. and Moore, I.C. (1985) Identifying and reducing losses in water distribution systems. Noyes Publication.

May, J. (1994) Pressure Dependent Leakage. World Water Environ. Eng., October, 10

Morris, R.E. (1967) Principal causes and remedies of water main breaks. Journal of the American Water Works Association, 54 (7), 782-798

Sullivan, J.P. (1982) Maintaining ageing systems- Boston's approach. Journal of the American Water Works Association, 74 (1), 554-559

GIS as an Active Management Tool in Water Distribution

D.F. Schindler *and* T.P. Garrard

ABSTRACT

Geographic Information Systems (GIS) are used by almost all Water Utilities in the UK - but their use has been largely restricted to recording the location of assets and producing drawings and reports. Most systems can be effectively described as CAD systems with links to databases, and as such they are of limited utility - particularly for operational purposes.

Recent advances (in both GIS technology and user implementation) treat the GIS as a true spatial modeller, in which the connectivity of the network, the hydraulic behaviour of components, and its relationship to consumers are paramount. Windows operating systems and inter-application communication encourage the use of GIS as an integrated component, rather than an isolated tool.

This paper describes how systems can be configured to reflect the overall pressure regime with clear benefits in such areas as maintaining the DG2 register (properties at risk of receiving low pressure). More importantly, the direct access to on-line field measurements, which is now possible, allows the system to operate in a predictive mode. Although traditional (and inbuilt) hydraulic models are employed, novel methods of applying the solver have been developed, with encouraging results.

The principal obstacle to widespread application of these techniques lies in the inadequate quality of network data held in most corporate GIS systems. The paper discusses the differences between the data-capture quality standards typically applied by IT departments and those required for an understanding of hydraulic behaviour, and then discusses a number of operationally useful procedures which become viable.

1   INTRODUCTORY COMMENTS

Throughout the last decade most Water Utilities invested heavily in GIS technology. The objectives, usually, were twofold. Firstly, to acquire and maintain an inventory of the distribution system assets. Secondly, to use the acquired data to achieve improved planning and operational efficiency.

To a large extent the first objective appears to have been met. Hand-drafted paper maps have been digitised and primary attributes of pipes and fittings recorded. Zonal and company-wide asset reports and analyses are readily generated, and the mapping is increasingly available to Desktop PC users or web browsers. A few systems export network data to hydraulic modelling packages (but substantial editing is normally required before the model can be run).

Almost no progress has been evident in the second area. It is conceptually easy to understand how a GIS could harness all the disparate spatial information sources relating to network operation, and use this data to assess and manage the process in a responsive fashion. Actually doing it has proved rather elusive! The difficulties lie in both software and data.

GIS software can be broadly generalised into two categories – those which are drafting (CAD) based, and those constructed around a spatial modeller. For cartographic data capture there is little to choose between the two approaches, but spatial modelling becomes essential to handle non-visual semantic issues.

Unfortunately, most current utility GIS datasets have been created using drafting-based systems, and are demonstrably unsuited for use in operational scenarios.

2   BASIC REQUIREMENTS OF AN OPERATIONALLY USEFUL GIS

Before a GIS system can become useful in an active (or responsive) mode, two requirements must be satisfied. The spatial datasets must represent, in a hydraulically sensible fashion, all elements of the distribution system; and the spatial 'engine' must be capable of interpreting the behaviour of the system in a way which improves operational decisions, particularly in emergency situations.

Typical data requirements are:

- The underground assets must be properly connected and the status of control devices must be correct. In particular, PRVs and NRVs must be correctly orientated, meter bypasses closed,etc.
- Above ground assets must be included, at least to the extent that, for instance, connectivity exists through service reservoirs.
- Customer locations must be known, together with consumption data.
- All *readings* whether from loggers, SCADA systems or intelligent valve keys must be accessible, and each channel must be referenced to the appropriate GIS asset.
- All *curves, profiles and controls* must be known (either from pre-definition but preferably via SCADA links) to the GIS to the extent that hydraulic simulations could be invoked directly.
- On-line links should exist to work management and customer contact systems.
- Although not essential, it is highly desirable to configure the system as a hierarchy of hydraulic zones. District metered areas (DMAs) are often the lowest members of such systems. (Although polygons provide a convenient means of displaying such zones in suburban and rural areas, they are less useful in complicated city areas. Zones should really be defined as connected sets of pipework delimited by zone meters and closed valves).

This (non-exhaustive) set of requirements may appear onerous, and it is – but without this base level of accuracy and integration the GIS will be little more than an electronic substitute for paper records. The good news is that once data is marshalled in an integrated framework, it becomes relatively easy to define routines to detect anomalies. As a trivial example, if pressures are available from loggers, most flow directions can be deduced - from which PRV orientation can be checked.

The spatial engine is the critical component. Different software developers will employ varying techniques, but to operate successfully in this environment they will need to offer, in addition to basic mapping functions, at least :

- A rule base allowing the definition of objects in terms of connectivity, attributes and symbology. This rule base should be extendable by end users.

- A relational data structure with in-built connectivity pointers

- Built-in and user-definable validation

- Good connectivity to external data sources

- Powerful, fast and user-controllable tracing features

- Ability to drive hydraulic and water quality models, and to handle the results

- Strong visualisation tools allowing, for instance, the animation of flow arrows in proportion to velocity.

- Ability for end users to create *procedures* which combine spatial, mathematical and query operations.

3   BEING REALISTIC

In developing a system of this sort it is important to recognise and accommodate certain realities.

Firstly, we are dealing largely with buried assets and there will always be a degree of uncertainty relating both to the configuration and condition of these assets.

Secondly, field measuring devices of a type affordable in the industry carry inherent inaccuracies both in the measurement and the timing which must always be considered, particularly in summing operations.

Thirdly, although the petrochemical and other industries have demonstrated that flow systems can be effectively regulated by integrated on-line monitoring, sheer economics preclude the wholescale adoption of such systems for water distribution.

The challenge is, essentially, to make the best possible practical use of available information and developing GIS technology.

4   A NEW APPROACH

A system deployed at South West Water Services Ltd in April 1999, which was originally designed to carry out a fairly limited range of tasks, has served to prove a number of new concepts. Whilst the original application was limited in scope, the systems integration and resulting data quality enhancements are now being used for a growing range of operational processes.

The initial requirement was to provide full time monitoring to identify customers at risk of low pressure (OFWAT's DG2 standard in the UK). The principle was that pressures and flows monitored at zone meters could be combined with the asset data held in the GIS, and customer locations from the corporate billing system and a hydraulic engine used to calculate the likely pressure at every customer connection point.

The basic system building block for this approach is the District Metered Area (DMA). Pipework and associated asset data for a DMA is obtained by tracing rather than using a polygon "cut". The boundaries of a DMA are defined by closed valves and features within the GIS are defined as DMA meters. The trace is controlled to stop at these points. DMA meters within the map carry an attribute that links the meter to pressure and flow readings that are collated in an Oracle database. This database currently acquires some 9 million readings a month and has been configured to act as a repository for everything monitored throughout the company.

It has been necessary to supplement the existing GIS data with additional hydraulic parameters. Automated techniques are used to generate pipe roughnesses and actual diameters. A field logging exercise has been used to obtain PRV settings. Once again, these values are now held centrally on the Oracle database and business systems are now in place to keep this maintained.

Automated network model building techniques are used to allocate properties to pipes. This process defines the distribution of measured flow within a DMA. Once each DMA is configured, a conventional hydraulic solver embedded in the GIS is used to calculate the headloss in every pipe. A single meter, recording pressure, is used to define the "driving head" for the DMA (any other meters are used as comparative checks for calculated pressures). As the elevation at the meter, the headloss in every pipe from the meter to a property and the property elevations are all known, the pressure at any property can be calculated. For simple single entry DMAs a pair of coefficients is output for each property. These can be recombined with future flow and pressure readings to calculate pressure without re-running the hydraulic solver. More complicated DMAs are run every time. These searches and runs are recorded (including the minimum property pressure for each DMA) in an ongoing system log. Processing errors are also recorded and brought to the user's attention.

Of primary concern is the fact that the network is a moving target. Several safeguards are in place to identify potential network changes and mark a DMA as due for reconfiguration. These include automatic comparison with the well maintained leakage system that holds the DMA and meter relationships in a database as well as comparison of calculated and logged pressures at "check" meters. Further up-to-date status information is obtained by accessing the boundary valve monitoring system.

5 PRAGMATIC SOLUTIONS

The approach adopted must be regarded as fundamentally pragmatic. The success of the system described above is directly related to the quality of the data upon which it operates. Every input has an accuracy limit and as these limits can be quantified, the total confidence in a particular output can also be determined. For this reason calculated pressures at customer locations are grouped into three bands (rather than being quoted in numeric terms). Red indicates that the pressure appears to be less than 15m, amber 15-20m and green anything greater than 20m. Customers found to have red or amber status are then subjected to more detailed investigation and data checking. Resources are therefore targeted at potentially problematic areas.

The system is not a pure modelling package and certainly not a traditional GIS. As a hybrid it appears (in many cases) to have combined the best of both worlds. Whilst the authors recognise that there are other approaches that could be considered, this one appears better suited to the water industry and its current practices because of its reliance on automation and realistic approach to results and error sources. On initial appraisal for example, detailed round the clock, real-time models appear to offer many benefits. In reality, however, few companies have online SCADA at anything lower than strategic/ key meter level. The resolution and density of the meters and transducers that are used in the water industry is significantly lower than those found in the petrochemical industries where such technology has been proven.

Establishing such online systems in trial zones is perfectly feasible. As a measure of how suitable they are for coverage throughout a large water company, the proportion of the organisation's network covered by traditional models known to be "in-calibration" and requiring no data updates should be a good guide.

6   BENEFITS

A number of benefits emerged from the process of configuring systems and data to be able to understand the operation of the network. The following were not part of the project scope but have already found uses elsewhere in the company.

–   Meters within the map now impart flow direction information. Combined with suitable tracing facilities this automates GIS data clean-up identifying pressure reducing (PRV) and non-return valves (NRV) that are incorrectly orientated or those that are shown with open bypasses.

–   Customers are linked to their source(s) of supply. In the event of a source outage the extents of its effect are automatically identified.

–   Customers influenced by pressure reducing/ sustaining valves/ pumps can be identified easily.

–   Desktop GIS users can "click" on a meter in the map and show flow and pressure data for a specified date range. An alternative "button" shows the water balance for an entire DMA (import flows minus export flows).

–   The system provides a quantifiable and repeatable method of focusing a modelling program. Many DMAs have been eliminated altogether from the modelling process, as a conventional model appears unlikely to improve on the configuration.

–   Identification of properties close to a water pipe *that are not included in the customer billing file.*

–   The ability to analyse either a DMA or an entire hydraulic system at any time in order to obtain time of flight/ velocity information on a single "button" push. This is currently being used in two ways: a) for incident management to assess the likely consequences of a contamination incident and b) by leakage teams to check the velocity changes to be expected from a step test program, hence determining the likelihood of dirty water problems.

- Many of the outputs from a conventional hydraulic model can be provided, in a controlled fashion, to a desktop user with no modelling experience.
- Pressure management prioritisation. Having highlighted low pressure areas, other parts of the network where pressure reduction is a possibility are easily identified. This has been extended to allow users to calculate the number of customers that will be affected, the expected leakage reduction, the expected revenue reduction from metered customers and hence a repeatable cost-benefit analysis for each valve location and setting.

This is not yet an on-line system, but it is increasingly active and responsive. As anomalies are resolved, more and more zones move to the 'calibrated' category. This is to say that the parameters held in the GIS are able to correctly predict the hydraulic effect of a network change without recourse to a modelling exercise.

7   CONCLUSIONS

GIS technology *can* be effectively employed in active management of water distribution systems, provided that appropriate measures are taken to ensure integrity of data, and that suitable software is selected.

A GIS based Dynamic Strategic Planning System

J. J. van der Walt *and* E. H. Johnson

ABSTRACT

Water and capital are very scarce resources in South Africa. Development of infrastructure should therefore be done in a responsible manner to ensure sustainability. Responsible development requires dynamic inputs from a wide range of normative aspects as well as a range of dynamic technical inputs.

This paper describes the development and application of a GIS based Dynamic Strategic Planning System (DSPS). The DSPS consists of a number of integrated modules that are used to evaluate the sustainability of a proposed development viz. Infrastructure-, Hydraulic-, Environmental-, Cost-, Financial- and Resource modules.

1   INTRODUCTION

During 1997 the possibility of supplying water to a new bulk consumer was investigated. It was realised that this would impact on a wide range of aspects such as purification, pumping, distribution and storage capacity. Raw water, human and financial resources would also be affected. It was therefore essential that the existing system be analysed to identify potential bottlenecks and spare capacity, as well as areas where new developments are planned.

An integrated dynamic strategic planning system (DSPS) evolved from the process of finding the optimal water supply system. The DSPS enabled Magalies Water to answer a range of 'what if' questions with relative ease and a fair level of confidence. It also paved the way for the evaluation of future projects and is presently expanded to an operational model. The structure of the DSPS is discussed in this paper followed by a generic description of the DSPS process and a case study to demonstrate the DSPS process.

2   STRUCTURE OF DYNAMIC STRATEGIC PLANNING SYSTEM

The structure of the DSPS is shown in Figure 1 and consists of a central GIS based database that is accessed by various modules. The DSPS integrates the functionality of a GIS with an Infrastructure Module, a Hydraulic Network Module, an Environmental Module, a Cost Module, a Financial Module and a Resources Module.

The **Infrastructure Module** uses ArcView [1] to present various types of GIS based topographic and demographic data which can impact on the current or proposed water supply system viz. pipelines, reservoirs, pump stations, dams, rivers, roads and railways. Demographic information on towns and major bulk consumers such as population, per capita water demand and demand patterns are used to link the Infrastructure Module with the Hydraulic Network Module.

The **Hydraulic Network Module** uses H2ONet [2] to simulate the bulk pipe network behaviour. The GIS based pipe network elements as well as consumption patterns are linked with the Infrastructure Module. The present system behaviour can be evaluated and the effect of new developments can be evaluated by adding pipe network elements. The effect of consumer demand patterns and unit consumption on the total demand can also be determined. First order system optimisation can be performed to suit operational requirements and specific consumer demand patterns. It is also used to determine the first order size of pumps, pipes and reservoirs.

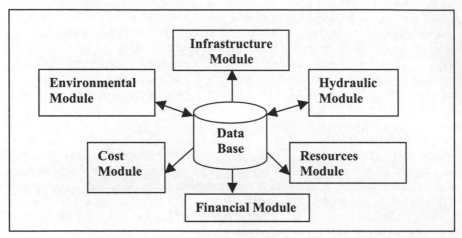

Fig. 1 – Structure of Dynamic Strategic Planning System

The GIS based **Environmental Module** is used to perform a first order check on the environmental impact. It can also give guidance in terms of flora, soil type, ground slope and erosion potential. A first order environmental impact assessment is established for each development scenario with the aid of a matrix, consisting of the possible actions causing the impact and the conditions of the environment. The magnitude and importance, determined for each development scenario, facilitates the selection of the infrastructure development option with the least detrimental environmental impact.

The **Cost Module** uses the data generated in the Hydraulic and Infrastructure Modules to calculate the cost of the proposed development. The module allows for estimation of pipelines, service reservoirs, pump stations, purification plants, and operational costs. The module can also be expanded to allow for other costs. The cost for the various items is calculated based on historic construction costs and projected inflation rates.

The costs of the various infrastructure components are based according to their size/capacity for a particular base year. Costs established are used for strategic planning purposes and are not expected to achieve the same accuracy normally associated with final design projects. However, total costs are estimated with this module in order to determine the financial module's expenditure component.

The **Financial Module** uses capital budgeting techniques to analyse the feasibility of the development. The Financial Module is presented in the form of an income statement that allows for:

- Potential income, determined from the future sales of water to the various types of consumers served by the proposed scheme. (Other income such as capital contribution and demand management charges are also allowed for.)
- Potential expenditure, such as operating expenses, capital investments, depreciation, renewal funds and levies.

In order to appraise the long-term feasibility of the project, it is necessary to use various economic analysis techniques. The techniques used by the Financial Module include payback period, net present value (NPV), internal rate of return (IRR) and return on investment (ROI).

The **Resources Module** focuses on the resources required to support the proposed infrastructure developments. The resources are grouped into four general categories viz. natural, production, human and financial. Performance indicators are established from historic data and with the aid of benchmarking the resource requirements for future development scenarios can also be established.

3   APPLYING THE DYNAMIC STRATEGIC PLANNING SYSTEM

3.1   Generic Description of DSPS Process

The DSPS process engages a number of modules that perform certain calculations. Check points are built in to determine if the DSPS process should continue or if the planner should revert to other options. The modules are computer based while the check points allow for human interface. Deliberate stops were introduced to abort the process if at any stage a non sustainable situation is encountered (Figure 2):

- The first check determines system *feasibility* by using the infrastructure and network modules.
- The second step checks the acceptability of environmental *impact*.
- The third step checks the system *viability* by applying the cost and financial modules.
- The last step checks the *sustainability* of the new development in terms of available and projected resources.

The processes that follow in between the check points constitute the various modules of the DSPS.

3.1.1   Step 1 - Demographic Data

The DSPS process commences with the identification of existing and possible future consumers. These consumers are bulk consumers such as large industries,

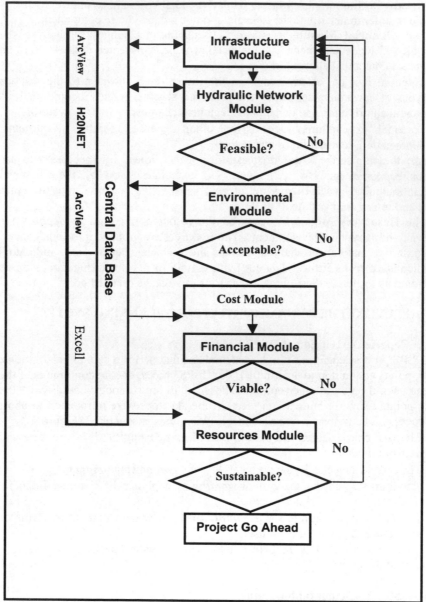

Fig. 2 – Dynamic Strategic Planning System process

mines and towns. The consumers, with all the relevant demographic data, are then displayed in the Infrastructure Module and superimposed on the existing bulk pipeline infrastructure.

3.1.2 Step 2 - Calculating the average demand

The next step involves the calculation of the average flows for domestic and industrial usage. The domestic consumption is calculated from the population data, the unit per capita consumption and a level of living index [3].

The level of living index is established from various criteria which establishes the unit water demands for a particular area. It takes into account a number of factors that were found to influence the water demand in especially rural villages. As average daily flows are the product of the projected population and the unit water demands for a particular scenario, the hourly flows still have to be modeled from previously established diurnal flow patterns. These hourly flows are used in the hydraulic model to simulate the load on various infrastructure components such as pipelines and pumps.

Industrial and mining water usage was established from actual data and projections based on the type of activity. Forecasts for the future water demands were made for various growth scenarios for the unit water demands (i.e. improvement in levels of service). A special script (Figure 3), developed in ArcView, was used to calculate the demands for each consumer growth rate scenario and design horizon and export it to the Hydraulic Network Module.

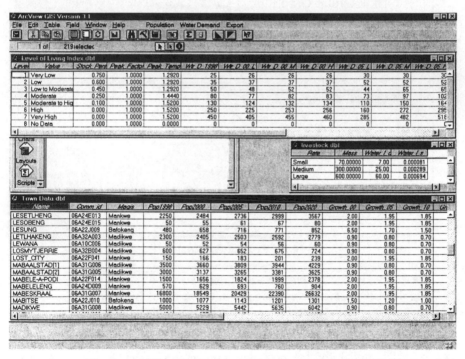

Fig. 3 – ArcView Script for calculation of consumer demands

3.1.3 Step 3 - Calculating the hydraulic performance

The Hydraulic Module used the demand data generated by the Infrastructure Module as well as the other pipe network data to simulate the network behaviour. Current demand sets were used to calibrate the model and future demand sets were used to determine bottlenecks. New infrastructure was added in areas where water shortages were predicted and the simulation was repeated until an optimal solution was obtained. Figure 4 shows typical reservoir levels of a number of reservoirs.

3.1.4 Step 4 - Checking the environmental impact

The environmental impact of the new development was checked with the Environmental Module. The ENPAT data set [4] was used to supplement the GIS data. If it was found that the new development was in an environmentally sensitive area some trade-offs were investigated. The dark areas (Figure 5) show, for instance, nature reserves in the Vaalkop area.

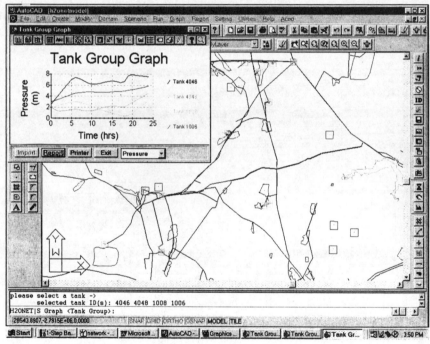

Fig. 4 - Network Module showing part of the Vaalkop distribution system as well as some simulated reservoir levels.

3.1.5 Step 5 - Calculating the capital cost

The capital required for the new development was calculated with the Cost Module. The pipe sizes and lengths as well as the reservoir storage and pump station capacity were obtained from the Hydraulic Network Module and the Infrastructure Module. Operational costs are estimated based on historic costs.

3.1.6 Step 6 – Financing the project

The capital investment required for the development as well as the current operating cost, water tariffs and water sales are used to determine the cost recovery strategy. This can provide first order estimates for water tariffs. A number of capital budgeting techniques [5] are used to determine the viability of the proposed project viz. payback period, net present value, internal rate of return and return on investment.

3.1.7 Step 7 - Check available resources

The last stage in the DSPS process is to check the available resources. The Resource Module checks if the new development can be supported by the current raw water, plant and human resources. This is achieved by comparing existing performance indicators against future requirements.

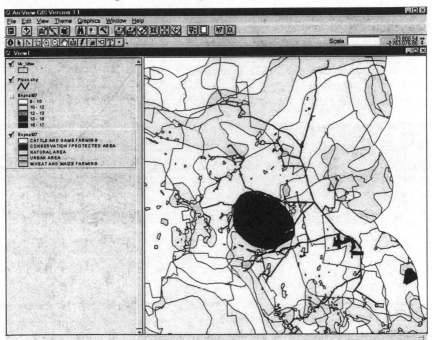

Fig. 5 - Environmental Module showing nature reserves in the Vaalkop area.

3.2 Applying the DSPS in the Vaalkop Supply Area

The Vaalkop water supply infrastructure covers an area of 12 000 km^2. The existing network includes 412 km of pipelines, 160 Ml reservoir storage capacity, treatment and pumping capacity of 120 Ml/day.

The effect of a new mine and urban development in the Rustenburg area as well as current shortages in the existing system were evaluated with the DSPS. The system was evaluated for the status quo, a 2005 scenario and a 2020 scenario. Each

110

scenario allowed for a low and high consumption by using the level of living index as a guideline. The analysis revealed that spare capacity is available in some sections of the network and confirmed the current shortages experienced in other areas.

3.2.1 Additional Infrastructure

Additional pipelines and reservoir storage are required for the mining and major township developments. The implementation of the new upgrades can be guided by the projected consumption. Project planning and implementation can be done in time to meet future demands. The total additional infrastructure required by 2020 involves about 190 km of pipelines, 95 Ml of reservoir storage and 90 Ml/d pumping and purification capacity.

3.2.2 Network analysis

Calibration of the Hydraulic/Network Model for the base year resulted in the theoretical demands being generally within 10 percent of those actually measured. The only exception was the rural villages. It was since discovered that high night flow rates contributed to the discrepancy between actual and predicted consumption. Operational measures are currently implemented to reduce night flow by repairing leaks and preventing unauthorised use.

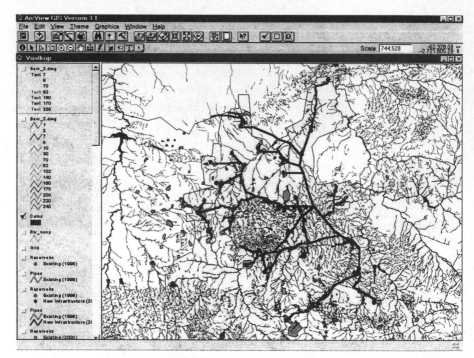

Fig. 6 – The Vaalkop supply system, indicating Rustenburg in the south and Thabazimbi in the north.

The DSPS also revealed that intelligent on/off and flow control systems on reservoir inlet valves can improve network efficiency and postpone network upgrades. This is achieved when for instance two or more reservoirs are prevented from opening and closing at the same time leaving the network under-utilised or overloaded for long periods.

3.2.3   Capital requirements

The Cost Module was used to estimate the total present day value of the investment required for the development scenarios. The cost of the required upgrades amounts to R34 Million (South African Rand) based on 1999 costs.

3.2.4   Funding the capital

If the capital investments required for the mine and the surrounding towns are funded through tariffs only, the marginal water tariff will increase by 5% each year from R0.13 per kiloliter to R0.36 per kiloliter. If the major consumers make capital contributions towards the investments, the tariffs will increase by a few cents in the long term. This increase also allows for the marginal increase/decrease in operational costs of the new development.

3.2.5   Resources

Applying the performance indicators established with the aid of the Resources Module for the base year to the future development scenarios, Magalies Water will need to increase its total staff from 54 to 81 for the Vaalkop System by the year 2020. Sufficient raw water and plant production capacity should be available to sustain the 2020 demand. If demands increase beyond the projections a shortage may be experienced from 2000 to 2004. This can, however, be contained through demand management on domestic and irrigation use. The increased demand can be accommodated without raw water augmentation schemes due to the increased return flows in the catchment areas. This is due to development in the Gauteng province.

4   FUTURE DSPS DEVELOPMENT

The Magalies Water water supply system spans two primary catchment areas and involves five distinct purification plants which feed into four distinct bulk distribution systems. The DSPS is currently applied to prepare a strategic plan for the whole Magalies Water system. Improvements on the DSPS include:

- Incorporating more detailed information in the Resource Module to indicate where, when and how much raw water will be available in the catchment areas for each of the four sub-systems.
- Limit the manual links between various modules.

The DSPS framework also lends itself for future expansion to an operational model which can in future include the linking of the:

- Hydraulic Module with actual network performance from SCADA to calibration Hydraulic Module on line.
- Network attributes to the network maintenance management system.
- Hydraulic Module with a water quality model of the distribution system.

- Hydraulic Module with the SCADA system to improve network usage through intelligent reservoir level control mechanisms.

5   CONCLUSIONS

The Dynamic Strategic Planning System provides an elegant, user friendly, dynamic and integrated tool to evaluate the sustainability of an existing and proposed new water supply scheme. The DSPS can be adapted for operational purposes to improve network efficiency and maximise its use. The DSPS provides a generic framework that can be populated with user specific data.

The DSPS not only provides a framework for strategic planning, but it can also be expanded for use as an operational tool. This is often essential to understand what happens in systems with numerous interconnected reservoirs, pipelines and pump stations.

REFERENCES

[1] Using Arcview GIS (1996) Environmental Systems Research Institute. Redlands. USA
[2] H20NET 2.0 User manual. (1997) MW Soft. Pasadena. USA
[3] VAN SCHALKWYK (1996) *Guidelines for the estimation of domestic water demand of developing communities in the Northern Transvaal.* Water Research Commission Report No. 480/1/96. South-Africa
[4] ENPAT Environmental Potential Atlas – Users Reference. (1997) Department of environmental affairs and tourism. South-Africa.
[5]   GITMAN L.J. (1994) *Principles of managerial finance.* 7th Edition. Harper Collins College Publishers.

Computer-aided Management of Water Resources for Small Island Communities: The Isles of Scilly, UK

D.C. Watkins

ABSTRACT

The Isles of Scilly lie 40km off the coast of Cornwall in SW England. Water resources assessments were conducted for two of the islands, St Martin's and St Agnes. These islands are small with land areas of around 2km^2 each and support populations of 70 to 100 people. Tourism causes the populations of the islands to increase by a factor of around 3 to 4 in summer, creating a seasonal stress on the limited water resources.

Water supplies are derived from a combination of rainwater collected from roof drainage and groundwater from small private wells. The groundwater is abstracted from shallow wells in blown sand deposits and deeper wells in a weathered granite aquifer. Wastewater is generally disposed of to septic tanks and soakaways.

A geographical information system and distributed numerical models were used to assess the water resources and to define source protection zones around wells. The integrated use of the two computer methods proved to be a useful combination of tools to aid in the assessment and management of the water resources of these small island communities.

1   THE ISLES OF SCILLY

Five small islands in the Isles of Scilly support permanent populations. Water is a critical resource on these islands and supplies are under a seasonal stress due to increased summer populations through tourism. Two of the islands, St Martin's and St Agnes, have recently been subjects of water resource assessment exercises.

1.1   General Information

The Isles of Scilly lie 40km west-south-west of the peninsular of Cornwall in the extreme south west of the UK, Fig 1. Of about 140 small rocky islands, 5 of the islands support a permanent population of about 2000 people but the population increases to about 5000 people with an influx of tourists in the summer months. About 80% of the residents live on St Mary's, the largest island and the administrative centre. The other four inhabited islands each support permanent populations of between about 50 and 120 residents. Other than tourism, the main industries are flower growing and fishing.

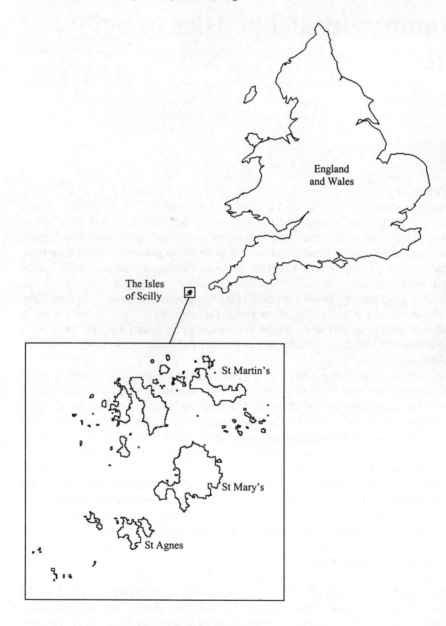

Fig 1   Location of the Isles of Scilly

The islands are small and low-lying. St Mary's, the largest island is about 4km by 3km in extent. The highest point on the islands is 50m above sea level and large areas are within 10m of sea level.

The islands of St Martin's and St Agnes have land areas of $2.2km^2$ and $1.8km^2$ and support permanent populations of 100 and 70 people, respectively. These populations increase to about 450 and 250 people, respectively, in summer.

The islands are a granite outcrop and are thought to be connected at depth, through the Cornubian Batholith, with granite outcrops that comprise the moors of Devon and Cornwall. The granite is Late Carboniferous to Permian in age, about 300 Ma. The granite is quite deeply weathered and the upper 2 to 3m have been highly weathered into a head deposit, a sandy matrix with shattered granite fragments, by periglacial actions. Most of the low-lying areas of the islands, below about 10m above sea level, are overlain by blown sand deposits of Holocene origin.

1.2 Water Supplies

1.2.1 St Mary's
The island of St Mary's derives its water principally from wells abstracting from the granite and from minor alluvial deposits. It has a central storage and treatment facility and a water distribution network. A water shortage crisis in 1992 resulted in the introduction of a small desalination plant. In times of high demand, brackish water is desalinated by reverse osmosis and blended with water derived from groundwater sources.

The other islands are more reliant on individual supplies for each property. These are derived from a combination of groundwater from wells and rainwater collected from roof drainage.

1.3 St Martin's and St Agnes
Up until about 1960, water was drawn by bucket and rope or by hand pump from shallow hand dug wells founded in the blown sand deposits or the granite head. In 1960-1985 a number of deeper wells were constructed. These were fitted with hand pumps or surface pumps with petrol driven generators.

In 1985, mains electricity arrived at the islands for the first time. This created an increase in the standard of service provided to the holiday industry and a corresponding increase in water demand. Automatic washing machines and dishwashers arrived. This event also led to the introduction of the submersible pump and its use in deeper wells. From about 1985 to present, a large number of wells were constructed, often in a rather haphazard fashion and with no regard to water resource issues such as the sustainability of the quantity and quality of supplies.

Rainwater collected from roofs has always been an integral part of water resources on these islands. Rain intercepted by the roof of a dwelling is fed by the guttering into a storage tank. From there it is pumped into a header tank in the roof space and gravity fed into the household pipework. The practice of utilising rainwater is fairly common on small islands where alternative supplies are limited

(Falkland 1991). In a few cases ultra-violet filters are used to treat the water but generally any form of treatment is rare. Commonly, households rely on rainwater storage for all domestic water supplies during the winter months and top up their storage tanks with groundwater abstracted from wells during the summer. A few properties are totally reliant on rainwater and a few rely totally on groundwater.

With the exception of a hotel, which has a small sewage treatment system prior to discharge to the sea, wastewater is disposed of to septic tanks with subsequent discharge of effluent to soakaways. This creates a potential conflict in the use of groundwater as a supply of drinking water and a receptor of wastewater, especially in the limited confines available to small island communities.

Some properties share wells with other properties and some small distribution networks exist, containing a few properties and a few wells. Some septic tanks serve more than one property.

1.3.1 Source Protection Zones

The Environment Agency of England and Wales (EA) currently have no jurisdiction in the Isles of Scilly. However, it was felt to be good practice to consider the principles of groundwater protection as applied by the EA, irrespective of the fact that they have no legal or regulatory standing on the Islands. The EA apply the concepts of vulnerability and risk to groundwater protection. Applying EA criteria, the granite aquifer from which water is abstracted in the Isles of Scilly is classified as a minor aquifer of high vulnerability.

The EA also set out Source Protection Zones (SPZs) around wells in which activities liable to cause pollution are limited or prohibited. Three SPZs are recognised:

- Zone 1 (inner SPZ) is defined as the larger of two criteria: 50m distance from the well or a groundwater travel time of up to 50 days.
- Zone 2 (outer SPZ) is a capture zone based on a travel time of 400 days.
- Zone 3 (catchment SPZ) is defined as the whole area within which groundwater eventually reaches the well.

2   GEOGRAPHICAL INFORMATION SYSTEM

As part of the management and analysis of water resources for the islands of St Martin's and St Agnes, a Geographical Information System (GIS) was utilised. The GIS used was Idrisi (Eastman 1997) which employs both raster and vector capabilities.

2.1   GIS as a database

The GIS was used as a database of information relevant to the water resources. Although a standard hierarchical type database would be sufficient for storing this information, the GIS proved to be a suitable platform for storing and cross-referencing information. It contains the added advantage of producing maps showing the distributions of information held.

Each well, and also each septic tank, was located and identified in a point vector file. Despite the small sizes of the islands, 42 wells and 42 septic tanks were located on St Martin's and 27 wells and 38 septic tanks were located on St Agnes, see Fig 2.

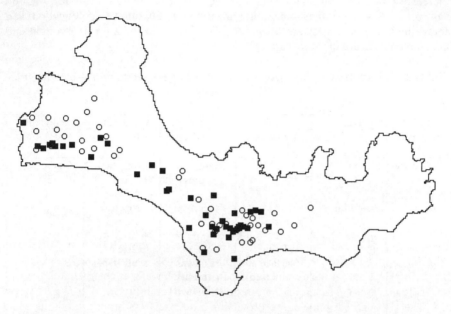

(a) St Martin's

(b) St Agnes

Fig 2   Locations of wells (indicated by open circles) and septic tanks (indicated by black squares) on St Martin's and St Agnes

Attributes files were used in the GIS to link the identification number of the wells with information on their properties. Attributes files were also used to cross-reference with other information such as the properties served. This information was further linked to files containing information on rainwater collection and storage and also on wastewater disposal. The categories of information held and linking attributes are listed in Table 1.

Table 1 GIS database files. Arrows indicate linkages used to cross-reference the data

GIS file	Wells	Welldata	Properties	Wastewater	Septictank
File type	*Point vector*	*Attribute*	*Attribute*	*Attribute*	*Point vector*
Data held	Well ID ◄────►	Well ID ►	Name and address	Septic tank ID ◄────►	Septic tank ID
	co-ordinates	Properties served ◄────	Property ID ────►	Properties served	co-ordinates
		Pump (surface or submersed)	No. of residents (summer and winter)	Size of septic tank	
		Type (hand dug or bored)	Details of rainwater collection system	Age of septic tank	
		Date drilled	Details of wastewater disposal system	How often emptied	
		Depth		When last emptied	
		Diameter			

2.2 GIS as an analytical tool

The power of GIS is in its ability to be applied analytically to geographically distributed problems, rather than in its use as a database. The GIS was used to construct digital elevation models of the islands. These played an important role in understanding the hydrogeology of the groundwater flow system. It was also used, in conjunction with numerical models, to delineate groundwater source protection zones around wells. The GIS was put to further use in integrating data with the numerical model in terms of processing both input and output.

The Digital Elevation Models (DEMs) of the islands were constructed using spot heights at 10m intervals supplied by the Ordnance Survey (OS 1992a, OS

1992b). The DEM is a raster coverage based on a regular 10m grid cell size. This required raster domains of 331 columns and 223 rows for St Martin's and 236 columns with 206 rows for St Agnes.

Not all of the cells are active; because of the irregular shape of the islands, many cells are located out to sea or over unconnected small islets. The DEM was therefore used to create a Boolean masking image, containing values of one over the island land area and zeros over the sea. Any information outside the areas of interest can be filtered out of any other coverages by multiplication with the Boolean masking coverage. Part of the resulting, filtered, DEM for St Martin's is shown in Fig 3, from which the relationship between the regions of steep granite and low lying blown sand can be seen. Note also the 10m grid spacing.

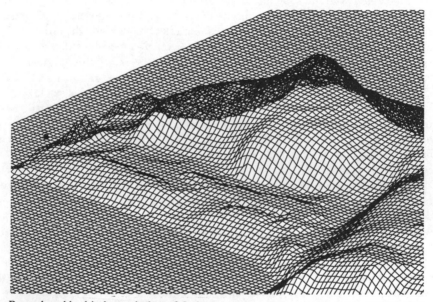

Reproduced by kind permission of Ordnance Survey © Crown Copyright MC/99-392

Fig 3   Digital Elevation Model of the western end of St Martin's as a perspective image (×3 vertical exaggeration)

One of the criteria for assigning source protection zones to wells is that the inner zone should be a distance of at least 50m from any well. This was calculated using the GIS. First the vector point file containing the well coordinates was mapped onto a raster file. The distance of any raster to the nearest raster containing a well was calculated. The resulting coverage was reclassified into a Boolean image using the criteria that distances less than 50m are allocated a one and distances greater than 50m become a zero. It was then possible to automatically test and identify which septic tanks were violating the inner source protection zones and which wells were affected.

A numerical model was run to simulate groundwater flows and to delineate the outer and catchment SPZs. This information was fed from the model back into the

GIS and the three zones were compiled for each well and mapped onto images of the islands. This allowed the SPZs to be correlated with a GIS coverage containing details of land use.

3 DISTRIBUTED NUMERICAL MODELLING

3.1 Groundwater Flow Modelling

Numerical models of the groundwater flow systems were constructed using aquifer properties derived from hydraulic tests conducted on wells. The finite difference code MODFLOW (MacDonald and Harbaugh 1984) was used for the flow modelling.

In keeping with the structure of the GIS raster coverages, a regular grid size of 10m and a block-centred finite difference scheme were chosen. This ensured compatibility in integrating the distributed numerical model with the GIS.

The islands were each modelled as single-layer, 2-dimensional horizontal flow systems. The inflow is rainfall recharge, taken to be 0.001 m/day, in line with values estimated previously and used in other studies (e.g. Burgess et al 1976). Outflow is to the coast where a constant head boundary was fixed at sea level.

Initial results using a single granite aquifer provided a poor correlation with field data. Water level monitoring of wells indicated that steep hydraulic gradients occur in the granite but shallow gradients are present when the water table intersects the blown sand, and possibly the head material. A second model was constructed whereby all land below 10m above sea level, as identified from the DEM using the GIS (see Fig 3), was deemed to have a hydraulic conductivity of 2 orders of magnitude greater than the granite, 4 m/day and 0.04 m/day. This was done to reflect the effect of the blown sand. This time a very good fit was found when compared to field measurements. A better approach would have been to construct a two-layer model with the higher permeability deposits overlying the main granite aquifer. However, it was felt that the extra complexity involved would not be justified in terms of increased model accuaracy and so all modelling was restricted to single-layer and 2-dimensional flow.

3.2 Groundwater Pathline Modelling

In order to simulate pathlines of potential contaminants in the groundwater system and also the groundwater travel time criteria for SPZs, a particle tracking routine was used. This was achieved using the code MODPATH (Pollock 1989) which was designed to interact with the MODFLOW code.

For each well, 80 virtual particles were placed around the grid cell containing the well. The code was then run as a backward-tracking routine to trace the paths of the particles and identify where they would have had to come from in order to end up in the well.

The routine was run three times for each island, once with no time limit which identified the total catchment zone for each well. The result for St Martin's is shown in Fig 4. The average time-span of particle, and therefore the mean age of the water, was found to be 4 years, a short time for groundwater. The model was run again with a time limit of 400 days to find the capture zone that defines the outer SPZ and once for 50 days to test the inner SPZ. In every case the 50m criterion for the inner SPZ was found to be greater than the 50 day time of travel criterion and in some cases greater than the 400 day capture zone. The results of the particle pathline tracking were captured graphically on screen as DFX images and converted by the GIS into raster images for SPZ mapping purposes.

Fig 4   Catchment zones (SPZ 3) for wells on the western end of St Martin's as found from numerical particle tracking techniques

The pathline tracking technique was run once more for each island. This time a single virtual particle was placed in each septic tank. The model was run as a forward tracking routine to test where the septic tank effluent could migrate to and which wells it could enter. A number of wells where identified as being at risk of contamination from septic tank effluent, including some which had already been abandoned for showing a regular presence of *E.Coli* bacteria.

3.3   Saline Intrusion

Being maritime islands, saline intrusion of seawater into the aquifer is a potential problem. This can result from wells being positioned too close to the coast, with their bases too deep or being pumped at too high a rate.

It is possible to numerically model the effects of saline intrusion but this would require a fully 3-dimensional, variable density fluid, advection-dispersion model, the complexity of which was considered to be outside the scope of the study. Instead, a simple analytical formula (Bear and Dagan 1964) was used to estimate the maximum yield that can be safely abstracted from each well, based on the well geometry, water level and aquifer properties. This was then compared with the maximum yield that can be abstracted from each well and the wells were classified in terms of high, medium or low vulnerability to contamination by saline intrusion.

4 CONCLUSIONS

As part of a study to assess the water resources and create a water management plan for two small island communities, GIS and numerical modelling were found to be applicable tools.

The GIS was used as a database to cross-reference information on wells, water use and wastewater disposal. It was also used to create a DEM which played an important role in understanding the hydrogeology of the islands. In conjunction with the numerical model, the GIS was used to identify and map the SPZs in accordance with Environment Agency recommendations.

Distributed numerical models were used to simulate groundwater flows and identify the impacts of well abstractions. Pathline tracking routines were used to identify the areas that contribute groundwater to each well and to delineate SPZs based on travel time capture zones. It was also used to identify potential pathways for contaminants leaving septic tanks.

By using the same regular grid for both the GIS and the numerical models, the two methods were integrated together and information was passed between the two systems for appropriate analyses, compilation and mapping of results. The two systems working together provide a valuable tool for the computer-aided management of water resources in small island communities.

REFERENCES

Bear, J. and Dagan, G., 1964. Moving Interface in Coastal Aquifers. *Proceedings of the Americal Society of Civil Engineers*, 99 (HY4).

Burgess, W.G., Clowes, U.R., Lloyd, J.W. and Marsh, J.M., 1976. Aspects of the Hydrogeology of St Mary's, Scilly Isles. *Proceedings of the Ussher Society* (3).

Eastman, R.J., 1997. Idrisi for Windows. Clark Labs for Cartographic Technology, Clark University, Ma., USA.

Falkland, A., 1991. *Hydrology and Water Resources of Small Islands: A Practical Guide*. UNESCO.

McDonald, M.G. and Harbaugh, A.W., 1984. *A Modular Three-Dimensional Finite-Difference Ground-Water Flow Model*. U.S. Geological Survey Open-File Report 83-875.

O.S., 1992a. Landform Profile digital topographic data for 1:10,000 sheet SV91NE. Ordnance Survey, Southampton.

O.S., 1992b. Landform Profile digital topographic data for 1:10,000 sheet SV80NE. Ordnance Survey, Southampton.

Pollock, D.W, 1989. *Documentation of Computer Programs to Compute and Display Pathlines Using Results from the U.S. Geological Survey Modular Three-Dimensional Finite-Difference Ground-Water Flow Model*. U.S. Geological Survey Open-File Report 89-381.

Zonal Network Management

R. Williams, M. Giel, M. Foulsham *and* D. Grimshaw

ABSTRACT
The day-to-day management of water networks demands that operators take account of many factors. They also need to assess the impact of their decisions (usually based on available records and local knowledge) on the rest of the network and on customers' levels of service. This is particularly true in emergency situations when decisions have to be made quickly and with a high degree of certainty. To support these operations, it is essential to have up-to-date mapping that illustrates key features of the network and accurate records showing the status of critical valves.

South East Water plc has taken the initiative to develop a Zonal Network Management System (ZNMS) following on from its decision to produce a prototype system. The company appointed Ewan Associates' specialist division, Technology Solutions, to carry out both stages of the development work. The ZNMS is integrated into the existing Geographic Information System (GIS), thus eliminating the need for duplicate data sets and reducing the costs associated with maintaining two systems.

The concept revolves around a hierarchical set of schematic network drawings – company level, management areas, water-into-supply zones and district meter areas – that provides operators with a method for viewing key network features. The system includes the status and recent operating history of key valves, as well as procedures for recording the status of critical valves.

The key feature of the system is the ability to link critical valves in the water network with the GIS. The system was developed in conjunction with Exeter University with the intention that it should be generic and combine GIS with advanced optimisation techniques. The schematics represent optimised views of the geographic network. The brief was also to update the GIS and "prove" data by close liaison with the water company's Operations and Leakage departments.

By providing operators with improved access to accurate data in the format required, the ZNMS will help ensure that supplies are safeguarded and customers' levels of service are maintained. It will also allow South East Water to respond effectively and quickly to emergency events.

1   INTRODUCTION
The efficient operational management of the water supply function necessitates a thorough understanding of the existing trunk and distribution mains network. The status, open, closed, or partially closed of specific valves within the system is critical to ensure good system management. Pressures from customers and

regulatory bodies are ever increasing and the potential for prosecution and compensation payment is increasingly real. A sound operational knowledge, available to all affected parties, ensures the effective and optimised operation of the distribution network, treatment works, pumping stations and reservoirs.

In South East Water, water is pumped and re-pumped over long distances. A high degree of interconnectivity exists when compared with other operational systems. This interconnectivity exists between supply zones and between the strategic transfer mains and the distribution system. This is particularly relevant in the Southern Area of the Hampshire and Surrey Region (formerly Mid Southern Water) where service storage is at a minimum. Due to the overall connectivity of the system, failure of Water Treatment Works, Booster Stations and Trunk Mains have a significant widespread and immediate affect, both on storage levels and distribution pressures and consequently levels of service. To manage these failures, strategic re-valving of the trunk and distribution mains is implemented. Consequently the network system can require daily adjustment in order to achieve efficient operation of the system.

Leakage control practice commenced in the late 1980's and relies heavily on good valve management. Due to the lack of culture regarding this management, significant difficulties have been experienced in producing reliable leakage information. With the completion of District Meter Areas (DMAs), for leakage control, valve management is critical to produce meaningful reliable information. Pressure control, implemented to reduce leakage also requires a culture of valve control and management.

During the high demand period of 1995, Crowthorne, Surrey Hill, Ewshot and Hogs Back Reservoirs became unacceptably low, several were millimetres deep. Similar problems were experienced in the Kent and Sussex Region (formerly South East Water Ltd.). Subsequent investigation revealed that the strategic control system in place had failed to maintain a current knowledge of the network. Control valves were found to be incorrectly open or closed. The network status had deteriorated to such an extent that the strategic movement of water could not be successfully achieved.

Further operational problems were experienced in 1997 with coliform failures. Uncontrolled valve operation gave rise to high chlorine residuals being experienced in the distribution areas fed from strategic mains. Changes in taste and odour resulted in customer complaints; effective valve management would avoid these.

In November 1997 the Drinking Water Inspectorate contacted all companies regarding discoloured water supplies and their intention to prosecute companies supplying water unfit for human consumption. The zoning of the distribution system and regular manipulation of this system greatly increases the risks of a discoloration event. Increased customer awareness can now transform an event into an incident more readily than in the past.

Both regions have a history of iron and manganese deposits within its mains. Recently water quality investigations revealed unacceptably high organic anaerobic deposits and water within dead legs of the District metered zones. The DWI has specifically stated that companies should maintain records, maps, diagrams including valve status and have in place procedures to manage operational activities.

Since 1990 the operational departments have retired at least thirty key members of staff. During recent operational incidents it has become evident that local knowledge no longer exists or is currently incorrect. The movement of operational staff has demonstrated the need for a readily transparent information system and procedures in order that the effective management of incidents may take place. Today's Managers have sound analytical abilities but lack specific local and strategic knowledge gained from practical experience.

This paper details the design and phased implementation of the zonal management system, based on the company's GIS, which will correct existing confusion of system operation and provide a tool to manage and operate the network.

2   THE ZONAL NETWORK MANAGEMENT TOOL

The efficient operational management of the water supply function necessitates a thorough understanding of the existing trunk and distribution network together with treatment work pumping stations and reservoirs. Currently this is gained from historical paper based schematics, paper based record logs and local knowledge.

The Zonal Network Management System is a software system based on a series of hierarchical operational schematics. The status of individual key boundary valves is displayed on the schematics and automatically linked to and created by the GIS Records Systems. The schematics are presented and used in a layered format to enable system operation. The hierarchy of schematics is regional, area, water into supply (WIS) zone and district meter area (DMA). A water into supply zone will typically have between 10,000 and 20,000 properties, a DMA between 500 and 2000 properties.

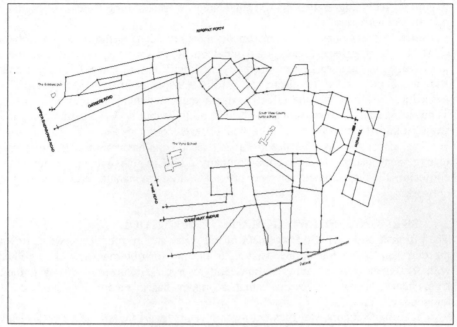

Figure 1 DMA Schematic Layout

The schematics also link to operational databases giving valve location, description and details. The status of specific operational control valves is updated by entering the GIS system, which will in turn revise the status on the operational schematics. Operational procedure manuals have been written for the updating of the valve status.

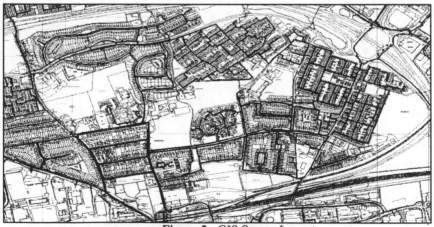

Figure 2 GIS Screen Layout

By using the schematics and operational procedure manuals, Duty Managers and other staff are able to gain an overview of the network operation, with the confidence that the records reflect the system. The success of the system depends on the accurate identification of strategic valves in the field. It is therefore necessary to erect valve marker posts with strategic ranking and procedural instructions.

Figure 3 Typical valve post and plate installation

In addition, the system holds CAD drawing schematics of individual works and Reservoirs. These can be displayed through the system.

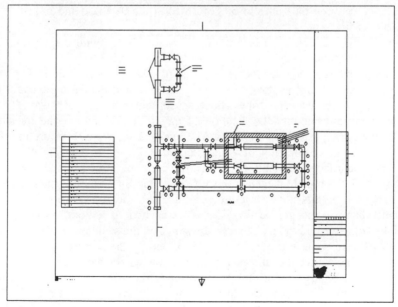

Figure 4 CAD Site Layout

3   TECHNICAL SPECIFICATION

GIS is a system to display, manage and integrate data in a geographically referenced environment. The Corporate GIS is a package called "GeoComm" based on the Oracle database and Microstation graphics engines. GeoComm was built around the use of Ordnance Survey 1:1250 and 1:2500 scale vector maps overlaid by vector lines and symbols representing the water network. All the programs and data display/management functions were designed to conform to this 'real world' view on life, i.e. a line on the GIS is directly related in terms of its position and length to the geographic location of the relevant asset.

The concept of Zonal Network Management required a very different approach and consequently presented some difficult technical questions.

- Schematic drawings have no direct geographic relationship and yet have to relate accurately to the real world; they are co-ordinates on paper but must occupy the same environment as existing geographic data and be displayed at the appropriate scale.
- System navigation functions within GIS were based on use of a street gazetteer or grid reference; zones were not managed / configured as separate entities for graphic display purposes.
- How do you geographically reference and control an object within a non geographic display?
- How do you navigate through a non geographic display whilst retaining the capability to return to a geographic view at the correct level and scale (and vice versa)?
- The use of physical zone boundaries for navigation purposes requires the relations between the zones to be described in the database and linked to the Gazetteer function.
- How do we maintain synchronisation between the graphic display and the gazetteers without restricting functionality?

The existing GIS user interface locates on road or grid reference; data is displayed according to scale and operator preference. Colours and database attributes are used to differentiate between zones; there is consequently no easy way for the operator to visually identify the location of any zone. To answer the questions posed by the concept of Zonal Network Management we were therefore required to design and build a completely new interface into GIS.

The new interface was built on a hierarchical zone structure starting at the Regional level through Management Area to Water Supply Zone then DMA / PRV. The use of this zone structure provided the method for integrating the Schematic data. A schematic drawing is associated to a zone or area, with hyperlinks being created to link between drawings on the same level. The physical boundary of each zone is generated in GIS providing the required focus for the display. A toggle function allows the user to switch between geographic and schematic modes at any level within the described structure.

Zones are linked using the Oracle Relational Data Base Management System, each zone is described in a database table by means of a unique key which by association describes all related zones. For example, management areas contain water into supply zones (WIS); therefore all the WIS zones in Management area 'A' will have the unique key ascribed to area 'A' as a foreign key on its table entry. The database tables are used to generate an enhanced gazetteer function used in system navigation. The physical boundaries used by the graphic display are linked via the relational structure to the gazetteer, which then allows the system to locate accurately on the selected zone. Due to the complexity and volatility of the network structure, particularly in the Hampshire and Surrey Region, it is very difficult to generate accurate boundaries; these will therefore be approximate and for navigation purposes only.

Navigation through the system has been enhanced. Window 1 has been configured as a navigation window with the display always maintained one level higher than the selected zone as displayed in the main data window. Selection can be either from the gazetteer or by selecting the required zone from the list provided and pointing in the navigation window using the mouse. Display control is now more exact and consistent; both scale and content is controlled by the system, and display management only applies within the current session.

Within the schematic environment valve status is identified by both colour and symbol type; green, yellow and red denote open, partially open and closed respectively. A new field, temporary valve status, has been created with a thematic session operating on it to generate the required colour. The content of this field can only be changed by an authorised user. Permanent valve status continues to be managed as before; changes can only be made by editors, following the creation of a red line file. A new and more comprehensive road gazetteer is being purchased from the Ordnance Survey, which will in time be integrated into the new structure.

In order to obtain a consistent and accurate picture of valve status we had to ensure that there was only ever one incidence of any valve. GIS provides all the tools and functions required to maintain the graphic elements, including valves; consequently the valve as described in GIS must be the master. We achieve coincidence across all displays and levels by relating all coincident valves to the same row in the database table. Rows in the valve table are created and maintained exclusively from within GIS. Only the temporary valve status field can be changed directly by operational staff, either from the schematic or geographic view; all other changes require the creation of a red line file.

For effective use of the system, particularly at operational level, valve location must be available to users from within the schematic mode. It has to be easy to relate from schematic mode to geographic mode and vice versa. To facilitate this a function has been incorporated that allows a third display window to be opened showing a geographic view centred on the valve selected from within a schematic drawing. Normal GIS functionality (Pan, Zoom etc.) is provided in this third window.

GeoComm manages graphic data by placing elements into a file covering a specified geographic area; this area can be of any size from an A4 piece of paper to the whole country. Each graphic file, whether schematic or geographic, contains all the elements relative to the display area it covers. Consequently a single physical element may exist in more than one file when both geographic and its schematic representation are managed within the same environment.

Graphic data elements are controlled by means of a graphic key, database table id + 'MSLINK'. This key is unique and can therefore be used to manage multiple display elements across a range of display types and levels, all coincident valves will have the same graphic key. There is unfortunately no easy solution to generating the links between the different graphic elements, tools configured within GIS can be used to make the correct association. There is, however, a considerable amount of data capture required to ensure that the correct graphic key has been ascertained and correctly allocated.

Data is the perennial problem of GIS both in terms of accuracy and completeness. For zonal management to be effective it is essential that data be consistent and accurate. The following data requirements are mandatory;

- Accurate and uncluttered schematic drawings.
- Complete and accurate graphic data. Valves must exist on GIS and be correctly placed and attributed.
- A well defined and maintainable key structure must be applied to all relevant valves.
- Effective QA procedures should be implemented.
- Data ownership must be assigned to specific roles with the operating companies.
- Information described on valve plates must relate to that held in the database.
- Updates to GIS data must be done as soon as possible after any change has been identified.

No system, however good technically or functionally, will succeed if the quality of data it holds is poor. An important part of the project will be the validation and consolidation exercise to be carried out on the GIS data.

4   COST BENEFITS

The cost benefits of a zonal management system are difficult to accurately establish. However, the DWI has effectively put the company on notice to establish procedures for managing operational activities to avoid discoloration.

Currently South East Water have 60 significant discoloration events per year which have the potential to be reported to the DWI. At least 35% of these could be avoided by improved system awareness, management and procedures. Currently two operational incidents are under review by the DWI and likely to result in prosecution.

South East Water has, on average, 12 significant operational incidents per month, which result in loss of supplies to properties. In 1997, Mid Southern paid approximately £45,000 in compensation costs. As customer awareness increases, compensation payments are likely to increase.

Total cost benefits for the two regions is therefore calculated very conservatively as £50,000 per year. This excludes direct cost attributed to the workload created by operational incidents.

5   PROGRESS TO DATE

An initial trial project was carried out during 1998 to prove the proposed system. This confirmed the viability, both technically and operationally. It involved producing and linking schematics for an area, a water into supply zone and a district meter area. The work included the erection of valve marker posts, production of operations manuals and updating procedures.

On completion a review was carried out, taking into account opinions from all effected parties. Based on the output from the trial, approval was given to complete the schematic production at all levels for the Hampshire and Surrey Region in 1999 and to produce regional and area schematics for the Kent and Sussex Region.

The company appointed Ewan Associates Ltd and its specialist operating division, Technology Solutions, to undertake development of the prototype system followed by implementation. The key stages within the project were:

- Development of specialist software to generate schematic diagrams from GIS. This was developed in conjunction with Exeter University with the intention that it should be generic and combine GIS with advanced optimisation techniques.

- Data Acquisition and Field Audits. This stage involved collation of all appropriate records, principally at DMA level, and discussions with company field staff to confirm operation of each part of the system. Field audits were carried out on all boundary valves and meters to confirm location and status.

- GIS Updates. Before the schematics generation software was applied it was crucial that all appropriate GIS data was error free and up to date. The updating process involved adding backlog data and including the findings of the valve and meter audits.

- Production of Schematics. Over 250 schematics have been produced and these were verified with operations staff before being integrated with the company's GIS.

At all stages throughout the project, QA procedures were implemented and management software introduced to track and facilitate quality control of the large volumes of data.

6   SCHEME COSTS

The cost of implementing a zonal management system, across both regions, in full is outlined in Table 1.

Table 1 Anticipated Cost of ZNMS

Activity	Proposed Expenditure by Year £'000s			
	1998	1999	2000	2001
Hampshire and Surrey Region Trial Western Area – Complete	19			
Production of Regional and Area Schematics		15		
Production of Water Into Supply Schematics		27		
Production of District Meter Area Schematics		20		
Production of Treatment Works and Reservoir Schematics			20	
Kent and Sussex Region Production of Regional and Area Schematics		15		
Production of Water Into Supply Schematics			27	
Production of District Meter Area Schematics			5	15
Production of Treatment Works and Reservoir Schematics				20
Total Expenditure	19	77	52	35

7 CONCLUSIONS

The ever increasing pressures on the water companies and the need for increased operational efficiency drives the need for new tools and systems. Fundamental to the water company is the management of the distribution network. This is supported by an increase in regulatory pressure and customer expectations.

The Zonal Management System is a tool which addresses these requirements, using the proven technical hardware and software. The system allows managers to very quickly understand network operation and technicians and inspectors to identify all key valves. The system utilises and develops existing data through the graphical information system. The work to date is considered to be the essential first stage. The potential for linking the system to telemetry systems, linking to network models and to provide a tool for accessing details of individual reservoir and treatment works has already been investigated.

The System is a good example of the increasing necessity for higher reliance on Information Technology as staff numbers and experience are lost from the Water Industry.

APPENDIX 1 - PROJECT SPECIFICATION

The Zonal Management Project falls into 6 key components, which are:-
1. Preparation of Operational schematics.
2. Verification of operational schematics by site audits and valve status checks.
3. Linking and developing GIS/Schematic interface.
4. Erection of valve marker posts.
5. Production of Operational Manuals.
6. System updating procedures.

The preparation of operational schematics, constructed in a hierarchical manner from the essential component of the project, the structure is :
1. Regional Area
2. Supply or Management Area
3. Water into supply
4. District Meter
5. Booster zones
6. PRV zones
7. Treatment works
8. Pumping Stations
9. Reservoir

All the above are produced on CAD with a consistent valve identification numbering system between each level. Specific data is available at each schematic level together with tabulated data specified by the user.

The following specification is used:-

1.   **Regional Schematic**
- All Trunk Mains and key distribution mains (typically to 8/6" diameter), with diameters.
- All Treatment Works with capacity, licensed peak day volumes, average day volumes with high and low lift pump details (average flows and total and available heads) with GIS links supporting A4 1:200 and 1:1250 location plans and A3 detailed site plan.
- All WIS/Reservoir Zone Valves with usual status, hand, and support A4 1:200 and 1:1250 location plans labelled to conform with agreed nomenclature.
- All Reservoirs with individual cells, capacity, twl, bwl and working depth with support A4 1:200 and 1:1250 location plans and A3 detailed site plan.
- All Boosters (average flows and total and available heads) with support A4 1:200 and 1:1250 location plans with A3 detailed site plan.
- All existing and proposed WIS/Reservoir zone meters, with support A4 1:200 and 1:1250 location plans, labelled to conform with agreed nomenclature.

- All Strategic Valves with type, usual status, hand and supporting A4 1:200 and 1:1250 location plans labelled to conform with agreed nomenclature.
- Place Names
- Area boundaries, meters with flows and boundary valves.
- Supporting data sheets for all of the above, as specified by user.

2. **Supply or Management Area**
- As per Regional schematics additional with all Trunk Mains and key distribution mains (typically to 6/4" diameter), with diameters, and flow direction.
- All WIS/Reservoir Zone Valves with usual status, hand and supporting A4 1:200 and 1:250 location plans labelled to conform with agreed nomenclature.

3. **WIS/Reservoir Zone Schematics**
- As per Regional and Area schematics additional with all trunk mains and key distribution mains (typically to 4/3" diameter), with diameters, and direction.
- All WIS/Reservoir Zone Valves with usual status, hand and supporting A4 1:200 and 1:250 location plans labelled to conform with agreed nomenclature.
- All district meters with supporting A4 1:200 and 1:1250 location plans labelled to conform with agreed nomenclature and average flows.
- All district boundary valves with usual status, hand, and supporting A4 1:200 and 1:1250 location plans labelled to conform with agreed nomenclature
- All pressure reducing valves with supporting A4 1:200 and 1:1250 location plans, labelled to conform with agreed nomenclature, with average flows, inlet and outlet pressures.
- Place names
- Supporting summary sheets for all of the above.

4/5/6. **District Meter/Booster Zones/PRV Zones**
- All key distribution mains (typically to 3" diameter), with diameters, average flow and direction.
- All Boosters (average flows and total and available heads) with supporting A4 1:200 and 1:1250 location plans with A3 detailed site plan.
- All district meters with supporting A4 1:200 and 1:1250 location plans labelled to conform with agreed nomenclature and average flows.

- All district boundary valves with usual status, hand and support A4 1:200 and 1:1250 location plans labelled to conform with agreed nomenclature.
- All pressure reducing valves with supporting A4 1:200 and 1:1250 location plans, labelled to conform with agreed nomenclature, with average flows, inlet and outlet pressures.
- Place names.
- Supporting sheets for all of the above.

7/8/9. Treatment works, Pumping Station, Reservoir

- All mains with diameter and flow directions.
- Regional Schematic
- All Trunk Mains and key distribution mains (typically to 8/6" diameter), with diameters, flow direction.
- All Treatment Works with capacity, licensed peak day volumes, average day volumes with high and low lift pump details (average flows and total and available heads) with GIS links supporting A4 1:200 and 1:1250 location plans and A3 detailed site plan.
- All Reservoirs with individual cells, capacity, twl, bwl and working depth with support A4 1:200 and 1:1250 location plans and A3 detailed site plan.
- All Boosters (average flows and total and available heads) with support A4 1:200 and 1:1250 location plans with A3 detailed site plan.

PART III

DATA MINING
AND INFORMATION

Real WWTP Influent Flow Data Analysis by Scale Extraction Using Wavelet Transform

J. Colprim, M. Rigola *and* M. Poch

ABSTRACT
Applying monitoring techniques to wastewater treatment plants (WWTP) requires the use of real data obtained from on-line sensors. The use of on-line data has problems such as the presence of some noise that may make data monitoring difficult. In order to avoid the confusion and obtain a filtered signal with the representation of the global trend as well as the local features, scale extraction techniques can be applied.

In this paper, a multiscale extraction with wavelets has been applied to the influent flow data of a real WWTP. Data studied have been obtained from two different places at the plant. The first from the raw influent and the second prior to the biological reactor where a maximum load device is installed.

A first visual analysis of raw data shows the presence of two different patterns corresponding to normal and overload influent days. Multiscale analysis has been applied, comparing the differences between the normal and overload days to select the correct maximum scale representation that gives a correct interpretation of the process pattern. The results obtained conclude that the use of multiscale analysis may help future control strategies to improve plant performance.

1   INTRODUCTION

The ever increasing technology applied to wastewater treatment processes results from the possibility of using on-line data obtained from plant sensors (i.e. flow meters, dissolved oxygen probes, pH and RedOx electrodes, respirometers...). When applying monitoring techniques, the experimental data obtained from on-line sensors are often masked by the presence of noise. Noise origin may be due to interfering physical or chemical processes, imperfections from sensors, or any other causes, which result in a random fluctuation of the true signal. In order to avoid the presence of this undesirable noise, the use of de-noising techniques has been widely applied. The main objective of any de-noising method is to remove the randomised behaviour of the signal without losing important information about local and global events.

Traditional signal processing methods such as Gaussian filtering or fast Fourier Transform (FFT) have been used with success. Nevertheless, the use of these techniques sometimes concludes with a de-noised signal with valuable

139

information being removed from the original observed signal. An extended application of de-noising techniques is the multiscale extraction of trends which represent the original signal at several scales. The scale of a process trend is a quantifiable parameter which characterises the resolution at which the underlying physico-chemical phenomena are being described by the trend. In particular, each scale specification is a real number which uniquely determines, and varies inversely with, the resolution of the trend (Cheung and Stephanopoulos, 1990).

Scale and resolution are common notions in mapping. Small scales give a local view of the map, while larger scale maps give a global view. Using the same area to represent the same map, large scale maps have a lower resolution than small scale. Scale extraction of process data allows one to represent the observed signal at several resolution levels. Correct interpretation of process trends by scale extraction requires a methodology capable of extracting the contributions made by the underlying processes as seasonal behaviours or random noise. Classical methods for multiscale analysis of process data such as Gaussian filtering (Witkin et al., 1983), Fourier analysis and windowed Fourier analysis have been traditionally applied. Nevertheless, the wavelet theory has been proved to be more efficient than windowed Fourier analysis, and combined ideas from scale-space filtering (Barclay, 1997; Dohan, 1997; Whitfield, 1997).

In this paper, we present the multiscale analysis of influent flow to wastewater treatment plants (WWTP) using wavelet transforms. The objective of this study is to differentiate the contribution of daily or local events and random noise from the original signal using multiscale wavelet analysis. This paper presents an application of the wavelet theory to a particular case study. Further readings about wavelet theory can be found in specialised papers (Bakshi and Stephanopoulos, 1994; Koornwinder, 1993 or Wickerhauser, 1994).

2 MULTISCALE WAVELET ANALYSIS

The wavelet transform is obtained by passing the original signal over low and high-pass filters, which result in approximate and detailed signals as depicted in figure 1. The wavelet filter is composed of four functions, the decomposing and reconstruction filters for the detailed and the approximate signals. The approximate signal is known as the signal at a scale increased by one with respect to the signal entering the filter. This filtering process can be consecutively repeated to obtain the approximate signal for successive representations at higher scales.

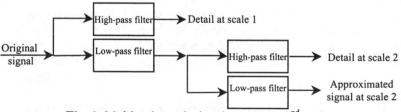

Fig. 1: Multiscale analysis scheme up to 2nd scale.

The number of coefficients representing the wavelet transform is halved at each scale applying the discretisation method. Therefore, with N discrete data, at scale j, the number of coefficients, of both the approximated and detailed signals, are $N/2^j$. The maximum available scale (S_{max}) is represented by one coefficient and is related with the number of available data as $2^{Smax}=N$.

The representation of the successive detailed and last approximated signals is named multiscale wavelet analysis ($\{S_o, D_1, D_2, ...,D_{Smax}, S_{Smax}\}$, where S_o is the original signal, D_j is the detailed signal obtained at scale j, D_{Smax} is the detailed signal at the maximum scale, and S_{Smax} the approximated signal at the maximum scale). The difference between the approximated signal at scales j and j-1 gives the detailed signal at scale j. Therefore, the multiscale analysis can be shown using only the approximated signals obtained at each scale ($\{S_o, S_1, S_2, ... , S_{Smax}\}$, where S_j is the approximated signal at scale j).

3   APPLICATION TO REAL WWTP INFLUENT FLOW DATA

The real WWTP has been designed to biologically treat up to 2200 m^3/h. Nevertheless, the plant accepts all the incoming water and passes it over the pre-treatment units. The surplus incoming water is by-passed after the pre-treatment and joined to the final biologically treated effluent (Fig. 2). Real data from on-line flow meters at the real WWTP were collected by the SCADA system. Flow data were obtained each minute from two different locations, at the raw influent and at the biological treatment inflow, during seven days (Fig. 3).

As shown in figure 3, the raw influent flow presents more randomised behaviour than the biological inflow. This is due to the homogenisation effect produced by the pre-treatment units. The by-pass effect can be observed with data from the last day, when incoming water flow reached up to 4000 m^3/h for four hours. This localised episode was related to a storm that occurred near the plant influence area.

The numbers of data items available are 9343 and 9374 for the raw and biological flow respectively. Thus, the maximum available scale, as mentioned above, for each signal is **13** (2^{13}=8192). We used this data to perform a multiscale analysis with wavelets to obtain a smoother profile for both the raw and biological flow.

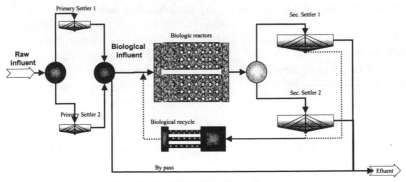

Fig. 2: Real WWTP flow diagram.

142

Using the free available Matlab toolbox developed by the *Universidad de Vigo* (UviWave 3.0), splines(3:3) wavelets were selected and the multiscale analysis was developed for the full signals. In spite of the signals having been treated as a unique data set, two amplified graphs are presented for the periods 2-4 and 5-7 days. This representation has been done to clearly observe the smoothing effect of the multiscale analysis.

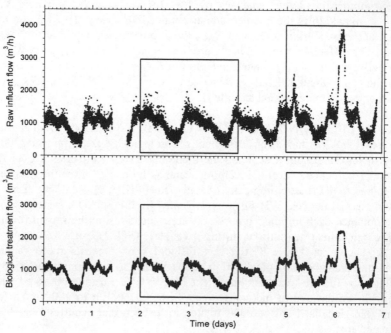

Fig. 3: Raw and biological influent flow at the WWTP.

3.1   Raw influent flow

Approximated signals for the raw influent flow at scales 5,6,7, and 9 are presented in figure 4. At scale 5, the signal presents a visual hourly profile with some random trends. The peak event is clearly defined and the random behaviour of the edge of the peak has been removed. The hourly profile observed between days 2 and 4 presents a more correct sharpness at scale 7, but the storm appears like a sinusoidal trend rather than a peak. Looking at scale 6 approximated signals, the peak appears clearly defined and the hourly profile easily identified. Finally, at scale 9, a sinusoidal curve is observed for each day, with a maximum load during the last day.

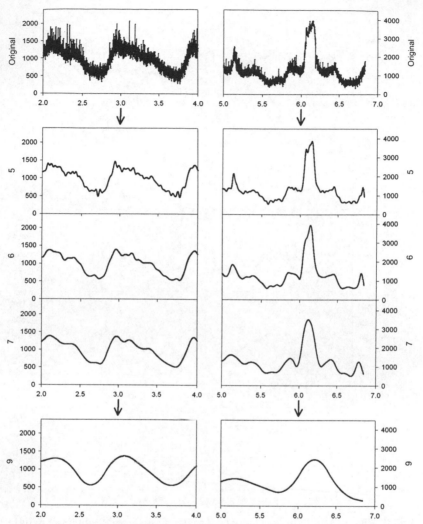

Fig. 4: Approximated signals at scales 5, 6, 7, and 9 obtained with the raw influent flow using splines(3:3) wavelets.

3.2 Biological treatment inflow

The behaviour of the biological treatment inflow data was similar to that shown by the raw influent data, despite the lowest random noise of the original data from the biological treatment inflow. In figure 5, approximated signals at scales 5, 6, 7, and 9 are presented. The 2-4 days profile in figure 5 presents a distinctive peak at a value higher than 2,500 m^3/h. This feature is due to the use of the original data obtained from the SCADA system, with missing values and errors. Under successive scales, this feature is easily removed from the original signal.

144

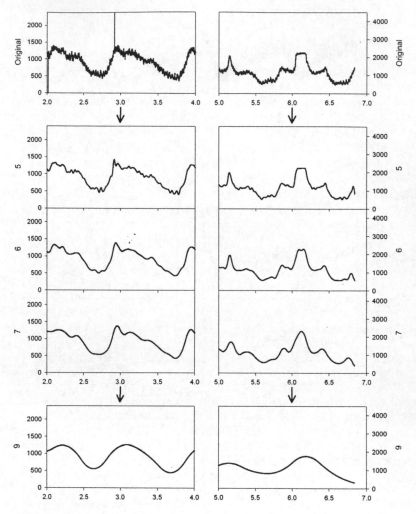

Fig. 5: Approximated signals at scales 5, 6, 7, and 9 obtained with the biological influent flow using spline (3:3) wavelets.

Again, as observed for the raw influent data, the hourly and daily profiles are obtained at scales 6, and 9, respectively. The by-pass effect can be observed during the 5-7 days profile. Up to scale 6, the by-pass is clearly identified as a profile cut-off. At scales higher than 6, the by-pass effect appears like a sinusoidal event. At scale 9, the daily average profile is observed as a sinusoidal trend for both normal and storm days.

4 CONCLUSIONS

Multiscale analysis for both raw and biological inflows has been easily obtained using spline (3:3) wavelets. In spite of the influent conditions (*e.g.*, baseflow or storm flow), an hourly profile without losing information has been obtained at scale 6. At scale 9, similar results are obtained for the daily average profile, showing a sinusoidal trend.

When monitoring WWTP performance, plant managers differentiate the global behaviour of the plant from localised events. The multiscale analysis requires an expert to decide which scale is most useful for each study. For instance, if we were planning to develop an event detection procedure, signal data at scale 6 would be used to develop further control strategies. As with data from flow meters, other data from on-line sensors can be obtained on real WWTP. Therefore, multiscale extraction can also be developed to identify other related WWTP problems using pH data, oxygen consumption (on-line respirometric tests), energy consumption, or data from other measured parameters.

REFERENCES

Bakshi B.R. and Stephanopoulos G., "*Representation of process trends - III. Multiscale extraction of trends from process data*", Computers chem. Engng. Vol. 18, No 4, pp. 267-302 (1994)

Barclay V. J. and Bonner R. F., "*Application of wavelet transforms to experimental spectra: smoothing, denoising, and data set compression*", Anal. Chem., Vol. 69, pp. 78-90 (1997)

Cheung J.T.Y. and Stephanopoulos G., " *Representation of process trends - Part II. The problem of scale and qualitative scaling*", Computers chem. Engng. Vol. 14, No, 4/5, pp. 511-539 (1990)

Dohan K. and Whitfield P. H.," *Identification and characterization of water quality transients using wavelet analysis. I. Wavelet analysis methodology*", Wat. Sci. Tech., Vol. 36, No. 5, pp. 325-335 (1997)

Koornwinder T. H., "*Wavelets: An elementary treatment of theory and applications*", Ed. World Scientific Publishing Co. Pte. Ltd., ISBN: 981-02-1388-3, (1993)

Uvi-Wave 3.0. Wavelet toolbox for use with Matlab. Grupo de teoría de la señal. Universidad de Vigo. http://www.tsc.uvigo.es/~wavelets/uvi_wave.html (1996)

Whitfield P. H. and Dohan K., "*Identification and characterization of water quality transients using wavelet analysis. II. Application to electronic water quality data*", Wat. Sci. Tech., Vol. 36, No. 5, pp. 337-348 (1997)

Wickerhauser M. V., "*Adapted wavelet analysis from theory to software*", Ed. A. K. Peters, Ltd., IEEE Press, P.O. Box 1331, 445 Hoes Lane, Piscataway, NJ 08855-1331, IEEE Order number: PC5656, (1994)

Witkin A., "*Scale space filtering*", Proc. 8th Int. Joint Conf. on artificial Intell., p. 1019 (1983)

Dissemination of Real Time and Operational Data on Intranet and Internet

B. Nguyen *and* F. Montiel

ABSTRACT

SAGEP (Société Anonyme de Gestion des Eaux de Paris) is responsible for supplying water to the City of Paris in accordance with strict criteria relating to pressure and water quality. In the performance of this task, operators working on a continuous basis and by remote control from Central Dispatching supervise, monitor and intervene with regard to the installations comprising the drinking water network. Central Dispatching initially computerised in 1992, is at the heart of the system that supplies drinking water to 2 million people. The unit processes in real time the 15 000 items of information that permit command and control of the water distribution facilities in Paris.

Since computerisation, the nature and mass information processed by this system have not ceased to evolve. The volume of data, which originally were essentially technical, has thus doubled over the last five years. In the latest version of SCADA, developed by SAGEP and which is currently in operation, information of a new type has emerged which can best be described as complex. Rationalisation of the operation of the installations and the need for water distribution of the quality beyond reproach have led to a gradual extension of the circle of users of this information to the managers of the production centres and the maintenance staff. Essentially designed at the outset for the operators of the command and control centre, it is now of interest to the entire company.

To meet these new requirements, it has proved necessary to design and set up a simple and effective system of communication, while at the same time retaining absolute control over security of access to the SCADA. Intranet has shown itself to be the most efficient method for accomplishing this task.

Today, however, this same demand for information is to be found among users. The opening up of Intranet to Internet appears to be increasingly translating itself into concrete terms.

1   INTRODUCTION

Dozens of new Internet sites appear every day on the Web. Some of these sites have clearly commercial objectives and seek to broaden their customer base using the new technologies. Others see in them a means to improve the service offered to a wider public. Others again have less pronounced motivations which can probably be situated between the need (or the desire) to make themselves better known and the wish not to miss out on the different stages in electronic progress.

The Intranet makes it possible to develop communication within the company and to give additional value to the information circulating therein. Intranet has become a means to create a genuine corporate culture. Internet or Intranet use the same principle, namely making available targeted information adapted to the user. In a manner of speaking, industrial supervision systems do the same thing; the only difference is the volume of information, and the mode and frequency of collection. In Paris, information concerning the state of the drinking water network is extracted from the SCADA system and subsequently disseminated by Intranet or Internet.

2 HISTORICAL BACKGROUND

The Paris water network has been telemonitored for over thirty years. Using a succession of different telemonitoring systems, the Dispatching operators are able to regulate pressure, supervise water levels in the reservoirs and monitor the quality of the water distributed. With the technological advances in process control and industrial data processing, the amount and quality of the information have substantially increased during the last decade. For obvious security reasons, the telemonitoring system remains strictly isolated from the world outside the control room.

In 1993, however, after installation of the first computer system, a special and particularly innovative interface made it possible to log on by telephone to a Minitel[†] server in which the main information concerning water distribution in Paris was updated every ten minutes:
- production from plants and springs,
- volume of water stored in the reservoirs,
- instant consumption of water,
- aperture angle of the telecontrolled valves,
- pumps in operation in the lift pumping stations,
- principal flows.

Limited to a dozen or so pages of text, this system of remote inquiry from a laptop computer was used and presented on different occasions outside Europe. Since then, the possibilities of communication and the organization of the water system in Paris have developed radically, but the basic principles governing the dissemination of information have remained the same.

3 CONTEXT

To properly understand the expectations and requirements of those involved with the water industry with regard to information, and also the means deployed to meet them, it is important to be familiar with the context of the water system in Paris. Société Anonyme de Gestion des Eaux de Paris (SAGEP) was established to

---

[†] Minitel is a French technology similar to Internet, dating from the 1980s. The information can be consulted from a specific videotext terminal.

produce, transport and distribute at the correct pressure the 630,000 m³/d required to meet the needs of two million Parisians.

The Operating Division at SAGEP is made up of Central Dispatching and seven production centres gathered together in three units. The centres are based on the production sites, some of which are at a distance of over a hundred kilometres from Paris. Central Dispatching issues the centres with instructions regarding production and manages the distribution of drinking water along with two private companies. The latter are responsible both for maintenance and intervention on the pipe network in Paris and relations with users.

3.1   Groups of users

Six groups of users of operational data have been listed. Each group brings together individuals or classes of individuals with expectations or requirements differing from the other groups. The six groups are as follows:

3.1.1   The operators from Central Dispatching

These ten people work in continuous relays in the control room from which they supervise and monitor by remote control the water supply to all of Paris. To this effect, they use a SCADA system which enables them to familiarise themselves with and regulate the water network at any given moment. The data they require are numerous and varied.

3.1.2   The production centre operators

These operators need to know the state of the installations for which they are responsible. As regards the plants, the important thing is the functioning of their water treatment process. They also make use of supervision software. They are increasingly interested by what affects directly or indirectly the downstream operation of their area of intervention (production from the other centres, water quality on arrival in Paris, etc.).

3.1.3   The SAGEP maintenance teams

These teams are kept fully up to date with the details of malfunctioning in the measuring and teletransmission instruments through interrogation of the lists of current and filed flaws. They forecast their maintenance programme as a function of operations.

3.1.4   The rest of SAGEP staff

The 530 SAGEP employees are distributed over the different departments and scattered geographically. Even if their job is not directly linked to production, they are likely to be concerned by the general functioning of the Paris water supply system and are anxious to enhance their expertise in this field. All of them wish to acquire a genuine corporate culture.

3.1.5   The distributing companies

Their statistics and distribution schedules by district are based on the indications provided by the flow meters installed on the pipes in Paris.

3.1.6 Parisian users

These are the subject of considerable efforts in terms of communication and are primarily concerned with the quality and type of water distributed to them (origin, characteristics).

As members of a group, users must have access to suitable information provided on an appropriate carrier, without throwing into question the mandatory safety rules governing water supply to a major city.

3.2 Information transmission carriers

The dissemination of information makes use of a transmission carrier. Several types of carrier exist at SAGEP: each one is characterised by an area of activity, controlled access and a transfer capacity. The transmission carriers are used by distributed applications such as in-company e-mail or accounting and by closed applications such as the SCADA at Central Dispatching. Three different information transmission carriers are available at SAGEP.

3.2.1 The industrial network at Central Dispatching

This network linking some one hundred local stations distributed mainly over Paris transmits information from the ground to the control room. The volume of data transferred at high speed by the industrial network is practically constant, that is to say approximately 15.000 items of information per second. This network together with SCADA (which makes use of it) are absolutely hermetic. One of the more distant production centres is not linked to the industrial network.

3.2.2 The in-company network

This is a conventional office-based network linking the company's micro-computers. Bridges make it possible to reach each of the company's remote sites including all the production centres. Access to this network is restricted to the staff of the company who are able to log on by personal or group access. SAGEP's Intranet makes use of this transmission carrier.

3.2.3 The Web

The Internet server for the www.sagep.com web site enables data to be disseminated to numerous potential users. Access is not controlled and the transfer capacity depends on the hook-up and the number of people logged on.

4 TYPES OF DATA

The types of data available on the SCADA in the control room fall into three very different categories: real-time data, calculated data and deferred-time data. Each of these three categories corresponds to a distinct requirement and is likely to be of interest to different groups of users.

4.1 Real-time data

These are the most numerous. Such information is of various kinds: it includes binary states (operation of the pumping equipment and state of the alarm systems) and telemetering, which represents by its quantified values the operating and distribution conditions (pressures, flows, water level in the reservoirs, water quality parameters, variable speed drive on the pumps, aperture angle of the motorized valves, etc.). The orders which are sent from the control room to the remotely-operated units (pumps, valves) also form part of the real-time data.

The designation "real-time" summarises the specific character of centralised supervision: these are raw data, frequently refreshed, often numerous, precise and overlapping, which aim to provide a general representation of the state of a system at any given moment.

Real-time data are known as "raw" because they are not validated by passage through a filter and consequently can be distorted by a certain number of factors linked directly or indirectly to the measuring instrument. Raw data can be incorrect and are therefore not always directly representative. They must be handled carefully, but on the other hand they benefit from highly frequent updating which gives them statistical validity.

Raw data would be ideal if they were not occasionally distorted. Analytical laboratories specialising in water quality are familiar with this problem given that they are confronted, on the one hand, with rigorous manual analyses using calibrated laboratory measuring devices and, on the other, by automated analyses for continuous supervision.

Despite the fact that, more often than not, technology imposes the choice between two methods of analysis, one can observe a logical development in raw continuous measurement. Precise and accurate manual measurement carried out at a given moment in the day cannot be representative of the possible changes in the parameters of water quality during that same day. Computerised analysis, like the real-time datum, enjoys the advantage of being more representative of the actual changes in the parameter measured but possesses a limited validity over time.

The real-time raw datum is not filtered; interpretation is therefore necessary to validate or invalidate this datum in the general representation of the state of the system. Such interpretation will be carried out by the operators posted at the control centres. When an anomaly appears, additional investigations can be launched together with the network operation or measuring instrument maintenance teams.

To sum up, real-time raw data are of great value for continuous supervision and monitoring of systems such as production and distribution of drinking water. These data form a whole that is frequently updated and which can only be used by qualified staff with a very considerable knowledge of the environment under observation. However, these same data lose some of their interest when they are incomplete, isolated or too localised; they are therefore not suited to wide dissemination by Intranet or Internet.

4.2 Calculated data

Certain physical quantities measured at given points benefit from being transformed or aggregated into new measurements. The new data thus generated improve the general representation of the system:

- A height of water in a reservoir of treated water gives by transformation a volume of water stored, taking into account the possible variations in the section of the reservoir as a function of the height.
- The sum of the volumes stored in the various tanks of one and the same reservoir gives the total volume available.
- The sum of the discharge flows through the outlet pipes of a plant gives the total production of the plant.
- An algebraic sum of flows entering and leaving a distribution sector give the amount for distribution in this sector.
- The pumps operating in a plant give the consumption of electricity of the plant.
- Lastly, a simple conversion makes it possible to change from a pressure to a piezometric level, or to change units of measurement: from bars to metres of water column, from litres per second to cubic metres per day.

The following data are calculated in Paris:
- filling percentage of the reservoirs,
- share of a given distribution sector in the total amount for instant distribution,
- percentage or volume of on-line water losses on an aqueduct,
- percentage or volume of margin available between current production and production capacity.

These calculated data are simpler to interpret than the real-time raw data; they provide a precise representation - but at a large scale - of the water supply system. They are not solely reserved for the control room specialists, but are also of interest to a wider public which is nevertheless capable of appreciating their value. The calculated data are well suited to dissemination on the company's Intranet.

4.3 Deferred-time data

There also exists another type of calculated data known as "deferred-time", with a lower calculation frequency; these are operation statistics. Generally based on differences in counters and volumes, they make it possible to calculate precisely the volumes distributed over a given period and not at a given moment. For example, one can calculate in this way the total volume distributed over the previous 24 hours and updated every 10 minutes. The history of this deferred-time data is especially interesting with regard to forecasting the production of water.

Similarly, the operation statistics are calculated once a day for the previous day. Other deferred-time data are the result of calculations and provide tendencies such as the age or the origin of the water distributed over a chosen sector. Deferred-time data are characteristically those of interest to a non-specialist public provided they are properly interpreted. They are well suited to dissemination on the SAGEP Internet site.

5   APPLICATION OF THE INTRANET SYSTEM

Setting up the information system is the result of a proper fit between the groups of users, the information they require and the physical carrier. Table 1 takes these parameters into account in order to establish the means of transmission of the information.

Table 1. Groups of users, information they require and physical carrier

Groups of users	Type of data	Frequency of updating	Information carrier
CCC	All	1 Second	SCADA
Connected Operators	Real Time + Calculated	10 Seconds	Automata
Non-connected operators	Real Time + Calculated	600 Seconds	Token-Ring
Maintenance	Real Time	1 Second	SCADA
SAGEP staff	Calculated + Deferred Time	600 Seconds	Intranet
CEP-SPE	Real Time + Calculated	10 Seconds	Automata
Users	Calculated	600 Seconds	Internet

The means of information transmission is represented in Figure 1.

Level 1, as shown in Figure 1, enables operation of the principal function of the system, which corresponds to the command and control of the installations and therefore to the supply of water to the City of Paris. This particularly sensitive system is isolated from level 2 by a unidirectional serial link between two computers connected to each of the data processing networks (SCADA real-time networks and Token-ring in-company network). The in-company network is rendered secure in relation to the Web by use of a Fire-Wall.

6 CONCLUSIONS

The emergence of new enhanced-performance SCADAs has permitted generalization of calculated data which the systems of ten years ago would have had difficulty in processing. Development, for example, of shifting consumption over the previous 24 hours could not be envisaged at that time. This new information presents an interest that goes beyond the restricted framework of the control room at Dispatching. Furthermore, management rules have also changed, with the production and maintenance departments no longer being mere executants. With the new information they receive, they are able to satisfy their legitimate needs to know what is happening beyond their area of intervention. The production centres thus optimize their availability during shutdown of the other centres. The maintenance teams are able to schedule their interventions more effectively. On the basis of the numerous raw data from the SCADA intended for a restricted number of users, one ends up, after several stages, with general data for the greatest number of users. The choice of intranet for the dissemination of this information became essential by reason of its user-friendliness and its degree of integration into the in-company network.

154

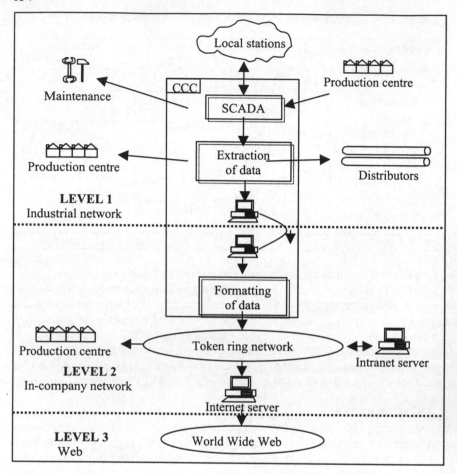

Figure 1. The means of information transmission

Data Mining and Knowledge Discovery for the Water Industry

D.A. Savic, J.W. Davidson *and* R.B. Davis

ABSTRACT
The water industry currently recognises that in addition to making data available across a company, it is equally important to be able to efficiently extract information from data, i.e. to have procedures for identifying logical, nontrivial, useful, and ultimately understandable patterns in data. Knowledge discovery is the process of identifying these useful patterns in data. These patterns can be relations, events or trends, and they can reveal both regularities and exceptions among data. The core of the knowledge discovery process is data mining, the automated analysis of large or complex data sets in order to discover significant patterns or trends that would otherwise go unrecognised. This paper describes basic principles of data mining technologies and aims to inform the water engineering community of the potential of these technologies for improving planning, decision making and performance. These methodologies, which have strong ancestry in the computational sciences and artificial intelligence, offer a way to provide decision support in water industry application. A short-duration study seeking to establish the benefits of data mining for Mid Kent Water plc illustrates the approach to data mining pioneered by the Centre for Water Systems at the University of Exeter. The study deals specifically with identifying the causes of mains bursts.

1   INTRODUCTION

Water utilities possess large quantities of data derived from different sources (SCADA, asset and customer databases, etc) and collected in different formats, which are stored but not archived properly or fully understood. The industry currently recognises that in addition to making data available across a company, it is equally important to be able to efficiently extract information from data, i.e. to have procedures for identifying logical, nontrivial, useful, and ultimately understandable patterns in data (Savic, 1998a). Knowledge discovery can be defined as the process of identifying these useful patterns in data. These patterns can be relations, events or trends, and they can reveal both regularities and exceptions among data. The core of the process is data mining, the automated analysis of large or complex data sets in order to discover *significant patterns or trends* that would otherwise go unrecognised. This type of analysis does not look for hard facts but for pointers that warrant further investigation. To make things more difficult, data mining appears under a multitude of names, which include knowledge discovery in databases, data or information harvesting, data archaeology, functional dependency analysis, knowledge extraction, and data pattern analysis.

The most commonly applied technique develops models that group data according to pre-classified examples. One of these techniques, decision tree induction, discovers a set of rules that can be applied to new (unclassified) data to predict which data records will have a given outcome. For example, when dealing with records reporting on pipe bursts, one would like to predict, based on the values of other attributes (age, quality, diameter, soil type, etc.) whether a pipe is likely to develop a break or not. It is the authors' view that data mining is just a step in the wider *Knowledge Discovery* (KD) *process*.

2 BASIC CONCEPTS OF DATA MINING

2.1 Classification

Some researchers make a distinction between technologies that discover understandable patterns in data (pure data mining) and those that discover a model that fit data (data modelling) regardless of whether the model is understandable or a black box. Data mining, taken here in a wider sense (i.e. embracing both technologies), can be divided into four major categories:

- *Undirected or pure data mining.* Here the data mining system is left relatively unconstrained to discover patterns in the data free of prejudices from the user (no hypothesis put forward and no interfering from the user).
- *Directed data mining.* Here the user will select a parameter, ask a question about it and the problem usually changes from a general pattern-detection problem to a better defined *induction* problem.
- *Detection of Anomalous Data and Patterns.* Here the user applies previous data mining results to analyse anomalous patterns and unusual data elements, i.e. that do not conform to the general patterns found.

- *Hypothesis testing and refinement.* The user presents a hypothesis to the system to evaluate it, but then - if the evidence for it is not strong - seeks to refine it.

Within these categories one can also identify another classification of data mining problems:

  (a) prediction
  (b) discovery
- *Prediction* incorporates classification tasks, regression and time series analysis.
- *Discovery* may include deviation detection, database segmentation, clustering, associations, rules, summarisation, visualisation and text mining.

2.2   The Data Mining Process

2.2.1   Data Screening
Almost any large database will have some problems with data errors and inconsistencies. This is especially true when data come from multiple sources that are maintained at different locations or by different administrators. Every effort should be made to minimise errors in data and missing values, because these can degrade the performance of the models obtained from the data. The basic checks should include type consistency, range validity, etc. A data mining system may be used for this purpose. Here, it can be used to search for general patterns and rules in databases and then identify anomalous data that do not conform to the rules found.

2.2.2   Selection of Training and Validation Datasets
To achieve robustness and generalisation of models data mining is commonly done using the *split record test* to develop a model and validate it. This method consists of splitting data into a training set and a set for validation. Only the training data are used to evaluate the fitness of the model developed (learned). The *training error* is calculated as the error between the modelled and target output for the training data. Similarly, the *test error* is calculated as the error between the modelled and target output for the validation data.

2.2.3   Selection of Relevant Parameters
It is important to include a number of parameters (data fields) that may have some relevance to the problem being studied. The data mining system will then discover which ones are the most useful and what is the relationship between the parameters. Omitting a highly relevant parameter from analysis will cause deterioration in prediction performance of the system.

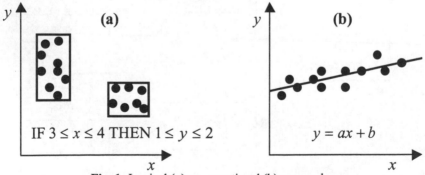

Fig. 1 Logical (a) vs. equational (b) approach

2.2.4   Knowledge Discovery and Encoding
This phase involves running the system, validating the patterns discovered and finally encoding the results of data mining in software that can be used for prediction or classification purposes in future.

3   DATA MINING TECHNOLOGIES

Based on the approach taken, data mining technologies can be classified into (a) logical and (b) equational. The difference between the two approaches can be clearly seen in Figure 1. Logical approaches usually employ the well-known conditional operators IF/THEN to represent knowledge. The logical approach in Figure 1(a) is best at dealing with crisp boxes while the equation in Figure 1(b) forms the basis of linear regression and has been widely used for statistical analysis. It is obviously good at representing linear patterns in numerical data while logic can deal with both numeric and non-numeric data.

3.1   Equational Approaches

Statistics and artificial neural networks (ANNs) are the best known equational approaches to data mining. These technologies almost always require the data set to be all numerical. Non-numeric data need to be coded into numbers in order to be used with equational approaches. This may cause a number of problems for data mining where non-numeric data are dominant. Regression analysis is a typical representative of the statistical approaches (Figure 2). It works well when the points to be approximated lie on a straight line or even if they approximate non-linear, smooth surfaces. However, their transparency (the ability of humans to understand the equation) decreases with the complexity of the equation found, in which case they become grey or black boxes. The greatest advantage of

Fig. 2 Data mining technologies

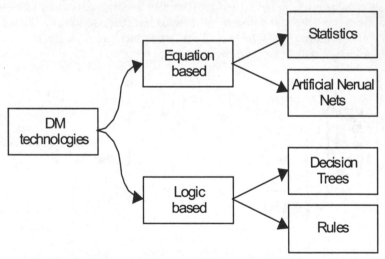

artificial neural networks over other modelling techniques is their capability to model complex, non-linear processes without having to assume the form of the relationship between input and output variables.

3.2   Logical Approaches

Decision trees and rule induction are related approaches to discovering logical patterns in large data sets (Figure 2). Although apparently similar, they differ in the way they discover information and more importantly in terms of their behaviour on new data items. The simplest types of rules express conditional or association relationships. Conditional relationships use so-called *conditional logic*, i.e. IF/THEN statements of the form "IF it is raining, THEN it is cloudy". Often some other form of logical approaches needs to be used when associations between data fields are expressed as statements. This is called *association logic*. For example, the fact that in 80% of cases when fault x occurs then fault y is also encountered can be expressed using an association statement. Decision trees also use IF/THEN statements to represent information, however, they may be viewed as a simplistic approach to rule discovery.   More on the equational and logical approaches may be found in (Savic, 1998b).

4   CASE STUDY – CAUSES OF MAINS BURSTS

Mid Kent Water plc (MKW) serves 230,000 properties, which equates roughly to 500,000 customers. The total length of main is 4,000 km. The company is split down into the hierarchy of WIS (water into supply) zones, water quality zones and DMA (district metering areas). The demand prediction for the company is flat with no future increase envisaged. The company therefore needs to convince the regulator that investment is required and identify where the investment is most beneficial.

Following a series of discussions on the benefits of data mining, a short-duration study seeking to establish the benefits of data mining was undertaken within the Centre for Water Systems at the University of Exeter. MKW were specifically interested in identifying the causes of mains bursts. It was anticipated at the time that a second phase of the project could be undertaken as a large-scale data mining exercise, if the apparent benefits observed from the trial study justify it.

The data mining procedure employed in this study was executed in three phases:
1. Data visualisation
2. Data mining
3. Statistical analysis

The objective of the first two phases was to discover significant patterns and trends and the objective of the third phase was to confirm or reject the conclusions of the previous phases using classical statistical analysis techniques. However, in this trial study these steps were carried out independently of each other to provide an impartial indication of the suitability of each to the MKW dataset.

4.1   Data

Data used in this study included data on mains performance (Halcrow, 1998) and more detailed data sets (GIS, customers, job management, water quality, etc.) for two water quality (WQ) zones. These zones were a WQ zone where water quality problems abound, and a WQ zone where a high incidence of bursts is the main

160

problem.

4.2   GIS and Data Visualisation

Data visualisation employs the simple techniques of GIS (geographic information systems), viewing and layering, that enable the visual determination of possible causal relationships. It should be noted that the standard of the data held at MKW was of a generally high standard. Specifically the spatial accuracy of the data sets was high which simplified the task of the spatial data compilation. This said a number of anomalies were identified during the data compilation, such as bursts more than 50 metres from the nearest pipe and rehabilitation updates not being recorded, which might need to be assessed by Mid Kent.

The analyses performed in this study indicate that *traffic loading* (an example given in Figure 3), *operational pressure range, water sources, pump locations and their regimes and soil type* could be the most significant factors affecting pipe burst rates.

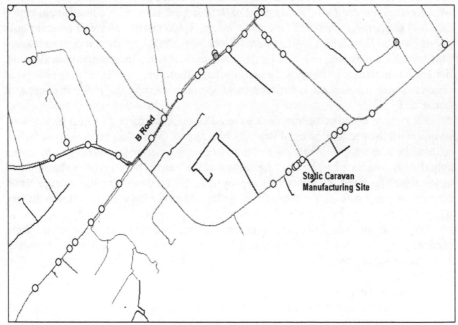

Fig. 3 Pipe bursts and local traffic

In the example given in Figure 3 it appears that where the traffic loading is substantially greater than the level for which the highway was constructed bursts are likely to occur. An example was identified in the sector where a B road is used as a route to a caravan park. The road has a significant number of bursts. These bursts when analysed primarily occur in the summer months. Note that the pipes on adjacent roads are of the same material, laid at the same time and in the same soil type but do not burst.

4.3   Decision Tree Techniques

The data mining phase uses specialised software to search for patterns in the data sets produced in the data processing phase. From the records in the data sets the software inductively constructs a classification model commonly represented as a decision tree, or often as a set of rules. The data set usually consists of a large number of records. Each record consists of several fields or *attributes* and every record is assigned a *class*. Classification models produced by the software are designed to predict a record's class on the basis of the attribute values.

The data sets used in the study consisted of records from the two water quality zones. Each record described a section of pipe with fields corresponding to *diameter, year laid, material, surface type, soil corrosivity,* and *number of bursts.* The objective of the classification model was to predict the number of bursts on a section of pipe given the values of the remaining fields. Therefore the number of bursts was selected as the record's class and the remaining fields as the attributes.

The software that constructs the decision trees is not specifically designed for classification of rare events. If a class appears relatively infrequently in the data set the misclassification of all cases of that class produces only a small increase in the percentage of error. Since pipe bursts are rare events, records with a number of bursts greater than zero occur relatively infrequently in the data sets. Therefore, the data sets had to be manipulated to force the software to classify records with more than zero bursts. To obtain a data set with the largest number of records with more than zero bursts the data from the two water quality zones were pooled to create a single data set. In the combined data set there were approximately 450 cases of pipes containing bursts. The number of records with zero bursts was artificially reduced from 11500 to 450 by the removal of randomly selected records. The number of bursts per pipe ranged from 1 to 9, with only 50 cases containing three or more bursts. The best results were achieved when the number of possible classes was restricted to two only, class 0 (for no bursts per pipe) and class 1 (for any number of bursts per pipe). However, trials were conducted in which further distinctions were made on the number of bursts per pipe.

The conclusions of this step indicate that the burst rate of a cast iron pipe appears to vary significantly with pipe *quality* (e.g. pipe material, ductile or brittle; standard of installation/ jointing etc) *and age.* There is also a strong indication that *smaller diameter pipes are more likely to have higher break rates.*

4.4   Statistical Analysis

The objective of this phase is to perform classical statistical analysis techniques to confirm or reject the conclusions of the previous phases of data visualisation and data mining. The analysis consists largely of two types of tests, analysis of variance and regression. Analysis of variance tests determine whether the factors identified in the data visualisation and mining phases significantly affect the pipe burst rate. The tests can determine whether the relationships cannot be explained by the random variation in the data alone. The other type of analysis, statistical regression, produces mathematical functions that are both predictive and descriptive. Mathematical functions of this type show the proportional effect of each of the identified contributing factors on the burst rate.

162

To be useful for planning and decision making the mathematical functions should predict the most likely outcome but must also provide accurate measures of confidence or certainty in the predicted result. Ideally the functions should provide all possible outcomes and their associated probabilities in the form of a probability distribution graph. An example of this type of graph is provided for a water quality zone (see Figure 4).

Probability of Cast Iron Bursts

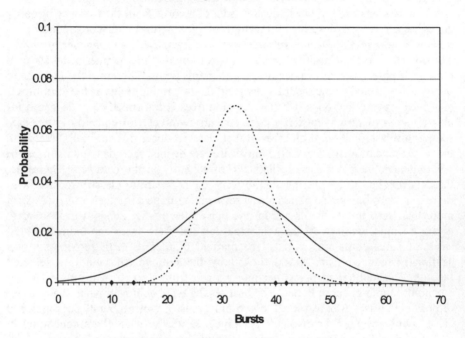

Fig. 4 Probability distribution graph for a water quality zone

The analyses were performed on AMP3 data and involved two types of tests: variance and regression. The study has confirmed the significance of some factors identified in the first two phases and has also found *a probability distribution that fits the random fluctuation in the burst rate of cast iron pipes*.

5 DISCUSION AND CONCLUSIONS

This feasibility study has demonstrated that data visualisation and data mining can identify significant patterns and trends in datasets. A statistical analysis has produced a probabilistic model that may be used as measure of a serviceability quotient and then applied to determine how standards can be imposed to improve the situation.

This study also indicates that the application of the techniques described do identify strong causal relationships for mains bursts. The application of the techniques would allow MKW to refine the analysis as to the nature and extent of problems giving rise to capital investment. Improved definition of problems at this early stage of the capital planning process would give rise to more efficient engineering solutions and capital savings.

Given the results obtained and the probabilistic model developed, recommended future work could involve:

- Design of a procedure which combines these techniques in the most effective way, based on observations made during the trial study;
- Broadening of the objectives to consider other causal relationships besides pipe failures;
- Application to the full range of MKW data, perhaps prioritised by zone according to perceived problems;
- Development of software which will enable MKW to interrogate and report on their data to satisfy ongoing asset management or reporting requirements;
- Interpretation of results to feed into the prioritisation process for the asset maintenance programme, potentially enhanced by combining with other optimisation techniques for determining cost-effective solutions.

REFERENCES

Savic, D.A. (1998a) Logisticians, Mine Your Own Business, *8th International MIRCE Symposium, System Operational Effectiveness*, Knezevic, J., Kumar, D. and Nicholas, C. (eds.), Exeter, UK, 191-203.

Savic, D.A. (1998b) Data Mining Technology For Logisticians: A Quantum Leap Into The Millennium, *14th International Logistics Congress*, Sun City, South Africa, November 24-27.

Halcrow, (1998) Mid Kent plc, AMP3 REPORT, Mains Performance Review, August.

Application of the *Group Method of Data Handling* for Managing the Water Quality of an Open-Cast Mining Affected River

F. Schlaeger *and* J. Köngeter

ABSTRACT

Brown coal open-cast mining results in several environmental problems, for example serious disturbances in the natural water balance or water quality. For controlling the complex relationship between water quantity and water quality in the Lusatian lignite mining district a decision support system has been developed. The foundation is a water balance model basing on a stochastic approach.

For coupling the water balance model with water quality, explicit algorithms which describe the transformation processes were implemented into the model. For most of the considered constituents, models have been developed which are based on physical approaches. For more complex systems like dissolved oxygen (DO) or the pH-value, the simulation results from physical models proved to be insufficient because of the lack of data. Hence, an alternative method, the self-organisation method *Group Method of Data Handling (GMDH)*, was applied. *GMDH*, which is based on the cybernetic principles of self-organization, is a powerful technique of data analysis for modelling and identification of non-linear systems through statistical analysis of input-output data. This particularly applies if only a small set of data is available.

In this paper the system structure and the development of physical algorithms is presented. Furthermore, the theoretical basics of *GMDH* and the development of *GMDH* algorithms are presented. Finally advantages and disadvantages of this method are discussed.

1   INTRODUCTION

Brown coal mining has a long tradition in Germany and covers large areas. There are three main districts of open-cast mining in Germany: the Rhineland district near Cologne, the mid-German district around Leipzig and the largest one, the Lusatian district.

The Lusatian open-cast mining area is located in the east of Germany about 80km from Berlin. Due to the one-sided energy politics of the former GDR the brown coal deposits were exploited without any consideration of ecological aspects. To enable brown coal quarrying the groundwater level had to be lowered by one or even more meters. The artificial lowering of the groundwater level affected an area of almost 2100 km². The drainage water had to be pumped into the receiving streams, in particular the River Spree. Hence, the discharge in the River Spree, which supplies the city of Berlin with drinking water, was strongly increased. As a consequence the induced drainage water diluted the municipal and industrial wastewater which, therefore, had only a small influence on the water quality. However, the concentration of mining induced parameters, such as iron and sulphate, was increased.

After the German reunion the exploitation of the brown coal open-cast mining was reduced because of economical aspects. As a result, the amount of mine drainage was reduced. Due to this the discharge of the River Spree decreased gradually after being raised over many years. Today, water is also diverted from the River Spree to flood the worked-out open cuts and to refill the drawdown cones of the groundwater. As the amount of mine drainage pumped into the River Spree is decreased, the diluting effect of wastewater diminishes. Thus, municipal and industrial sewage influences the water quality considerably. In addition, the mining induced parameters continue to have an impact on the water quality of the River Spree and its tributaries.

2 DECISION SUPPORT SYSTEM

2.1 Project

Since open-cast mining will end approximately in the year 2030 the complex relationship between water supply and water quality requires a rigorous water management for the next three decades. Therefore, the ministry of Saxony concluded the development of a decision support system for the River Spree and its tributaries. This water quality decision support system is coupled to the water balance model GRMDYN which was developed by WASY GmbH (Berlin). GRMDYN is based on mass-balance principles. In time steps of one month, discharges and run-off are calculated by stochastic simulation.

The main components of the water quality model are explicit and deterministic algorithms which describe transformation processes. They have to be coupled to the water balance simulation. Alongside the mining induced parameters of iron, sulphate and pH-value, the components DO, COD, nitrogen, phosphate, chloride and water-temperature have also to be implemented. Since the simulation of water quality in steady-state models based on mass-balance is subject to a coarser consideration than in dynamic models, it is not necessary to simulate all processes in their entire complexity. The goal is to develop explicit algorithms which describe the essential processes with sufficient precision that a simple but realistic model is obtained.

As water quality is coupled on water quantity the results are stochastic. By simulation of different scenarios and the statistical analysis of the simulated results, assumptions about the impact of future management policies on the water quality can be made.

2.2 System Structure

The system structure covers approximately 115 km of the river Spree and 110 km of its tributaries. The rivers are subdivided into reaches averaging 15 km. The reaches are limited by nodes which match the water quality gauging stations. At all nodes the water quality is balanced. Because of different objectives the locations of nodes for balancing water quality and balancing water quantity differ. Within this structure all relevant intakes which affect water quality are considered. In **Figure 1** the northern part of the system structure for balancing water quality is shown.

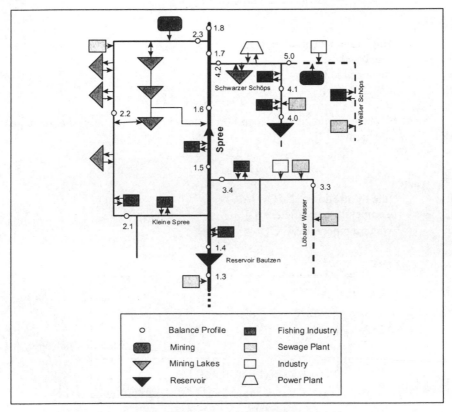

Figure 1: System Structure

3 ALGORITHMS

3.1 Physical Models

The majority of substances in water are subject to changes in concentration due to physical, chemical and biological processes. Physical models describe these processes with mathematical equations. All relationships in these models are rigidly determined and fixed [6]. The development of physical models is described in the context of modelling and simulating the parameters sulphate, ammonia and nitrate.

168

In natural rivers sulphate is present only in small amounts (10-30 mg/l). Exceptions are rivers which are affected by lignite mining. In some River Spree tributaries the concentration of sulphate exceeds 800 mg/l. Sulphate is a reasonably conservative substance. It is only subject to biochemical degradation by assimilation [1] or coprecipitation with Fe and Al phases [7]. Further interactions between sulphate and other constituents can be neglected. Therefore, a first order kinetic model (1) was chosen [11]:

$$u \cdot \frac{dc(SO_4)}{dx} = c_0(SO_4) \cdot k_{SO4} \tag{1}$$

where:

$c(SO_4)$: sulphate-concentration, [mg/l]
k_{SO4}: rate of reaction, [1/d]
u: mean velocity, [m/d]
x: distance, [m]

Since biochemical processes depend on water temperature the *Arrhenius* equation (2) can be applied [2]:

$$k_{SO4} = k_{20} \cdot \theta^{(T-20)} \tag{2}$$

where:

k_{20}: rate of reaction at 20°C, [1/d]
θ: Temperature coefficient, [ ]
T: water temperature, [°C]

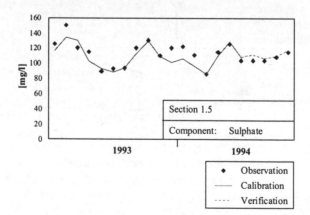

Figure 2: Simulation results of sulphate

Figure 2 depicts the results of the calibration and verification of the chosen model for section 1.5 of the River Spree. This section is characterised by the tributary Löbauer Wasser which is strongly affected by municipal sewage. Because of the lack of data, calibration and verification had to be carried out with small sets of data. The measured values are represented by dots. For better clarity, the simulated values are represented by lines. After the calibration the model is verified against an independent set of data (10-15 % of the observed values, dotted line). The quality of verification is estimated by the relative error between measured values and verification results [9]. It can be shown that the results are sufficiently accurate. The relative error averages 2.8 %.

In the model the nitrogen components ammonia and nitrate are considered. Nitrogen dynamics have to be modelled in a considerably more complex manner than sulphate because of their substantial biochemical role, the important oxidation-reduction reactions, and because other important water quality variables such as DO are affected by nitrogen. The main transformation process is nitrification. Nitrification can be simulated by first order kinetics as a one-stage process with oxidation of ammonia to nitrate directly. Further sink or source processes like resuspension of ammonia or accumulation of nitrate are approximated by constants [3]:

$$\text{Ammonia:} \quad u \cdot \frac{dc(NH_4)}{dx} = -c(NH_4) \cdot k_{NH4} + k_r \tag{3}$$

$$\text{Nitrate:} \quad u \cdot \frac{dc(NO_3)}{dx} = c(NH_4) \cdot k_{NH4} - k_a \tag{4}$$

where:

$c(NH_4)$: ammonia-concentration, [mg/l]
$c(NO_3)$: nitrate-concentration, [mg/l]
k_{NH4}: rate of reaction (temperature dependend), [1/d]
k_r: resuspension of ammonia, [mg/l/d]
k_a: accumulation of nitrate, [mg/l/d]
u: mean velocity, [m/d]
x: distance, [m]

Figure 3 depicts the results of calibration and verification for ammonia and nitrate at section 1.5. It is obvious that the results are less accurate than the results of the sulphate simulation. The relative error between measured values and verification results are 33.2 % (NH_4) and 8.9 % (NO_3).

The reason for these results is the lack of data on the one hand and the more compound transformation processes on the other hand. Hence, for more complex systems like dissolved oxygen (DO) and the pH-value an alternative approach, the self-organisation technique *Group Method of Data Handling (GMDH)*, was applied.

170

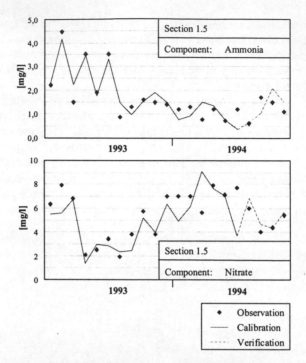

Figure 3: Simulation results of nitrogen

3.2   *GMDH* Models

3.2.1   Characteristics

Like Artificial Neural Networks (*ANN*) the *GMDH* approach is based on the connectionism as a representation of complex functions through networks of elementary functions. *GMDH* has implemented a stronger power than *ANN* due to the principle of induction. This principle consists of self-organisation as an adaptive creation of a network without subjective points given and of external complement enabling an objective selection of a model of optimal complexity. Another reason for application of *GMDH* is that the resulting models are explicit whereas the results of *ANN* are implicit models, which can be changed into explicit models only approximately and with much effort [10].

The quality of the resulting model depends on the range of the data set used for calibration. It is a fundamental problem that every modelling process is based on a finite amount of data whereas the reality is infinite. In this way the modeller is responsible for the selection of data which is used for calibration or reserved for verification, respectively. In the revised *GMDH* using the *PESS-Criterion* the modeller does not have to determine a data ranking list, since each data is used as learning data as well as checking data. This allows an objective system analysis to be performed.

3.2.2 Theoretical background

This section gives a brief description of the *GMDH* algorithm. The *GMDH* algorithm is a procedure for constructing a high-order polynomial, named the *Kolmogorov-Gabor* polynomial (5):

$$y = a_0 + \sum_{i=1}^{m} a_i x_i + \sum_{i=1}^{m}\sum_{j=1}^{m} a_{ij} x_i x_j + \sum_{i=1}^{m}\sum_{j=1}^{m}\sum_{k=1}^{m} a_{ijk} x_i x_j x_k + \dots \quad (5)$$

where:

y :   Output (dependent) variable
x :   Input (independent) variables
m :   Number of input variables
a :   Polynomial parameter

    The basic information which is necessary to construct this polynomial is a set of *n* observations. In the basic *GMDH* the available data (*n* data) are subdivided into two sets, the training set and the checking set [4]. In the revised *GMDH* the data are subdivided dynamically (cross-validation principle), thereby avoiding heuristics. This implies that all data are subdivided for all combinations into a set of (*n*-1) training data and one checking value [8].

    The first step to construct the revised *GMDH* algorithm is to take all independent variables x_1, x_2, ... x_m two at a time and to generate optimal basic functions. To develop these optimal basic functions the revised GMDH algorithm starts with second order polynomials in which each term (6 terms) is considered to be an inner variable. By using a selection criterion (see below) each of these inner variables is selected only if the quality of forecasting is improved. For each of these optimal functions the unknown polynomial parameters A_i (maximum of 6) have to be determined by using the training set so as to minimise the mean-square error between the output *y* of the training set and the evaluated optimal function. In this way $m \cdot (m-1)/2$ regression surfaces z_{ij} like those illustrated in *Figure 4* are developed.

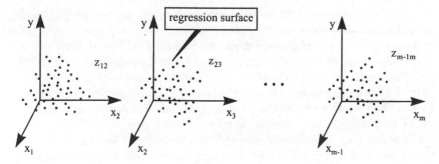

Figure 4:   Construction of polynomials

    By evaluating the constructed polynomials at the *n* data points new variables (vector Z_i) are obtained (**Figure 5**). These new variables are polynomials in the original variables x_1, x_2, ... x_m.

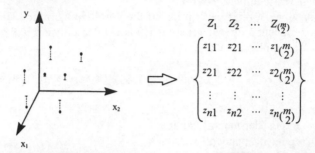

Figure 5: Construction of new variables

The next step is now to determine the columns of Z which best predict the observed values of y by applying a selection criterion. The old variables are replaced by those columns of Z. As selection criterion the *Prediction Error Sum of Squares Criterion* (*PESS-Criterion*) is applied (6). *PESS* is the sum of errors between each observed value y_t and its appropriate simulated value $f()$ divided through the number of observations n. Each simulated value is a function of all appropriate independent variables (vector X_t) and the parameters determined with the training data without the tth data set (vector A_t):

$$PESS = \frac{1}{n} \cdot \sum_{t=1}^{n} \left(y_t - f\left(X_t, A_t \right) \right)^2 \tag{6}$$

where:

y_t : t^{th} observed value of output variable
X_t : Vector of tth value of all independent variables (checking data)
A_t : Vector of parameters which were determined with training data without t^{th} data set
n : Number of observations

In this manner each data is once used for checking the constructed model. The advantage of the *PESS-Criterion* compared with other selection criteria is that by applying the cross-validation principle the selection is free of heuristics and the knowledge incorporated in the data can be revealed optimally [5].

The values of Z's which satisfy the criterion with $PESS < L$ are the new variables for the next iteration. The other values of Z's are screened out. The limit L is a prescribed value which has to be chosen by the modeller. In this way he influences the number of new variables in every iteration step.

The last step in the development of the model is the test for convergence. At the second and all the subsequent iterations the smallest error or the minimum of the *PESS-Criterion* is compared to the minimum of the preceding iteration. If the value is smaller than the previous one, the first two steps are repeated. If the value is greater, the process is stopped. In the latter case the polynomial of optimal complexity was constructed in the preceding iterations. This implies that the complexity of the model is increased as long as the forecasting quality of the model improves. For the example in **Figure 6** this process would be stopped after the fifth iteration and the result from the fourth iteration would be chosen.

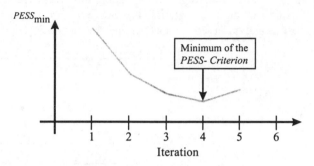

Figure 6: Test for convergence

3.2.3   Results and Discussion

In the following section the results of simulations using *GMDH*-models for DO and the pH-value are presented. For the verification process (not the checking process within the *GMDH* algorithm) 15-20 % of the available data was used. The data was within the range of training and checking data. The selected, dependent variables and the networks between input and output are presented in **Figure 7**.

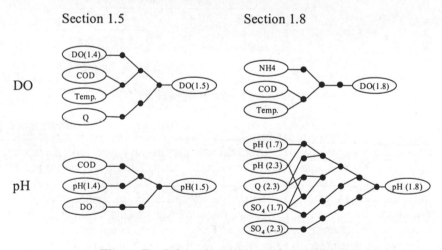

Figure 7: Selected variables and networks

Figure 8 shows the results of calibration and verification of a DO model for two sections of the River Spree. The first section (1.5) is the same as described above. The second section (1.8) is influenced significantly by the tributary Kleine Spree. This tributary is heavily affected by mine drainage.

The variables identified for the DO-models are DO (at the upstream profile), water temperature, ammonia and COD (see **Figure 7**, top). The latter two variables are indicators for municipal sewage, which causes an oxygen demand by nitrification or degradation of organic material, respectively. The relative error between verification results and observed values averages 4.8 % or 3.7 %, respectively.

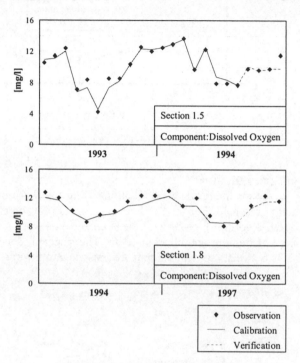

Figure 8:   GMDH results of DO

Figure 9 depicts the results of a pH-value model for the sections described above. The relative error between verification results and observed values averages 0.7 % and 2.6 %, respectively. The variables identified are discharge, pH-values, and sulphate concentrations at the upstream profile as well as at the profile of the tributary (see **Figure 7**, bottom). The identification of sulphate is physically meaningful because the acidity of the mine drainage results from the disintegration of the sulphur bearing mineral pyrite (FeS_2), which predominates in coal seams. When pyrite is exposed to air or water it oxidises to sulphate and releases dissolved iron and acidity into the water [12]. Therefore, high sulphate concentrations indicate low pH-values in the tributary as well as in the Spree section considered.

The analysis of uncertainty of these models was performed by sensitivity analysis for the selected variables and by Monte Carlo simulations to identify the parameter variance-covariance structure. It can be shown that the models are applicable when the input data does not exceed the range of the data applied for calibration. However, input values which exceed the field of the data which were employed for calibration can lead to insufficient results. Hence, a *GMDH* model should always be controlled and if necessary updated, if new data is available.

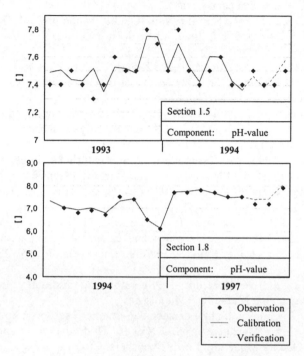

Figure 9: GMDH results of pH-value

4 CONCLUSIONS

In this paper a heuristics free self-organising technique, the *Group Method of Data Handling (GMDH)*, for constructing water quality models for a river affected by lignite mining is presented. Because the amount of available data is small, physical modelling cannot provide a prediction model for complex, non-linear systems like DO or pH-value. Therefore, a *GMDH* algorithm is applied to develop prediction models. The results indicate a high forecasting quality for the constructed models.

However, compared to physical models, forecasting with *GMDH* models could fail when input-data exceeds the range of data used for learning and checking. Obviously, methods of experimental system analysis, like *GMDH*, cannot replace the necessary analysis of causes of events, but appropriate application of these tools may reveal the secrets of a system-structure. If the physical descriptive model is not known explicitly and if the characteristics of the structure are well incorporated in the observed data, a physically meaningful structure may be identified by using *GMDH*.

REFERENCES

[1] Brehm, J.; Meijering, P.D. (1996): Fließgewässerkunde; Quelle & Meyer Verlag, Wiesbaden (in German)

[2] Chapra, S.C. (1997): Surface Water-Quality Modelling; McGraw-Hill Series in Water Resources and Environmental Engineering; R.R. Donnelley & Sons Company

[3] EPA (1985): Rates, Constants, and Kinetics – Formulations

[4] Farlow, S.J. (1981): The *GMDH* Algorithm of Ivakhnenko; The American Statistician, Vol. 35, No. 4, 210 pp

[5] Ivakhnenko, A.G.; Müller, J.A. (1992): Parametric and Nonparametric Selection Procedures in Experimental Systems Analysis; System Analysis, Modelling and Simulation (SAMS), Vol. 9, 157 pp

[6] Jokiel, C. (1995): Gewässergütesimulation natürlicher Fließgewässer; Mitteilungen des Instituts für Wasserbau und Wasserwirtschaft der RWTH Aachen; Köngeter, J. (publisher); Bd. 103; Academia Verlag, St. Augustin (in German)

[7] Kimball, B. A.; Broshears, R. E.; Bencala, K. E.; McKnight, D.M.(1994): Coupling of Hydrologic Transport and Chemical Reactions in a Stream Affected by Acid Mine Drainage; Environ. Sci. Technol., Vol. 28, No. 12, 2065 pp

[8] Lembke, F. (1996/ Dezember 1998): Knowledge Extraction from Data Using Self-Organizing Modeling Technologies; http://www.informatik.htw-dresden.de/~muellerj/ self-orgn.htm

[9] Martin, J. L.; McCutcheon, S. T. (1999): Hydrodynamics and Transport for Water Quality Modeling; Lewis Publishers, CRC Press

[10] Otto, P. (1994): Identifikation nichtlinearer statischer und dynamischer Systeme mit Künstlichen neuronalen Netzen; 39. IWK Band 3; Ilmenau (in German)

[11] Schlaeger, F.; Schramm, J.; Köngeter, J. (1998): Development of a Coupled Water Management-Quality Model to Control the Water Regime of the River Spree; Proc. 11th IAHR APD Conference, Yogyakarta, September 7.-9.,1998, 213 pp

[12] Stumm, W.; Morgan, J.J. (1996): Aquatic Chemistry, Chemical Equilibria and Rates in Natural Waters, 3rd edition; John Wiley & Sons, New York

PART IV

RELIABILITY AND FAILURES

Network Asset Strategy Modelling[*]

D. Burnell *and* J. Race

ABSTRACT
This paper outlines a strategic framework to help develop leakage policy within Thames Water. The framework has been developed at the request of directors following the 1997 Water Summit. Two key elements are described:

- a model to analyse mains repair location at the street level. The model takes account of all London mains repairs since 1990. A novel clustering method has been developed to group together similar streets so as to reduce "noise" in the data. Repairs are found to be strongly linked to asset age (inferred from the date when the street first appeared on a London map), soil "fractivity", and pipe diameter. There is a seasonal pattern for transverse fractures - the most common failure on 3" and 4" mains - which matches soil-moisture deficit. Most failures are prompted by ground movement as the underlying clay dries out or becomes moist, exacerbated by extreme winter weather. The model provides a basis for projecting future repair rates and where renewal will be most cost-effective.

- a model of leakage dynamics. Leakage is a result of asset failure. The model links daily data on mains and service repair activity with daily leakage levels (derived via Minimum Night Line). The model is first calibrated on the past, to estimate seasonal and cold-weather recurrence. Future leakage scenarios are then constructed and compared, adjusting detection effort and weather. The model can also be used to estimate the economics of leakage by looking at the equilibrium leakage level eventually reached, for a range of detection effort. The costs and benefits of infrastructural renewal can also be explored.

Together with other studies, the models help provide a map of key leakage choices and focus attention on areas where more data is still required.

1   INTRODUCTION

Following the 1997 Water Summit, Thames Water's in-house Corporate Modelling consultancy group was asked to develop a framework for assessing asset performance to help with strategic decision-making on infrastructural investment and leakage control. Part of the response to this remit (Crerar et al. 1999) was to initiate new studies to collect more data on asset condition and usage patterns. This paper outlines key features of the framework developed to exploit this data and provide a high-level 'map' of leakage issues.

The suite of decision-support models were written in-house and made use of corporate data assembled on mains and service-pipe repairs since 1991, zonal Minimum Night Lines and new data sets on traffic loading, soil and street age. The models address the following key questions:

---

[*] © Thames Water

179

- what can be learnt from the location of mains repairs and how can this be used to prioritise future repair and renewal effort?

- what do the various daily time-series on leakage, mains repairs and weather imply about the dynamics of leakage? Can we account for the past and how does this help plan for the future?

Along with other analysis, aspects of this work have been used to support Thames Water's Strategic Business Plan for the 1999 OFWAT Periodic Review. The aim of this paper is not to present these results but to set out the broad approach, which could be applied elsewhere given similar base data.

2   MODELLING ASSET CONDITION

2.1   Modelling Requirement

Like any other asset, network infrastructure presents management with a choice between repair and renewal. For example, knowing when it is cost-effective to replace a vehicle may be obvious given a steadily rising garage bill or a mechanic's detailed appraisal. The difficulty with distribution mains, on current technology, is in assessing condition - as a basis for making this choice - on a dispersed and largely invisible asset.

One readily available piece of information on a main's condition is its repair history, assuming this has been reliably kept for a number of years. The challenge is to organise this data so as to (a) make sense of past repairs and (b) project likely repair rates into the future. If this can be done for every pipe in the network, we can then start to judge how far it is possible to target renewal so as to maximise the benefits of the investment.

2.2   Streets as a basis of location analysis

Location analysis of repair history needs to be comprehensive. If all repairs are known to be included, areas with low rates are as instructive as high ones. To achieve this means working to the best common accuracy possible. Whilst other location details may be available, repairs within London are always logged by street and locality. Within a built-up urban area the street is a natural index of location. By assembling all data by street, we can carry out location analysis at the level of accuracy which the historic record supports.

Indexing on streets brings a number of challenges. Street-names are ambiguous, easy to miss-spell; some streets are long. There are 147 'Albert Road's in London and many ways of writing St John's Wood. To meet these difficulties a set of algorithms have been developed to convert repair addresses into locations. The algorithms use AddressPoint, the Post Office/OS database of every deliverable address in the Thames Region.

From AddressPoint we have extracted the grid-reference, "outcode" (e.g. NW10) and house number of each end of each street (up to 2 bends per street). AddressPoint gives the same location to every flat within the same building, so one can count buildings as well as households and so infer the connection density for each street.

From this master list, each repair address in the corporate repair database has been matched by street. Where no immediate match is possible, short-lists are constructed of near-miss spellings and other locality data used to pick from this. Once the street is matched, a grid-reference is constructed by interpolating along the street using its "outside house" number.

This process on its own provides many insights. In particular the schematised streets and associated repairs provide a visual impression of where the activity has taken place, and the extent to which repairs are clustered on particular streets. The display below shows how the repairs have been distributed across one London Zone over the past 8 years.

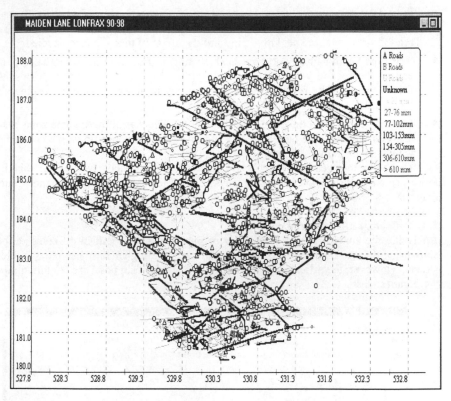

Fig. 1 Distribution of repairs for a London zone

Because the address-matching process is fully automated, it can be applied to every kind of corporate data where the address is known e.g. services and valves, meter installations, older repair data sets.

In order to develop an understanding of where repairs are needed, many other data sources have been brought into the model. The aim is to include all likely explanatory variables so their importance can be assessed via statistical analysis. Key data sets include:

- Ages of street, based on the ever-widening 'sprawl' of London as derived from Ordnance Survey maps from 1822 onwards. In London, original street age is a good indicator of pipe age. This is supplemented with design drawing-record data where available.

182

- Soil 'corrosivity' and 'fractivity' polygons. These measure the tendency of soil to corrode, or to induce fracture through changes in moisture content. This data was obtained by Thames Water R&D.

- Main-road traffic density estimates, via Thames Water R&D from the London Research Centre;

- Pipe data downloaded from the company's Geographical Information System. Some 18 million pipe segments have been indexed by street. This allows repair rates to be normalised per km of pipe and the effect of pipe diameter studied.

- War-time bomb-damage locations have been identified from maps produced at the end of the second world war and held at the Metropolitan Library, to test the theory that wartime repairs (and materials) may represent a point of weakness in the network.

- Each street has been linked to a Pressure Zone. This has allowed some zone-based analysis, although the large size of London's pressure zones limits the conclusion which can be drawn.

The model thus acts as a street-based gazetteer of repair activity. It allows this data to be depicted and analysed in many ways, including maps, histograms and scatter-diagrams.

2.3   Factors affecting repair rates

Analysing the annual repair rates per km of pipe for streets built in different periods shows that (as one might expect) mains repair rates are higher for those pipes in older streets. However, the difference is not as big as one might suppose: the Victorian pipes were built to last!

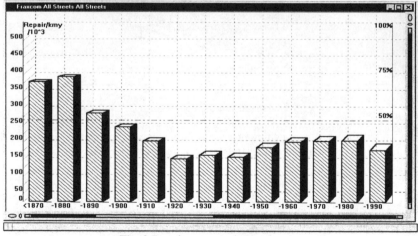

Fig. 2 Repair rates against street age

The post World War 2 expansion probably used poorer quality materials; later on there is an overlap of materials which is more difficult to interpret. Similar analysis has shown that repair rates tend to be higher for :

- those pipes in soils which have a higher fractivity i.e. exert greater ground movement forces.
- smaller diameter pipes, repair rates fall off steadily as pipe diameter increases. There is a greater length of smaller diameter pipes, which means that the vast majority of repairs are arising on smaller pipes (<=6")

Repairs (especially transverse fractures) show a marked seasonal pattern, which is similar in shape to soil moisture deficit patterns and also to ground temperature changes. Extreme weather in Feb 1991 and Jan 1997, whilst visible as spikes, cannot explain all of the winter rise.

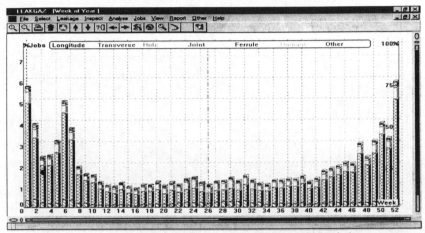

Fig. 3 London Mains Repairs 1990-98 - Seasonal Variation

Analysis confirms that repairs do not occur randomly across the London network. Some 55% of streets have not required any mains repair since 1990 and 20% of streets account for 80% of repairs. These findings suggest that careful targeting of repair or renewal efforts is well worth while.

2.4 Statistical Analysis of repair Locations

To analyse repair location, we need to allow for interactions between the various explanatory factors. For example, London has sprawled out from its centre over time, so the oldest streets will tend also to be the most congested. Multi-factor analysis is needed to avoid spurious conclusions.

Even with several years' data, there is too much 'noise' to do this at the street level. A 100m street with no bursts in five years has a burst rate of 0, whilst one with 2 bursts has a rate of 4.0 bursts /km/yr. Both are far from the overall rate of around 0.2 burst/km/yr.

To deal with this we have developed an innovative method of "clustering" streets which are similar from an asset-performance viewpoint, then examining their composite repair record. We group together streets into 'clusters' which are similar in terms of:

- street age bands
- soil type
- road classification

184

- proportion of small diameter pipes

For example, a cluster might comprise all streets built in the 1870s which are B-class roads and build on high-fractivity soil.

Some 220 London street-clusters were identified, and their repair rates and associated data fed into the analysis. The model explains over 75% of variation in cluster mains repair rates across London, with soil type and street age being amongst the explanatory factors for the best fit model. The predicted and actual repair rate for each cluster are as shown.

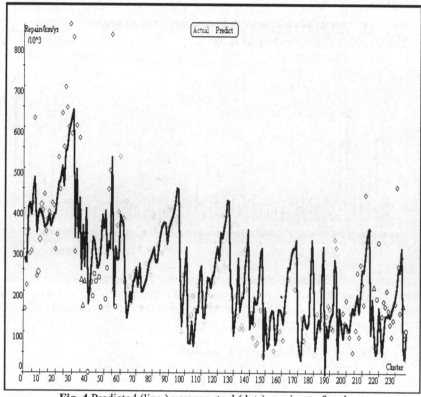

Fig. 4 Predicted (line) versus actual (dots) repair rate for clusters

The saw-tooth reflects the way the clusters have been ordered. The make-up of the predicted repair rate is indicated below, with the different shaded areas showing the contribution of different factors to the cluster repair rate.

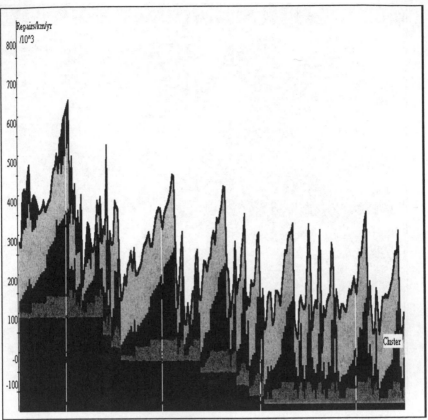

Fig. 5 Cluster Repair Rate Model - Contribution of Different Factors

The model provides a basis for many investigations. For example,

- Project repair rates into the future under different renewal assumptions. We can extrapolate repairs as streets 'age' over time, increasing the age contribution to the repair rates until the pipe is renewed.

- the co-incidence of repairs on a street can be examined (e.g. to explore the number of repeat repair jobs);

- repairs can be grouped at the zonal or km-square level as well as by street or cluster;

- step a repair-map though time to see the build-up of a winter event;

The model has been developed as a high-level strategic tool. However, the data and approach could help with targetting renewal at the planning level. The map below (with the river Lea coming in from the North and Regents Park on the left) help pin-point high-repair clusters within Central London.

186

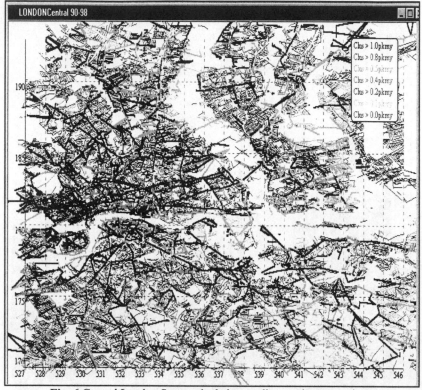

Fig. 6 Central London Streets shaded according to cluster repair rates

3   LEAKAGE DYNAMICS AND ECONOMICS

3.1   Key Questions on Leakage Dynamics
Leakage is the result of asset failure. Key questions are:
- what is the make-up of leakage e.g. customers supply pipes versus company assets?
- what are the costs of meeting different leakage targets with the current infrastructure?
- what is the economics of leakage, and how is this affected by different renewal policies ?
- what is the effect of weather on leakage and on the economics of leakage?

These issues are fundamental to answering strategic questions :
- what is the best mix of Opex (repairing current assets) versus Capex (renewal of selected pipes) to meet anticipated future leakage targets?
- what is the best way to spend repair budgets to maximise leakage savings?

The LEAKPLAN model has been developed to address these questions. It focuses on *the timing of repairs and their collective impact on leakage.* The model ties together
- daily data since 1995 on "find and fix" activity on mains, valves, services and customer supply pipes;

- daily leakage rates based on Minimum Night Lines. (So far the model has been applied to North and South London and Provinces.)

Whilst still in its early days the model is already contributing to the understanding of the economics of leakage and forecasting of future activity.

3.2 Modelling the mechanics of leakage

The model seeks to account for past trends. Leakage is assumed to be the result of cumulative recurrence (of mains and services) less cumulative repairs. The difficulty is that we know the net result of these changes on the Minimum Night Line, but not how they were made up. Calibration requires estimates of

- the leakage-gain for each type of repair;
- the underlying level of recurrence and growth;
- the impact of severe winter weather;
- the make-up of leakage, notably the amount of leakage which is undetectable background losses.

The weather impact is modelled first, using just the visible mains "finds" over the past few winters. It is reasonable to assume that such jobs are reported quickly, so their find date is known and can be tied in to daily weather factors.

The weather modelling is done via multiple regression. The explanatory variables are various weather parameters, including

- max and min daily temperature;
- derived parameters such as the number of days since a freezing-spell began;
- threshold parameters, such as degrees below zero (or zero, if the temperature is above zero), which only impact in cold weather.

Whilst it is relatively easy to account (in retrospect) for a particular winter, it is hard to find factors which are stable from one winter to the next. This has led to the introduction of a long-term weather 'memory' factor. Physically, this may reside in the network or the soil. A severe winter, which triggers many bursts, will tend to relieve the stress in the system so the following year will see below-average levels of bursts. This effect can be seen in various long-term repair time-series for London (one going back to 1920) which we have analysed.

The weather model is assumed to apply equally (but less obviously) in triggering other jobs. Once a weather sub-model has been calibrated, underlying recurrence is modelled by a separate regression to link together cumulative repairs and daily Minimum Night Line.

More elaborate models are possible. In particular, a view needs to be taken on the extent to which recurrence grows because the same leaks are getting bigger, as opposed to new leaks arising. Assumptions are also needed on whether detection rates will be higher for larger leaks than for small ones. Data from other studies now underway within Thames Water (notably the Water balance Study) will be used to calibrate more elaborate models in due course.

Once factors have been found which give a credible account of the past, this gives a basis for forward projection of different operating strategies and weather scenarios, assuming the infrastructure remains unchanged.

188

3.3 Sensitivity Analysis

The strategic virtue of such a model is not that it gives a rock-solid "central case" but that it allows alternative assumptions to be rapidly explored. The graph below shows how projected leakage (under one, purely illustrative, set of assumptions on background losses and future detection policy) might vary under different weather scenarios (namely repeats of the 1994/5, 95/6 and 96/7 winters). The model suggests that a severe winter will take a long time to work through: the three scenarios only converge at year end.

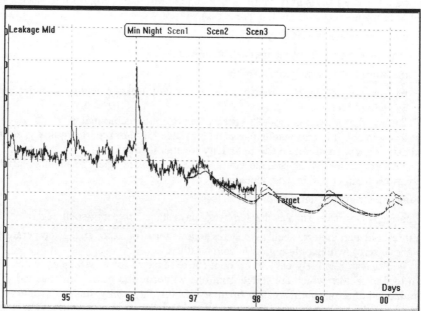

Fig. 7 Dotted lines are projected leakage for different weather scenarios under illustrative assumptions about background losses etc

3.4 Estimating the Economics of Leakage

The scenarios above show how, for a given level of detection, leakage will tend towards an equilibrium level at which the detection rate is exactly holding in check the overall leakage recurrence. The long-term equilibrium is also of interest for economic assessment.

LEAKPLAN uses this to estimate the economics of leakage by finding the leakage level at which overall costs (cost of supply + detection costs + repair costs) are minimised. This approach runs through a range of operating policies, extrapolating what has happened in the past. It is not an idealised equilibrium, but takes account of the effect of winter weather. As already noted, a bad winter will perturb leakage upwards and the system will take months to work this through.

The economics of leakage depend crucially on the level of Background Losses - an assumed floor on leakage with current detection methods. LEAKPLAN is now being further extended to allow asset renewal to be included in the calculations. New assets, replacing the worst pipes with ones with minimal failure rates, will add capital costs but lead to lower detection and repair costs. This element will allow the trade-off between

Capex and Opex in reducing leakage to be explored. LEAKPLAN is only as good a predictor as its modelling assumptions allow. Potentially, though, it gives crucial insights on a complex dynamic process.

4   CONCLUDING OBSERVATIONS

This paper has outlined two major decision-support models to help strategic leakage thinking. Both make use of repair data systematically collected since 1991:

- One model, LEAKGAZ, focuses on mains repair location, and shows there is a strong pattern to this once age and soil condition are taken into account. This provides a basis for evaluating alternative renewal strategies, and for targeting repair and renewal activity.

- The other model, LEAKPLAN, concentrates on repair timing, and the impact of repairs (and cold weather) on leakage. Making sense of the past is a starting-point for projecting the future and exploring the impact of different weather and detection scenarios.

Leakage is too complex a phenomenon to be reduced to simple models, but interactive models with easily-modified assumptions at least allow these uncertainties to be explored.

The results presented in this paper are based on analysis of Thames Water assets, and will not necessarily hold for other parts of the UK or elsewhere, because different local conditions and pipe types will apply. However, the analytical approaches developed are generalised and could be applied to other datasets.

Over the coming year Thames Water expects to make substantial advances on District Metering. The Framework now established will be enhanced by data and insights from these DMA's, as well as from special study areas.

ACKNOWLEDGEMENTS

We would like to thank Thames management for their support and encouragement during this project. This project has drawn on discussions with colleagues in Engineering, Customer Field Services and R&D. Thanks are also due to R&D and computer systems experts for their help in providing data. The views are those of the Authors.

REFERENCES

Crerar, Race, Burnell and Barton, 1999, Water Balance Study – Microscope on Leakage and Demand, *Water Industry Systems,* vol. 2, Research Studies Press Ltd, UK.

Probabilistic Forecast of Failure Occurrences in Water Networks with Short Maintenance Records

P. Eisenbeis *and* Y. Le Gat

ABSTRACT
Managers of drinking water utilities need technical decision tools for network maintenance. One of these tools, used at Cemagref, allows for probabilistic forecasting of failures. This is a parametric proportional hazard model with covariates, assuming a Weibull distribution for the inter-arrival failure times. It has already been shown to be effective with complete maintenance historic records despite the fact that parameterisation cannot be transposed from one service to another.

At present, most of the services have only short maintenance records (less than 10 years). In order to assess the usefulness of the model for short records, we have processed data of the rural service of the Charente-Maritime area (Atlantic Coast of France). These networks involve pipes made of different materials: PVC, asbestos-cement, cast and ductile iron, with maintenance records available since 1988. A difficulty could arise a priori from the lack of data over the period running from the date of laying of the main to the beginning of the observations. Counting the first inter-arrival times from the starting date of observations does not handicap the model; the method accounts for right censoring of the data.

Firstly, using the failure records from 1988 to 1996, we parameterise a specific model for each material (the set of relevant covariates varies according to the material). Secondly, we forecast the number of failures which may occur in the year 1997, using a Monte-Carlo method. We are thus able to compute the differences between actual and forecasted numbers of failures, and then assess their statistical significance by a Chi-square test. This has led to encouraging results. The model is effective even for networks with short maintenance records, and therefore should be integrated in Geographical Information Systems.

1   INTRODUCTION
The statistical analysis of failures is one of the ways to assess the state of a drinking water network and, more precisely, of a pipe [1]. These statistical methods are either classical methods or specific methods, like survival analysis.

Since 1994, Cemagref [2] has been using survival analysis and, particularly, the Weibull Proportional Hazard Model (WPHM) on data coming from water services with long maintenance records (from 20 to 50 years). However, most of French water services have been keeping these data for 5 to 10 years.

The aim of this study was to examine if the WPHM could be used on such services. Data coming from the rural water service of "Charente-Maritime" has been used. In spite of only 10 years of maintenance data, the results are encouraging.

We present the statistical methods used to assess the influen1cing factors, to compute the survival function and to forecast the number of failures. Secondly we briefly present the data. Lastly, we present the results and some comparisons between forecasted and actual failures.

2 STATISTICAL SURVIVAL ANALYSIS APPLIED TO DRINKING WATER PIPE FAILURES

Statistical survival analyses are specific methods, used generally for epidemiological studies. In particular they allow the use of "censored data", i.e. data for which a time period is missing.

2.1 The Weibull Proportional Hazard Model [3]

In the Weibull accelerated life model (and proportional hazard model), it is possible to link explanatory variables to the lifetime of the pipe. These explanatory variables (also called covariates) might influence the time to failure and the failure rate. The lifetime T is defined as the interarrival failure time. The lifetime used in the analysis might be either observed or a right-censored failure time, i.e. pipes that are still intact at the end of the observation period.

Accelerated lifetime models assume that $\ln T$ (natural logarithm of the lifetime T) is related to the p covariates X_j via a linear model :

$$Y = \ln T = X\beta + \sigma W$$

where Y is the $n*1$ vector of the log of the n observed lifetimes, X is the $n*(p+1)$ matrix of covariates ($X=[1,X_1,X_2,..,X_p]$). The $(p+1)*1$ vector β ($\beta=[\beta_0,\beta_1,\beta_2,..\beta_p]$) and the scalar σ are unknown regression parameters. β_0 and σ are respectively called "intercept" and "scale", and W is a random $n*1$ vector of errors supposed to follow the extreme value distribution.

For a given pipe subject to the stress x (the vector of covariates values for this pipe) the survival function for the Weibull accelerated model for each individual pipe as a function of time is:

$$S(t, \beta, x) = \exp\left[-\exp\left(\frac{\ln t - x'\beta}{\sigma}\right)\right] = \exp\left[-t^{1/\sigma}\exp\left(\frac{-x'\beta}{\sigma}\right)\right]$$

The method to estimate the vector β and the scale σ uses the maximization of the log-likelihood function, which is the log-transform of the joint density of probability of the observations. The right-censored data take the value of their survival function at the censored time. This analysis is performed with the statistical software SAS.

2.2 Estimation of forecasted number failures

Based on these parameterized survival functions, the forecasted number of failures for each pipe is computed using the vectors β and x. Monte Carlo simulation is then used to forecast the expected number of failures within a given time horizon [4]. The recourse to Monte Carlo simulation is necessary because the number of previous failures (NOPF) is a highly significant covariate that complicates the theoretical calculation of the distribution of the number of future failures.

Solving the survival function as a function of t, the failure time corresponding to a given survival probability S is obtained:

$$t = \left(\ln\left(\frac{1}{S}\right) \exp\left(\frac{x'\beta}{\sigma}\right) \right)^{\sigma}$$

The Monte Carlo simulations are carried out in the following way:

- A random number (0,1) is chosen. The corresponding failure time for the given survival function is calculated (see Figure 1).
- If the failure time is shorter than the defined time horizon, a new failure time is calculated using an updated version of the survival function (the NOPF increases by one unit). This is repeated until the chosen time horizon is reached. The accumulated number of failures within the time horizon is calculated.
- For each pipe, this elementary scheme is repeated 1000 times and the mean value of the 1000 simulations is an estimator of the expected number of failures within the time horizon. The upper and lower confidence limits and the standard error can also be estimated.

194

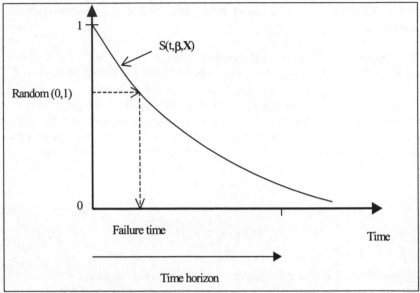

Figure 1 Forecast of failures based on survival functions

It is then possible to classify the pipes according to the number of failures, or the probability of failures, or to classify them, group by group, according to the influencing factors.

3   PRESENTATION OF THE DATA [4]

"Syndicat des Eaux de Charente-Maritime" (Atlantic Coast of France) is in charge of several small rural drinking water networks. These networks were previously independent and managed by the municipalities. Therefore, their characteristics are different from network to network. Nine of these networks have been chosen to represent the variability of the factors (see Table 1).

Overall, 1243 km have been observed over a period of nine years, that is to say 1212 links of pipes and 735 failures. The distribution of the pipes by material is shown in Figure 2.

The factors taken into account, due to their variability, are the length, the age of the pipe (taken at the beginning of observation for pipes without any failure observed inside the observational window, or otherwise at the date of the previous observed failure), the diameter, the kind of joint, the type of soil (acid, humid, stony), the level of traffic and the kind of supply (gravity feed or pumping).

The failures observed (i.e. a leak or a break having had a repair) were reported on paper documents that gave information about the date of occurrence, the kind of repair and, sometimes, the cause of the failure.

Table 1 Characteristics of the chosen networks

Networks	Urban or Rural	Modal laying date	Modal material (s)
Ecoyeux	Rural	1960	Asbestos cement
Marans	Urban	1935	Cast iron
Port d'Envaux	Rural	1966	Steel
Tonnay-Charente	Urban	1905	Cast-iron, PVC
Charente-Seudre	Rural	1965	Asbestos cement, PVC
Montguyon-Montlieu	Rural	1940	Cast-iron, PVC
Néré	Rural	1950	Cast iron
Saint Bris des Bois	Rural	1950	Asbestos cement
Saint Saturnin des Bois	Rural	1950	Cast iron

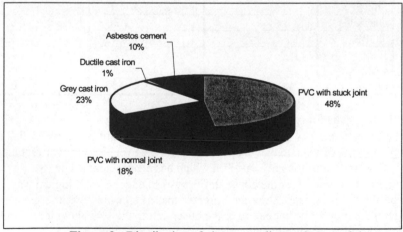

Figure 2 : Distribution of pipes according to the material

Because of the short period of observation, the first failure observed is not the actual first failure due to the presence of "left-censorship", meaning that the failure records between the laying of the pipe and the beginning of observation are missing.

4 THE RESULTS [4]

In order to compute the influence of the different factors, it was necessary to split the pipes into six strata according to the material and the number of previous failures (NOPF). Actually there is an interaction between influential factors and the material of the pipes (e.g. soil humidity will not act on grey cast iron and PVC in the same way). Moreover pipes made of ductile cast iron were not studied because of the small number of pipes.

Table 2 shows the results, giving the influencing factors and the values of the parameters β (see 2.1).

Table 2 Influencing factors and the computed parameters *(in grey, non influencing factors according to the Wald test)*

		Asbestos cement		Grey Cast Iron		PVC	
		NOPF= 0	NOPF>0	NOPF= 0	NOPF>0	NOPF= 0	NOPF>0
	Intercept	14.32	8.84	4.92	7.59	12.89	5.79
	Scale	1.26	1.90	0.71	1.74	1.15	1.37
Parameters β	Gravity feed						0.85
	Age left	-1.81		0.47		-1.65	1.99
	Traffic			-0.78			
	Humidity	-1.66	-1.45		-2.25		
	Acidity	2.35		-0.75	-1.87		
	Stones				-2.93		
	Diameter			0.52	1.38		
	NOPF						-1.59
	Joint or stuck					1.74	-1.60
	Length				-0.53		-0.29

This table calls for some comments:
- The scale parameter is generally greater than 1, indicating a survival function with a negative exponential shape; this is not true for the survival function of the first stratum of the grey cast iron which has a sigmoidal shape ;
- For grey cast iron pipes the covariates behave in a technically logical way. The data for the parameter "Age-left" greater than 0 signifies that the greater the age of the pipe at the beginning of the observation, the less likely it is to fail;
- For asbestos-cement pipes covariates are less significant. The covariate "age-left" acts differently from grey cast iron pipes;
- For PVC pipes more covariates are significant, particularly the type of supply (by pumping or gravity) and the type of joint, which act differently between the two strata;

It is important to establish if this model gives satisfactory predictions. With this aim we attempted to forecast failures for the year 1997, using the method presented in the section 2.2 and compared the forecasts with the actual failures that occurred in 1997.

Table 3 shows that for each stratum the global number of forecasted failures is very close to the actual number. The little difference is due to the variation of the actual number from one year to another. A comparison over a period of 5 years would have been more interesting but the maintenance records were too short. Moreover, the number of actual failures (5 failures) was too weak to make a statistical comparison for asbestos-cement pipes.

Table 3 Comparison of the forecasted number of failures and the actual number of failures that occurred in 1997 *(for chi-square test, the probability value, related to 3 degrees of freedom, is reported in parentheses)*

	Asbestos cement		Grey cast iron		PVC	
	Forecasted failures	Actual failures	Forecasted failures	Actual failures	Forecasted failures	Actual failures
Total	7.3	5	26.8	20	50.9	47
1st quartile	3.4	2	13.5	14	26.6	18
2nd quartile	1.9	1	6.6	2	10.5	12
3rd quartile	1.3	1	4.2	4	8.1	10
4th quartile	0.7	1	2.5	0	5.6	7
Chi square	1.337 (0.953)		4.066 (0.254)		1.942 (0.585)	

This table shows also the number of failures by quartile and the results of a chi-square test on these quartiles. In spite of the short forecast period of one year, the results of the chi-square test give a value greater than 0.05. Therefore, the short maintenance records do not prevent making a good forecast.

5   CONCLUSIONS

The Weibull model presented in this paper was used earlier on water networks with long maintenance records. This study shows nevertheless that it is also efficient in services with short maintenance records. It also shows that the factors act differently according to the material and the number of previous failures of the pipe.

The calculated parameters are quite different from the value estimated on the previous studies. Each water network contains specific features from a failure viewpoint and a general model cannot be applied. Results recently obtained in Norway in the city of Trondheim [5] show the same trends.

Water services are then well advised to create databases containing the pipes, their characteristics and their maintenance, in order to use them to make a diagnosis of the state of their network and to anticipate its deterioration. Associated with a Geographical Information System, this tool will facilitate decision making with the objective of rehabilitation, with the complementary analysis of other criteria such as the deterioration of water quality and the increase of head losses or of water losses.

REFERENCES

[1] ANDREOU S. (1987), Maintenance decisions for deteriorating water pipelines, Journal of Pipelines, 1987;7, pp. 21-31.

[2] EISENBEIS P. (1994), Modélisation statistique de la prévision des défaillances sur les conduites d'eau potable, PhD thesis University Louis Pasteur of Strasbourg, Collection Etudes Cemagref n°17, 190 p. + appendix.

[3] KALBFLEISCH J.D., PRENTICE R.L. (1980), The statistical analysis of failure time data, John Wiley & Sons, 1980, 321 p.

[4] PALLOIS F. (1998), Le vieillissement des canalisations d'eau potable- prévision statistique sur un réseau à faible historique de maintenance, School Engineer report, Cemagref Bordeaux – ENGEES, 96 p.

[5] RØSTUM J; DÖREN L; SCHILLING W (1997), Deterioration of the built Environment: Buildings, roads and water systems. Proc. 10th European Junior Scientist Workshop, Island of Tautra, Norway, 24-28 May 1997. Norwegian University of Science and Technology, IVB-report B2-1997-2, ISBN 82-7598-040-2

Risk Assessment Tool for Water Supply Network

K. Odeh, F. Fotoohi, R. Kora, J-P. Feuardent *and* P. Feron

ABSTRACT

In the scope of water supply network operating safety, Lyonnaise des Eaux is expected to perform reliability measures in order to control and improve quality and security services. This paper proposes a risk assessment study for water supply networks using a reliability assessment tool, *SAEP* (Sûreté d'Approvisionnement en Eau Potable). The authors evaluate quantitative and qualitative measures in order to compare the safety of different investment scenarios. The authors present their tool via a study on South Paris water network (ESP).

For many risk assessment studies, networks have become widely used for modelling complex systems, which are subject to component failures [10]. A water distribution network is an assembly of various components such as pumps, pipes, plants, boreholes. The above assembly can be modelled as a stochastic network. Three types of reliability problem need to be investigated to perform the reliability analysis for a water distribution network: a) mechanical reliability, which studies the probability that sources can reach all the demand points; b) flow reliability, which studies the probability that a water distribution system can meet a required water flow level at each demand point; c) hydraulic reliability, which studies the same problem as flow reliability, but takes into account the pressure at each demand point. The investigation will be focused on the flow reliability problem using Shannon factorisation [8] and proposing new quantitative measures.

1   INTRODUCTION

The hydraulic reliability assessment for a water distribution network estimates the probability that each and every demand node of the network receives water at the predetermined pressure. Obviously, hydraulic reliability [12] is a subset of the flow reliability, which is a subset of the mechanical reliability [3]. In the estimation of hydraulic reliability, minimal trees of pipelines connecting source to demand nodes are generated then reduced into minimal trees satisfying water quantity node demand. Secondly, only trees satisfying demand at the required pressure are maintained. These minimal trees, whose generation is a NP-hard problem [6], are the pathset usually used in a risk assessment study. The validation of each tree in terms of hydraulic functionality is a laborious task and requires pressure calculation procedures. The difference between the estimation of flow reliability and hydraulic reliability is generally overshadowed by the likely errors that are incorporated in the analysis due to poor component data availability. Our choice of flow reliability

200

is an intermediary approach between mechanical and hydraulic reliability. Anyway, cutsets and pathsets calculated by *SAEP*, could be subsequently validated for hydraulic function.

Most flow reliability system studies have a two step approach: a) generating minimal pathset or cutset [1], [5]. b) evaluating availability or other quantitative measures using minimal sets generated in step a. Network simplification, like parallel and serial reduction [7], could be applied. But in flow reliability, reductions generate multistate components, which are not adapted to our binary method.

Many approximation reliability evaluation techniques can be applied to solve large network problems: a) probabilistic cut-off procedures during the cutset enumeration process based on qualitative and quantitative criteria [2]. b) taking into account first terms order in the inclusion-exclusion formula of availability calculation [4].

2   MODEL & METHOD

In a water distribution system, components are interconnected to form a network of arcs and nodes. Arcs, described in the original network in
Figure 1, are pumps, valves and pipes, while nodes are consumption areas and water plants. Having a canonical description of the network will simplify the reliability calculation algorithm. A typical network, source to single terminal, would include the source node corresponding to the supply source(s), demand node corresponding to consumption areas and arcs corresponding to the rest of the components. Arc(s) starting from the source node correspond to supply source(s) and those ending at the demand node correspond to consumption areas. Each arc has the same capacity as the corresponding component. We consider that consumption arcs are perfect (cannot fail).

The availability of this network is the probability that a tree connects the source node to the demand node with the flow in each consumption arc equal to its capacity.

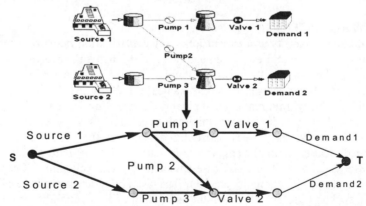

Figure 1. **Network modelling in *source to single terminal network (s-t)***

The second step is the application, on the generated source to single terminal network, of an algorithm based on Shannon factorisation. This method, applied in electrical domain [10], generates cutset and evaluates the availability in a single step approach. The Principe of factorisation is: given a Boolean function φ, for every variable X_i on which φ depends, we can write:

$$\varphi(X_1,...,X_n) = X_i\varphi(X_1,...,X_{i-1},1,X_{i+1},...,X_n) \vee \overline{X}_i\varphi(X_1,...,X_{i-1},0,X_{i+1},...,X_n)$$

In our case, φ is the structure function of the network. φ takes the value 1 if the system is operating and 0 if the system is in breakdown (water lack). $X_1,...,X_n$ are variables describing the state of network components. Applying the mean expectation on the two sides of the last formula, we obtain the system availability. The development of the Shannon formula generates a binary tree where depth is the number of components and paths from root to zero leafs correspond to cutsets. For each network configuration corresponding to a node of the Shannon tree, *SAEP* uses a simple simulator to verify network demand satisfaction. Simulator complexity is a linear function of network component number.

The maximum node number of Shannon expansion tree is $2^{n+1}-1$ where n is the number of components network. Exploring the whole Shannon tree for a large network becomes intractable. To resolve this constraint, two strategies have been applied: a) heuristic concerning variable order in Shannon factorisation has been adopted to reduce the number of tree nodes: variables corresponding to component having an important capacity are chosen first. b) truncation of Shannon tree [9]: an upper bound A' surpassing the real availability A, by at most ε, is calculated.

New quantitative measures have been developed in order to quantify damage. Using minimal cutset, we calculate a *Mean Water Lack* indicator:

$$ML = \sum_{k_i \in K} P(k_i)L(k_i)$$

K Minimal cutset

$P(k_i)$ Occurrence probability of the minima cut k_i

$L(k_i)$ Water lack engendered by k_i cut

This measure corresponds to the network expectation water lack in cubic meter per day.

This paper assumes that component availabilities for pumps, valves and plants have been previously determined. Only pipe availability is calculated by *SAEP* using the formula developed in [11].

3   *SAEP* TOOL & EXAMPLE

SAEP is a PC software product developed under the Windows 95 environment and uses an algorithm and model described in the previous paragraph. Figure 2 describes the *SAEP* functional schema.

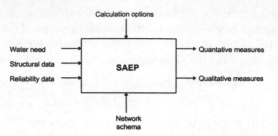

Figure 2.   *SAEP* tool functional schema

The network schema is described via a user-friendly editor. Each element is characterised by structural data (tank volume, pipe capacity and length, ...), and reliability data (failure probability). Failure probabilities of pipelines are calculated by using their structural data.

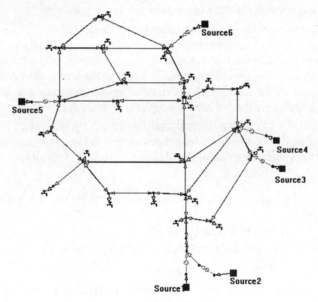

Figure 3.   ESP network

Applying *SAEP* on the network in Figure 3 (45 components), the evaluated availability was 0.98075729 with a maximum error of 1.86E-8. Other measures have been evaluated: minimal cutset and *mean water lack*. Time execution was 70 seconds on Pentium pro 200MHZ.

In order to compare two investment scenarios in terms of safety, we show in Figure 4, per scenario, the *mean water lack* in future years. Scenario 1 corresponds to the construction of a new plant (source6) while scenario 2 corresponds to the doubling of source5 capacity. Results show that scenario 1 is more reliable than scenario 2, which is logical because of the decentralisation of sources.

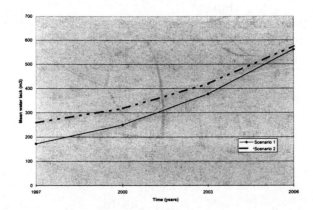

Figure 4.   Reliability comparison for investment scenarios

4   CONCLUSION

The preceding example demonstrates how our flow reliability method can be applied on a real water distribution network with an acceptable execution time. New measure quantifying damage (water lack) has been proposed. *SAEP* tool integrates the whole approach developed in this paper and allows, via its friendly interface, to perform a risk assessment study for water supply network.

REFERENCES

[1] Aggarwal K.K., Chopra. Y.C., Bajwa. J.S. (1982). Capacity consideration in reliability analysis of communication systems, *IEEE Trans. Reliab.* Vol. R-31, N°2, Jun., pp. 177-181.

[2] Hughes R.P., (1987) Fault tree truncation error bounds, *IEEE Trans. Reliab.* Vol. 17, N°2, Jun., pp. 34-46.

[3] Kansal. M.L, Kumar A., Sharma. P.B. (1995) Reliability analysis of water distribution systems under uncertainty, *Reliability Engineering.* Vol. 50, pp.51-59.

[4] Limnios N. (1981). Fault Trees *(in French). Hermès edition.*

[5] Misra K.B., Prasad K.K. (1982). Reliability analysis of networks with capacity constraints and failures at branches and nodes, *IEEE Trans. Reliab.* Vol. R-31, N°2, Jun., pp. 174-176.

[6] Odeh K. (1995). New algorithms for probabilistic and logic treatment of fault-trees *(in French).* PhD thesis, Université de Technologie de Compiègne.

[7] Rushdi A. (1987). Capacity function-preserving star-delta transformation in flow networks, *Reliability Engineering.* Vol. 19, pp. 49-58.

[8] Shneeweiss W.G (1984). Disjoint Boolean product via Shannon's expansion, *IEEE Trans. Reliab.* Vol. R-33, N°4, Oct., pp. 239-332.

[9] Shneeweiss W.G (1992). Approximation fault-tree analysis without cut sets. *In : Proceedings Annual Reliability and Maintainability Symposium*, pp. 370-375.

[10]   Theologou 0. (1990). Contribution to reliability network evaluation *(in French)*. PhD thesis, Université de Technologie de Compiègne.

[11]   Walski T.M., Pelliccia. (1982). Economic analysis of water main breaks, *J. Am. Water Works Assoc.,* pp. 140-147.

[12]   Wu. S-J., Yoon J-H., Quimpo R.G. (1993). Capacity-weighted water distribution system reliability, *Reliability Engineering.* Vol. 42, pp. 39-46.

The Effect of Feeder Pipe Configuration on the Reliability of Bulk Water Supply Systems

J.E. van Zyl *and* J. Haarhoff

ABSTRACT

A stochastic analysis technique was used to calculate and compare the reliability of bulk water supply systems with different feeder pipe configurations. The technique is based on an extended time Monte Carlo simulation of the system and incorporates both deterministic and stochastic behaviour parameters, including water consumption and pipe failure events. In the past, this technique was applied to the analysis of a bulk water supply system with a single feeder pipe between the source and the storage reservoir. As a result, the supply flow rate to the storage tank could only be 100 % (pipe operating) or 0 % (pipe failing) of the feeder pipe's capacity. In this study more complex and realistic feeder pipe configurations were considered, including different numbers of feeder pipes in parallel and interconnections between two parallel pipes. The results show that the feeder pipe configuration has a major effect on the reliability of bulk water supply systems. The method may be used to design optimal new bulk water supply systems for a given level of reliability, or to analyse and improve the reliability of existing systems

1   INTRODUCTION

A stochastic analysis technique for estimating the reliability of bulk water supply systems was introduced in 1996 by Nel and Haarhoff. The technique is based on an extended time Monte Carlo simulation of the system and incorporates both deterministic and stochastic behaviour parameters, i.e. water consumption, pipe reliability and fire events. Nel and Haarhoff carried out the analysis on a bulk water supply system consisting of a single feeder pipe between the source and the storage reservoir. As a result, the flow rate could only be 100 % (pipe operating) or 0 % (pipe failing) of its capacity. In this study more complex and realistic feeder pipe configurations were considered in the analysis of an existing bulk water supply system.

In its most simple form, a bulk water supply system may be described in terms of a source, which is connected to a service reservoir with a feeder pipe or pipes. The users of the system are supplied via the service reservoir. For clarity, the terms *source, feeder pipe(s)* and *reservoir* will be used throughout this paper to denote the corresponding elements in the bulk supply system as described above.

The *reliability* of a bulk water supply system may be defined in terms of its reservoir's failure behaviour, i.e. the frequency and length characteristics of failure events.

Various feeder pipe configurations were analysed in this study, including a single pipe, two or more pipes in parallel, and short interconnecting pipes between the parallel feeder pipes. The results show that the feeder pipe configuration has a major effect on the reliability of bulk water supply systems.

The method described may be used to design optimal new bulk water supply systems for a given level of reliability, or to analyse and improve the reliability of existing systems.

2 BACKGROUND

It can be shown that there is a certain interaction, indicated in Figure 1, between the configuration of the feeder pipe(s), the reservoir capacity, and the reliability of a bulk water supply system. In principle, any two of these three parameters can be specified, and the third calculated as a dependent variable. Design guidelines often do not exploit this inherent flexibility, but prefer to define rules for sizing the reservoir and feeder pipe characteristics. In South Africa, for example, storage reservoirs are sized for 24 to 48 hours of the annual average daily demand, with additional measure required for fire water. The feeder pipe is simply sized for a capacity of 1.5 times the average demand.

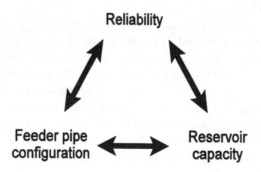

Fig. 1   Interrelationship between bulk water supply variables

In this study, a probabilistic method (Nel & Haarhoff, 1996) was used for relating the reservoir capacity, feeder pipe(s) properties and system reliability. The method is based on a Monte Carlo simulation, in which stochastic functions are used to describe consumer demand, fire water demand and pipeline failures.

Consumer demand is described in terms of deterministic and stochastic components. The deterministic component consists of long term, seasonal, weekly and diurnal patterns, and the stochastic component of serial correlation and "white noise". Fire water demand is described in terms of the probability of a fire occurring within a time step, and fire flow rate and -duration distribution functions. Pipe failures are described in terms of the probability of an independent failure event occurring and a repair time distribution function for the duration of the event.

Functions for describing demands and pipe failure behaviour were developed to handle the characteristics which could be extracted from the available data. Various data sets were analysed in a number of previous studies. Certain inherent assumptions are made (e.g. independence of pipe failure events) as a result of limitations in the available data. Like any mathematical model, the reliability of the stochastic analysis results is dependent upon the quality of the input data. The stochastic analysis technique may be extended to include more detailed descriptions of failure and other processes if the relevant data is available.

The development of this method has led to the analysis of a number of urban and rural bulk water supply systems, resulting in greater insight into the behaviour of these systems. The most important conclusions of these studies have been that:

- fire demand has very little effect on the reliability of bulk water supply systems,
- the greatest single determinant of the system's reliability is the reliability of the feeder pipe(s),
- the empirical design rule of feeder pipe design capacity of 1.5 times the annual average daily demand is appropriate for most urban systems analysed.

One of the biggest problems in the application of stochastic analysis to bulk water supply systems is the lack of clear guidelines as to the acceptable levels of reservoir failure rates. Kwietniewski and Roman (1997) suggested some reliability requirements for water supply systems, based on a wide number of social and technical considerations, but a great deal more research is needed in this field.

3 METHODOLOGY

The test system used in the analysis is based on an existing rural bulk water supply system in the Republic of South Africa. It consists of a source in the form of a large storage reservoir complex feeding water to a 3.5 Ml reservoir through a 58.8 km long, 350 mm diameter feeder pipe with a capacity of 25 l/s.

Using historical pipe failure data, a conservative pipe failure rate of 0.1 failures/km/a was selected. The data also revealed that failure events have an average duration of 16.8 hours, with a normal distribution and a standard deviation of 4 hours.

The community served by the reservoir has an average annual daily demand of 15.52 l/s. Water is mainly supplied through stand pipes, resulting in a demand pattern with negligible seasonal variations, but a weekly and diurnal demand pattern and an auto correlation coefficient of 0.67. Source failures and fire demands were not considered in the analysis.

The analysis was performed in two parts. In the first, the number of parallel feeder pipes between the source and the reservoir were increased from one to four to determine the effect this has on the system's reliability. The parallel pipes are assumed to have identical diameters, which will of course be smaller than the original pipe's diameter to maintain the same flow rate. For the purpose of the analysis, it was assumed that the available pipe diameters are continuous, i.e. any pipe diameter can be used, whether commercially available or not. Another assumption made was that the failure characteristics of the feeder pipes are independent of pipe diameter.

In the second part of the analysis, two parallel pipes were interconnected at an increasing number of evenly distributed points along their lengths. The interconnections were assumed to be supplied with an adequate isolation valve configuration so that any section of pipe may be isolated. In this way the effect of a failure event, e.g. a pipe burst, could be limited to the section of pipe where it occurred.

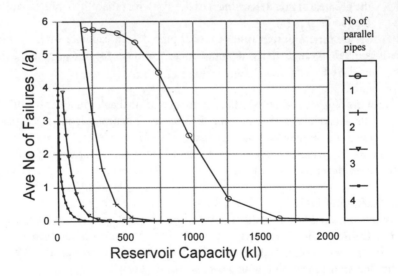

Fig. 2   The annual average number of reservoir failures for different numbers of parallel feeder pipes.

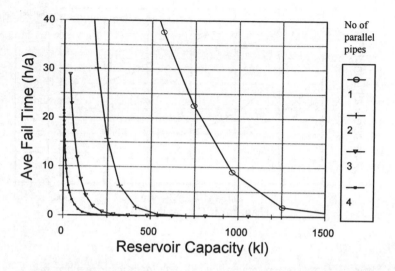

Fig. 3   The annual average reservoir fail time for different numbers of
parallel feeder pipes.

4   RESULTS AND DISCUSSION

4.1 Parallel feeder pipes

In the first part of the analysis the number of parallel feeder pipes was increased from
one to four. The results are shown in Figures 2 to 5 in terms of the following
reservoir failure parameters:

-    the *Average Number of Failures*, or the number of times the reservoir failed
     on average in a year (Figure 2),
-    the *Average Fail Time*, or the average time the reservoir failed in a year
     (Figure 3),
-    the *Average Failure Length*, or the average length of failure events
     (Figure 4),
-    the *Failure Length Standard Deviation*, or the standard deviation of all the
     failure events (Figure 5).

The first two figures (Figures 2 and 3) show that the results for the Average
Number of Failures and the Average Fail Time are very similar in form. These graphs
provide some useful insights into the behaviour of the system:

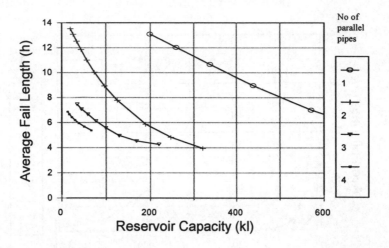

Fig. 4   The average reservoir failure length for different numbers of parallel feeder pipes.

- For any feeder pipe configuration there are characteristic curves for the Average Number of Failures and the Average Fail Time as a function of reservoir capacity. These curves may be used to compare the reliability of different systems.
- An increase in the number of parallel feeder pipes greatly increases the reliability of the system in terms of both parameters. As an example, the Average Number of Failures for a 750 kl reservoir reduces from approximately 4.5 failures/a to less than one failure in 15 years when the number of parallel feeder pipes is increased from one to two.
- These results may be expressed in another way by calculating the minimum reservoir capacity required for a given level of reliability. If, for example, an Average Fail Time of 30 minutes/a is taken as an acceptable level of reliability, the minimum required reservoir capacity would be 1.5 Ml for the case of a single pipe, and 0.5 Ml, or one third, if two parallel pipes are used.
- It is interesting to note that for the case of a single feeder pipe, the Average Number of Failures does not exceed 5.9, even for very small reservoir capacities. The reason for this behaviour is that the feeder pipe has an average failure rate of 5.9 failures/a, limiting the number of possible reservoir failures. For more than one feeder pipe, the total pipe length, and thus the upper limit for the number of failures is increased.

Figures 4 and 5 show the reservoir's failure behaviour in terms of the Average Failure Length and Failure Length Standard Deviation for different numbers of parallel feeder pipes. As can be expected, the Average Failure Length generally reduces with increasing reservoir capacity. The same holds true for the Failure Length Standard Deviation.

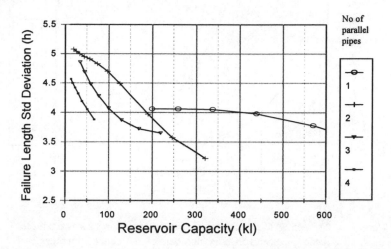

Fig. 5   The reservoir failure length standard deviation for different numbers of parallel feeder pipes.

4.2 Interconnecting pipes

The effect of interconnecting pipes on the reliability of the system was determined for two parallel feeder pipes between the source and reservoir. The number of interconnection pipes, evenly spaced over the length of the parallel pipes, was increased from zero to three. Figures 6 and 7 show the results for the Average Number of Failures and the Average Failure Time respectively for different numbers of interconnecting pipes.

It is clear from the figures that the effect that increasing the number of interconnection pipes has on the reliability of the system is similar in order of magnitude to that of increasing the number of parallel pipes. If, for example, an Average Failure Time of 30 minutes/a is taken as acceptable, the required reservoir capacities for the cases of no interconnecting pipe and one interconnecting pipe are 500 kl and 245 kl respectively. The equivalent reservoir capacity requirement for an extra parallel pipe is 250 kl, which represents virtually the same reduction in the required capacity.

The Average Failure Length, Failure Standard Deviation and Maximum Failure Duration results for the analysis using interconnecting pipes are not shown. Suffice

212

to report that the behaviour is similar to that of pipes in parallel, although the curves tend to be closer together.

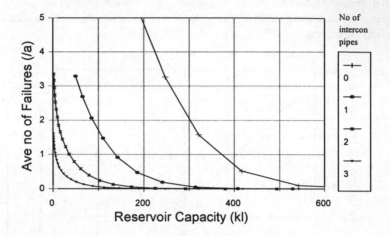

Fig. 6   The annual average number of reservoir failures for two parallel feeder pipes and different numbers of interconnecting pipes.

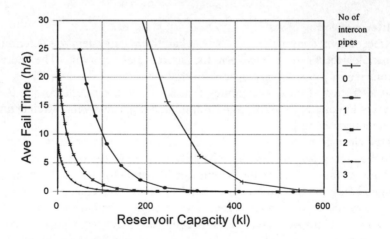

Fig. 7   The annual average reservoir fail time for two parallel feeder pipes and different numbers of interconnecting pipes.

4.3 Comparing different options

The various results may be compared by fixing the required reliability and then extracting the minimum reservoir capacities for the different pipe configurations under consideration. Three arbitrary reliability values were selected, namely Average Fail Times of 30 min/a, 2.5 h/a and 24 h/a. The results for these reliability values are shown in Figure 8. The great variation in required reservoir capacities for different bulk pipe configurations is evident from the figure.

It may also be observed that an increase in the number interconnecting pipes results in an improvement in the system reliability comparable in order size to a corresponding increase in the number of parallel pipes. It is, of course, a much cheaper option to increase the number of interconnecting pipes. This observation has great significance for the design and improvement of bulk water supply systems. In many cases, the use of two parallel pipes, interconnected at a number of points along their length, may prove to be the optimal feeder pipe configuration, providing the required reliability at minimum cost and with good possibilities for optimal future improvements in system reliability. This is, however, not true universally, but is dependent upon the specifics of the system under consideration.

The 95 percentile Failure Lengths, or the failure duration which will not be exceeded in 95 % of failure events, were also calculated and are shown in Figure 9. It is interesting to note that the 95 percentile Failure Length generally increases with decreasing reservoir capacity (and corresponding increasing pipe system reliability), although some exceptions to this rule may be observed.

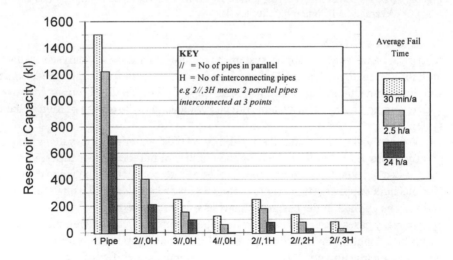

Fig. 8   Reservoir capacities required for different pipe configurations to achieve Average Fail Time reliability levels of 30 min/a, 2.5 h/a and 24 h/a.

214

A further step in the procedure would be to compare the costs of the different alternatives in order to select the optimal feeder pipe and reservoir configuration. The high cost of duplicating the entire feeder pipe over its long length prohibits a parallel pipe for the system under consideration, but for a short enough feeder pipe length the two parallel and interconnected pipes option becomes very attractive.

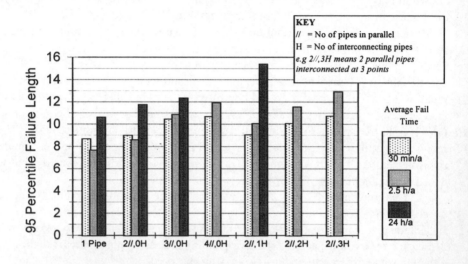

Fig. 9 The 95 percentile reservoir failure length for different pipe configurations and Average Fail Time reliability levels.

5 CONCLUSIONS

A typical rural water supply system in South Africa was analysed using a probabilistic method. The method was extended to account for parallel feeder pipes and interconnections between the feeder pipes. The results brought improved understanding of why rural systems fail and could lead to the design of more reliable systems.

The analysis showed that the reliability of bulk water supply systems may be greatly improved by either increasing the number of parallel feeder pipes, or in the case of two or more feeder pipes, the number of interconnections between the pipes. Various bulk supply system failure parameters may be calculated using the method and may be applied in the design or analysis of these systems.

In many cases the use of two parallel pipes, interconnected at a number of points along their length, may prove to be the optimal feeder pipe configuration. This is, however, not true universally, but is dependent upon the specifics of the system under consideration.

The improved reliability brought by parallel pipes and their interconnections should be balanced against their additional cost, an aspect not addressed in this paper.

To conduct such an analysis, estimates of acceptable levels of reliability, as well as the cost of noncompliance, are needed. This is an area where little information is currently available.

ACKNOWLEDGMENT

The research presented in this paper was funded by the South African Water Research Commission.

REFERENCES

Edwards GT. (1996) *Betroubaarheid van waternetwerke.* Unpublished, Department of Civil and Urban Engineering, Rand Afrikaans University, Johannesburg, South Africa

Kwietniewski M, and Roman M. (1997) Establishing performance criteria of water supply systems reliability, *Aqua* **46** (3) 181-184

Nel D, Haarhoff . (1996) Sizing municipal water storage tanks with Monte Carlo simulation. *Aqua* **45** (4) 203-212

Nel DT, Haarhoff J, Engelbrecht RJ. (1996) A probabilistic technique for sizing municipal water storage tanks. *Computer Methods and Water Resources III*, Abousleiman Y et al (Ed), 145-152

PART V

QUALITY MANAGEMENT AND CONTROL

A Methodology for Assessing the Risk of Trunk Mains Failure[*]

N. R. Cooper, G. M. Blakey, C. Sherwin, T. Ta, J.T. Whiter *and*
C.A. Woodward

ABSTRACT
A probability based model has been developed using GIS to estimate the risk of
failure of water mains greater than 300mm in diameter. Data from past fractures
has been analysed to determine a probability score for each water main in Thames
Water's London Water Supply Region. This is combined with spatial data on the
consequences of water mains failure to develop a broader risk of failure score.
This model can be used to aid operation and planning functions, enabling:

- Prioritisation of Investment
- Optimisation of maintenance and service recovery
- Further analysis into causes of water mains failure
- Proactive planning to reduce burst frequency

1   INTRODUCTION

Thames Water Utilities Limited (TWUL) are responsible for the operation and
maintenance of approximately 30,000km water mains. Most of these mains were
installed by a variety of predecessor organisations and some are over 100 years
old. As the network has expanded over the last 150 years, different types of main
have been laid as manufacturing methods have changed and new materials become
available.

The majority of mains now owned by TWUL are either cast, ductile iron or
steel. These pipes vary not only in the quality and thickness of metal used but also
in the jointing systems employed and the internal and external linings used to
prevent corrosion. These pipes were laid to a variety of standards in a wide range
of different soil types and conditions to convey waters with a range of corrosivities
(Ford, et al 1996).

In terms of Trunk Mains failure (those greater than 300mm in diameter), the
potential threat to public safety and the costs of repair, reinstatement and
compensation makes the study of mains fractures essential.

This paper details a methodology developed by a team in Thames Water
Research & Development for assessing the risk of trunk mains failure. A model
has been developed which aims to answer the following questions:

---

[*] © Thames Water

219

- How likely is it that a burst will occur ?
- How much water will be released ?
- Where will the water go ?
- What are the potential consequences ?

In trying to understand why pipes fail, two distinct approaches can be adopted. A mechanistic approach will attempt to establish and analyse quantifiable relationships between factors which influence pipeline failure. By measuring or estimating these factors relationships are applied which will predict failure under specified sets of conditions.

Alternatively, a probabilistic approach will look at a set of failures and attempt to explain their distribution through analysis of common factors or conditions. This approach explains the distribution of pipe failures through a limited number of factors, and although it does not provide an absolute measurement of failure probability, it can provide a robust measure of the likelihood of one pipe or set of pipes failing relative to others. It is this second approach that Thames Water has adopted in its Trunk Main Burst Risk Model (TMBRM).

2   THE PROBABILITY MODEL
A dataset of approximately 700 bursts was available for an 8 year period between 1986 and 1994. Variables were selected according to what factors might be important in influencing mains failure, but also what data was available. The trunk mains within the London Water Supply Area were divided up into 50m lengths. The variable is examined included pressure, traffic load and ground elevation, however, ultimately it was possible to explain the distribution of mains failures using 4 variables. These were:

- No. of Buses (per hour)
- Pipe Diameter
- Soil Corrosivity Class
- Pipe Density Function

Mains age was not included. Primarily this was because no direct and reliable data on trunk mains age is available. Urban age mapping has been undertaken on the London Water Supply Zone using building age as a surrogate for mains age, however, this technique is more reliable for smaller distribution mains where the link between urban development and mains age is strongest. Trunk mains are often laid in anticipation of development or to provide a strategic link, and in this case building age does not provide a good estimate of mains age. In addition, operational records and anecdotal information indicate that the relationship between mains age and failure rate is not a simple one. Evidence suggests that a number of failures in rapid succession is not necessarily indicative of the future failure rate or mains condition. For these reasons mains age was discarded from the assessment, although the urban age dataset was retained for future enhancement.

3   CONSEQUENCES

The main driver behind understanding and assessing the risk of trunk mains failure was the potential consequences of a serious main burst. These consequences were categorised under the following headings:

- Domestic properties
- Commercial properties
- Repair and reinstatement costs
- Additional Factors

Information held about the different categories of consequences were stored using a common unit of the postal incode (e.g. W10 8QT). Some data such as sum insured and rateable value only had mean values at the postal outcode level (e.g. W10) but to ensure continuity, these values were allocated to the smaller incode 'cells'.

3.1   Domestic Consequences

Estimation of the consequences from flooding domestic properties is based on a number of parameters, as detailed below.

3.1.1   Rateable Value

The rateable value element is used as an estimator of potential property damage. This assumes higher value properties have higher rateable values and that the consequences of damaging these properties would be greater. This information provides a method for estimating the consequences of a mains burst on buildings of different values.

3.1.2   Mean Sum Insured

Mean Sum insured data is used to refine the estimate of domestic consequences further. Available from the insurance industry, this gives an indication of the value of buildings and contents insurance, and thus the likely costs associated with damage to the fabric of customers' properties or their contents. This helps to highlight areas where property and contents insurance claims may be higher or lower than suggested by rateable value alone.

3.1.3   Basement Probability

Because of the high incidence of residential basement properties in the central London area it was decided that an estimate of basement probability was required. This would primarily enable an assessment of the potential injury risk to occupants of properties below street level. In addition, the consequences of increased damage to basement is estimated.

Basement probability is assessed primarily on building age. This was determined using a combination of two methods; a commercially available database and GIS based property agebands using historical Ordnance Survey mapping from 1840-1990.

3.2 Commercial Consequences

In terms of commercial properties, unsuccessful attempts were made to determine a relationship between a commercial basement and either the location or the age of the property. Instead, a commercial property database was searched for references to basements and different floors in the address field. In estimating the consequences of flooding a commercial property, a measure was developed which identified the consequences of both physical damage to property and loss of business suffered as a result of flooding or disruption.

Estimation of the consequences from flooding commercial properties is based on the parameters outlined below.

3.2.1 Mean Rateable Value for Each Floor

A ratings list database was searched to look for reference to a basement and different floors in the commercial property address field. This enabled mean rateable values to be calculated for each floor.

3.2.2 Property Damage

Rateable value was summed per postcode for basements and ground floor properties to obtain the property damage score per postcode unit. This gives an estimate of the likely damage to commercial buildings on ground floor and basement levels.

The sum of rateable values per postcode unit for basements and ground floor properties produced the Property Damage Reserve per postcode unit.

3.2.3 Loss of Business Factor

Rateable value was summed per postcode for basement, ground and upper floors and multiplied by a 'Loss of Business' factor to get a loss of business score per postcode unit.

The property damage score and the loss of business score were summed to give the Commercial Consequence Score for each postcode unit. This makes the assumption that a loss of business claim will be directly related to the rateable value. This may not be a particularly robust assumption, however, it does provide a basis for assessing the broad costs of a mains failure on a relative basis between different commercial properties.

3.3 Additional Consequences Factors

The Domestic and Commercial Consequences Scores outlined above provide the basis of a quantifiable consequences calculation. However, there are other factors which are included in the model which, although they do not contribute directly to the consequences score, do give added insight into the effect of trunk mains failure. These factors include:

- Sensitive customers
- Dialysis patients
- Hospitals

- Key Customers
- Tourist attractions
- Red routes (main arterial roads)

Currently, these factors do not have scores. As the system is developed it should be possible to add scores to these so the effects of trunk mains failure upon traffic disruption, media attention and broader customer issues can be quantified.

4 FLOODABLE AREA

Determining the extent of flooding from a main has two uses within the model. Firstly, it enables a consequences score for a particular burst main to be calculated from the domestic and commercial consequences scores for each postcode unit. Figure 1 shows the intersection of postcode polygons with a number of floodable areas. The model identifies where a predicted floodable area overlaps a postcode polygon. The percentage area of the postcode affected by the burst determines the proportion of the consequences score which is allocated to a specific event. In this way the total domestic and commercial damage from a burst main is calculated. Secondly, local level sensitivity analysis can indicate where floodwater is likely to breach the road from a single burst point.

This functionality is highlighted in figure 2. A single, primary burst point is identified (marked B1). Secondary flooding points highlight where water is channelled along the road surface, breaches the curb and has the potential to impact properties.

224

Figure 1. Floodable area overlaid by postcode polygons

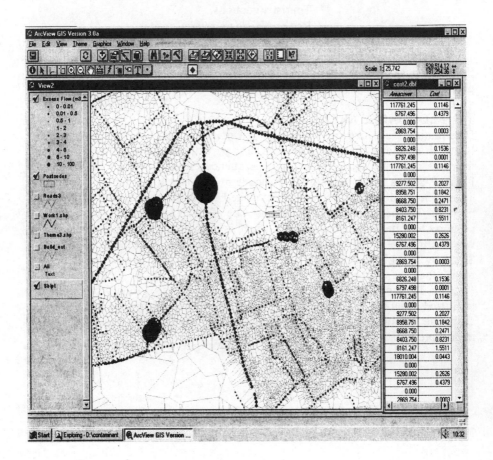

Figure 2. Identification of secondary flood locations

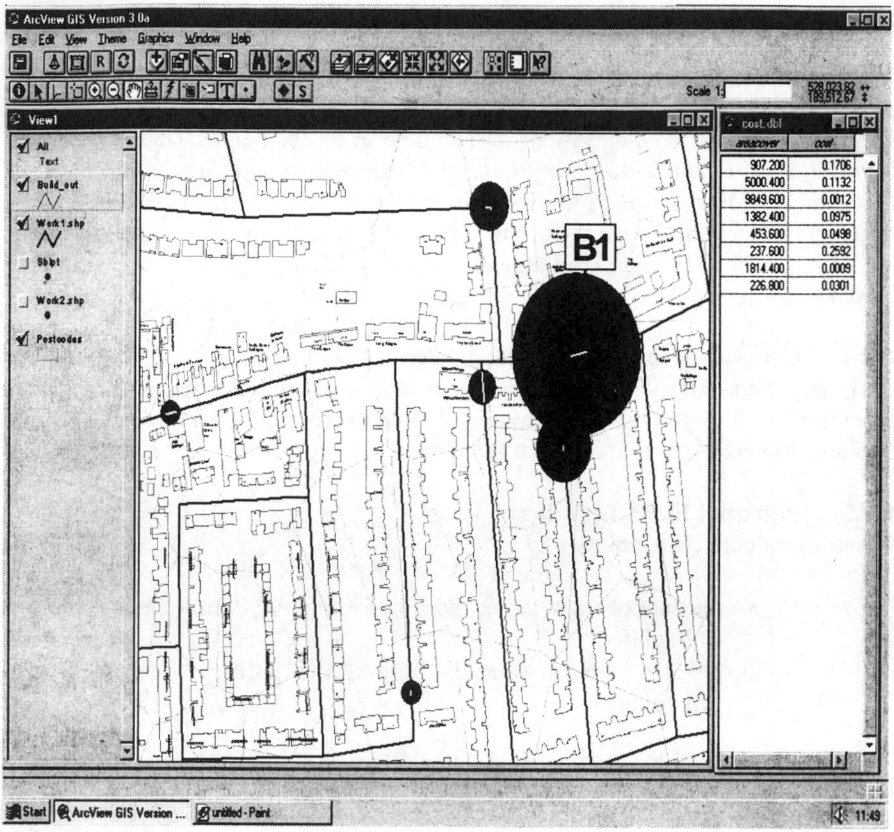

4.1 Methodology

4.1.1 Flood Water Volume

Mains bursts were simulated along each length of trunk main in the Network. The burst flow was assumed to run at 100% flow for three hours - a three hour shutdown time assumed. This value can be altered by the user on an individual burst by burst basis to make the result mode site specific. In addition, road gullies and drains were assumed to be blocked. This was done to reflect a worse case scenario, and also the fact that drains will quickly block due to silt washed out with the floodwater.

4.1.2 Floodable Areas

In calculating the floodable area from a burst the road networks were treated as a series of open channels analogous to a drainage network. The model calculates at what point the network reaches 'bank full' state and calculates what volume of water will flood over the curb. This gives an approximate area affected by flood water. A number of different elements combine to form this area, as detailed below.

4.1.3   Three Dimensional Road Network

This was determined by using the Ordnance Survey Digital terrain Model and Thames Water manhole elevations to calculate grids for height, slope and aspect covering the London Water Supply Region. These grids are intersected with the Meridian road line coverage to give a master Meridian road network with height, slope and aspect values attached. The road width around the meridian road line is determined from Meridian road feature classes and London Research Centre road classifications.

4.1.4   Cross Sectional Area of Road Network

Curb height was always assumed to be 0.1m and the camber of the road assumed to be 0.075m. Road width was determined from Meridian road feature classes and London Research Centre road classifications.

4.1.5   Potential Flood Locations

These are calculated as points where:

- the rate of burst water discharge $(m^3.s^{-1})$ at 3am >channel full volume
- the rate of burst water discharge $(m^3.s^{-1})$ at 8am >channel full volume

Figure 3 highlights the estimated floodable volume (volume>channel full) from each main. A more detailed description of this methodology can be found in Blakey and Ta (1999).

Figure 3. Estimated floodable volume

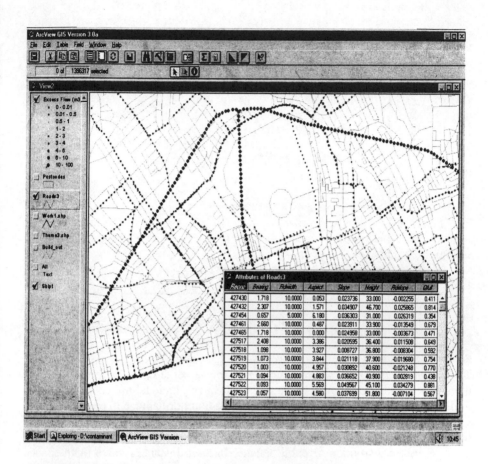

5 USES OF THE TRUNK MAINS BURST RISK MODEL

5.1 Investment Planning

Where a fixed amount of funding is available, or a fixed length of main needs to be rehabilitated or replaced, the Model provides a robust and defensible methodology for prioritising investment.

In terms of prioritising mains the matrix shown in figure 4 can be used. Plotting the probability score against the consequences score for each main within a sample produces a risk matrix. Those mains located towards the upper right hand corner should be identified as high priority. These are the mains which pose the greatest risk of failure and highest consequences score.

Figure 4. Risk Matrix of Trunk Mains Failure

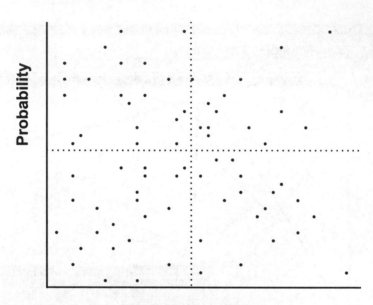

Consequences

5.2 Contingency Planning

There may be instances where mains may not be candidates for replacement as there is no public safety or strategic threat. However, their failure could still result in high compensation claims, operational expenditure or damage to public image.

The Trunk Main Burst Risk Model enables analysis of risk values which reveal those mains which may produce high consequences in terms of property damage, traffic disruption or public image. In these instances valve maintenance programmes can be targeted and vital spares can be kept in stock. This ensures that if a failure does occur the main can be put back into service as quickly as possible, with the minimum of disruption.

5.3 Event Response

Thames Water's Event Support Group manage events such as burst trunk mains, ensuring that disruption is minimised and customers service levels are maintained. Work by this group has identified that there is a 'Golden Hour' at the beginning of most events - if actions can be taken to minimise consequences during this vital period, damage and disruption can be greatly reduced.

By ensuring that contingency plans are in place this 'Golden Hour' is not wasted. For instance by establishing and communicating shut-in, the amount of damage can be reduced significantly. Additionally, contacts for local transport companies, sensitive customers and commercial customers are recorded as attributes within the model, enabling a rapid response to stakeholders' needs.

6 SUMMARY

The large cost and potential safety issues arising from trunk mains failure have highlighted the need to understand the probability of trunk mains failure and the associated consequences.

A GIS based model has been developed which estimates the risk of trunk mains failure. Data from past fractures was been analysed to determine a probability score for each water main in Thames Water's London Water Supply Region. This was combined with spatial data on the consequences of water mains failure to develop a broader risk of failure score.

Allocating risk to individual burst events was achieved using a terrain model developed to calculated the area potentially affected by a main failure.

The model highlights potentially high priority mains and enables ancillary information associated with individual mains to be accessed rapidly. In this way the model provides an excellent tool to assist investment planning, contingency planning and event response functions.

REFERENCES

Ford, R., Whiter, J., & Woodward C.A. Burst water mains and how to predict them AGI'96. 24-26 September 1996. Birmingham, UK

Blakey, G.M. & Ta, T. Mapping road elevation and its use in the prediction of water mains burst flooding. ESRI UK User Conference, Keele, 28-29 April 1999.

Assessing Water Quality Issues in Water Distribution Systems from Source to Demand

C. Fernandes *and* B. W. Karney

ABSTRACT
The quality of the water delivered by a distribution network may degrade for many reasons. The goal of this paper is to review water quality issues in distribution systems considering both external actions in the watershed and internal mechanisms in the supply system. For example, the external influence of urbanization may degrade water resource systems and thus severely interfere with the characteristics of the water used for human consumption. Moreover, internal mechanisms such as hydraulic transients, especially through the presence of leaks and the role of system start-ups, can also cause water quality degradation. To deal with the complexity of these interrelated challenges, an integrated approach for the analysis and design of water distribution systems is required. To be comprehensive, a water quality model must account for both the role of different flow characteristics and the various mechanisms of constituent/disinfectant transport.

1   INTRODUCTION

The environmental degradation of urbanized watersheds is a difficult and complex problem. In particular, the compromising of watersheds whose water is used for human consumption, in terms of both quantity and quality, interferes with both water treatment and water delivery. The increasing effort to operate the whole system to achieve specified water quality criteria requires an integrated approach to deal with new threats and challenges to delivery.

This paper overviews the main water quality issues in water distribution systems that are required to consolidate the integrated approach. The goal is to evaluate, based on an understanding of hydraulic and transport models, the consequences of the current modelling approaches on the watershed-treatment-distribution system. Several specific issues are identified: the interference of external actions to the water distribution system, the changes to water quality components, and the role of the modelling approach itself. An integrated framework is proposed to complement and consolidate the research to date.

2   SIGNIFICANCE OF EXTERNAL ACTIONS

The problem of water quality in a distribution system is often viewed as detached or separate from the larger issues of environmental degradation. Indeed, on one level, the treatment plant is designed to achieve this separation. However,

231

degradation processes resulting from human development often exceed natural limits, and thus strongly affect natural cycles. For example, the process of urbanization, particularly when linked with inappropriate soil and land use may seriously impair the ability of water supply systems to supply high quality water.

However, important differences exist between developed and developing countries. In developed countries the impact of environmental degradation has been at least partly controlled. The combination of integrated water management projects allied to socio-cultural improvement, economical stabilization, resolution of conflicts, and public involvement is responsible for a new environmental attitude. In developing countries this level of technical consciousness is not evident and environmental problems are common and serious. Degradation of water-supply watersheds, the increasing risk of flooding within the watershed and inappropriate final disposal of wastewater and solid waste are common environmental concerns. In many cases, water quality degradation is observed in the intake and there is an increasing risk of transient flow effects in the distribution system. It is distressingly common to interrupt water treatment or delivery due to deterioration of raw water or due to a lack of electrical power. In such a context the usual rules for assessing the topological reliability or flexibility of the distribution system are inadequate.

In terms of modelling, most water quality models assume as boundary conditions supplies and demands that are, if not constant, at least well-behaved and predictable. In fact, the influence of external actions on water distribution systems is often ignored. Assuming, for example, water quality degradation at the intake and the flexibility of the treatment system to guarantee the required water quality criteria, it is possible to cause significant changes to constituent/disinfectant concentrations, changes that may be of greater significance than many transport mechanisms. Since these changes may also create a different hydraulic scenario, the final effect can be magnified.

3   WATER QUALITY MODELLING APPROACH

Water quality modelling in water distribution systems is a tool that has been used to predict the transport and fate of dissolved substances. These models are gaining importance for many reasons, including increasing governmental regulations, changing customer expectations, the growing number of reported water quality incidents and a heightened consciousness brought about by the use of monitoring and management techniques for water quality purposes (Rossman and Boulos, 1996). This section comments on the modelling process and briefly identifies where improvements are necessary.

A water quality model is the product of two basic modules. The first determines hydraulic conditions based on one-dimensional equations of momentum and continuity (Karney, 1996) and, in most cases, a certain hydraulic scenario is assumed (e.g., steady, unsteady incompressible, or unsteady compressible). The second module, related to water quality, is dependent on the hydraulic model to evaluate flow paths in the pipeline system, the mixing from different sources, the dilution of contaminants and the travel/detention times of the water (Elton et al. 1995).

The basis of any water quality model is an equation describing the transport of both water and contaminant/disinfectant. The one-dimensional advection-dispersion equation with reaction (accounting for contaminant/disinfectant decay) represents the basic physical transport mechanisms. It considers advection by bulk transport (velocity) and dispersion due to pipe wall effects on the axial spreading of a constituent as

$$\frac{\partial C_i(x,t)}{\partial t} + u_i \frac{\partial C_i(x,t)}{\partial t} + D \frac{\partial^2 C_i(x,t)}{\partial x^2} + KC_i(x,t) = 0 \tag{1}$$

where $C_i(x,t)$ = constituent concentration, x = longitudinal space dimension; t = time; u_i = mean velocity of flow in pipe segment "i"; K = coefficient of concentration decay rate; and D = coefficient of longitudinal dispersion.

Computer implementations of this concept have been developed to simulate the water quality behavior of individual constituents/disinfectants in water distribution systems based on steady state or dynamic conditions (Islam and Chaudhry, 1998). Most commonly, the reaction of chlorine or related compounds in the pipe system (Rossman, 1993) is analyzed, a logical first-step considering chlorine's wide use as a disinfectant in drinking-water systems. Chlorine influences DBP formation, bacterial regrowth, corrosion of pipe materials and release of dissolved metals into the bulk water. In most cases, advection effects are responsible for the main mechanisms of mixing throughout the network (Rossmann, 1993), and many models (EPANET, PICCOLO, NET and DWQM) neglect the contribution of axial dispersion.

The focus of a water quality model is to adequately approximate the governing equations by controlling or reducing numerical errors. Case studies have been presented using assumed (i.e., fictitious) data or using limited field data. Unfortunately, only a few detailed field studies have been reported in the literature. Although a logical beginning, concerns with this approach relate to the identification of the primary mechanisms of transport in the bulk flow since this strongly influences the concentration of a constituent/disinfectant. The classical one-dimensional advection equation with reaction, under most circumstances, represents the main transport mechanism in water distribution systems. This assumption also implies plug-flow for the velocity profile and does not directly account for the shear stress at the boundaries. However, in real flow conditions, the shear stress at the boundary of the pipe has an effect on the velocity distribution across the pipe (Axworthy and Karney, 1996). For example, Axworthy and Karney (1996) show that under periods of low flow, dispersive effects may become important, causing significant differences when comparing the advective and the advective-dispersion models. In the case of hydraulic transient effects, few results have been reported. The impact of velocity profile and shear stress conditions may be strong enough that the advection consideration alone is insufficient to guarantee the accuracy of water quality predictions. Complimentary studies are required to investigate the effects over the transport conditions in the bulk flow and its effects over the transport of constituents/disinfectants.

4 WATER QUALITY IN WATER DISTRIBUTIONS SYSTEMS

Water quality problems in distribution systems are strongly influenced by many different factors. These include the decay of chlorine-based residual, bacterial regrowth, predictions of disinfectant residuals and the presence of assimilable organic carbon (AOC) (LeChevallier et al., 1994). Other factors include biofilm formation on pipe walls which can induce depletion of chlorine residuals (Wable et al. 1994), as well as chemically and microbiologically induced internal corrosion of the pipe wall, long detention times in storage tanks and the growth and decay of disinfection-by-products (DBP) (Romain, 1996).

The quality of water can also be degraded by transient effects, not only by increasing the risk of pipe failure, but also by making biofilms, corrosion and/or tuberculation on the pipe wall more susceptible to transport by high velocities and shear stresses (Karney et al., 1994; Karney and Brunone, 1999). The role of the velocity profile and shear stress in water quality problems due to the nature of transient effects associated with turbulent flow is a potentially important transport mechanism. In essence, transient flow can influence the overall mechanism of transport of mass in the system (Brunone et al., 1999).

As mentioned previously, most water quality models incorporate a chlorine decay model with first-order decay rate (Rossman 1993; Islam and Chaudhry, 1998). In most cases, the water quality model has been successfully applied to case studies which have compared theoretical results to those obtained from field measurements. However, dynamic conditions are generally represented by slowly changing conditions (quasi-steady state) and few results relate to the influence of pipe material and diameter.

Nevertheless, the problem of modelling bacterial regrowth and the effects of biofilm are not necessarily related only to the interpretation of the mechanisms for the disinfectant-residual decay. Other important factors that may be ignored in the modelling process include internal corrosion and DBP formation. Bacterial regrowth in distribution systems is a common phenomenon and, in general, occurs when organic materials and sediment accumulate in distribution pipes, chlorine residuals dissipate and when water temperatures increase. The presence of nutrients in water, corrosion, sediment accumulation and hydraulic are other factors (LeChevallier, 1990).

Biofilm formation on pipe walls is a possible source of increasing coliform bacteria counts in distribution systems (LeChevallier, 1990). This growth can deplete chlorine residuals at higher rates than would be expected in their absence. The mechanism of disinfectant diffusion in the biofilm not only inactivates bacteria but also induces movement to the bulk water. The influence of alternative disinfectants to chlorine (e.g., monochloramine) and the pipe material have been reported as critical factors in determining the concentrations of a given bacterium in the distribution systems (LeChevallier, 1990).

Internal corrosion of the pipe wall, due to chemical reactions between the pipe wall and the bulk water or microbially induced corrosion, clearly influences water quality. The pH and alkalinity of final water and the influence of higher temperatures and higher chlorine residuals combined with increased flow rates produce increases in alkalinity and concentration of dissolved metals in the bulk water.

The increasing DBP concentrations observed in water distribution systems due to the longer contact time between precursors (humic and fulvic acids) and oxidants is another concern. As a result, a balance between DBP formation and microbial protection is a focus of intense research. The existing numerical models for predicting DBP formation/decay are similar to those developed for prediction in a treatment plant (Singer, 1994). The batch-scale experiments do not account for hydraulic conditions, biofilm formation, mixing conditions and pipe wall reactions and most of those effects are ignored by existing water quality models.

The combination of all these water quality characteristics indicates the complexity of the modelling problem, even considering only steady boundary conditions. In a more comprehensive analysis, Lu et al. (1995) present a model that accounts for the simultaneous transport of substrates, disinfectants and microorganisms considering the interaction/reaction of the components with each other and reactions with the pipe wall and biofilm. Despite the consideration of diffusion in the radial direction combined with advective transport in the axial, the analysis does not directly consider the influence of different flow velocities.

Besides the interaction/reaction of the water quality components, it is necessary to evaluate those effects under different flow conditions. It is expected that the characteristics of the velocity profile and shear stresses may have significant influence over the mechanisms of biofilm formation, bacterial regrowth and internal corrosion; however, few water quality models consider such factors. To illustrate, models of disinfectant consumption at pipe wall usually assume no chlorine is returned to the bulk water. Rossman et al. (1994) assume that the axial diffusion is responsible for chlorine transported to the pipe wall with no effect over the transport at the concentration front. Thus, the model may be unable to account directly for varying levels of biofilm throughout the system that will influence chlorine consumption at the pipe wall.

5   HYDRAULIC MODELLING

Hydraulic modelling of water distribution systems is traditionally performed using steady or 'quasi' steady models. Even if transient effects are considered, they are seldom considered dominant. Despite this, actions external to the water distribution system (e.g., decreased water quality at the intake or power failure events), may cause a higher number of transient events. Even though time-limited, these effects may create new hydraulic characteristics such as more extreme pressures and velocities in the distribution network. New hydraulic conditions such as start-up and shut-down events may deteriorate water quality not only due to direct effects on the transport mechanism, but also by enhancing the risk of pipe breaks and distributed leaks within the system. Leaks are a particular concern since they generally allow flow in both directions, thus also permitting constituents to enter the network, particularly under transient (low-pressure) conditions. Hydraulic conditions often require a more careful assessment than is attempted.

One solution may be to use a continuum approach to hydraulic modeling both for the mathematical description of flow characteristics and also for the appropriate numerical model (Axworthy, 1997). More specifically, the unsteady compressible/incompressible and uniform/non-uniform, quasi-steady and steady flow conditions are all special cases of the full one-dimensional hyperbolic partial differential equations of momentum and continuity. However, the computational procedure for solving the various flow equations has traditionally varied. The method of characteristics is appropriate for the unsteady compressible case (Wylie and Streeter, 1993), the modified Euler technique for unsteady incompressible flow (Chaudhry et al. 1985), and the backward Euler method and the lumped-system approach for quasi-steady flows (Islam and Chaudhry, 1998). The possibility of different numerical procedures solving specific transient flow characteristics has led to numerical criteria for representing the transitional flows (Axworthy, 1997; Axworthy and Karney, 1996; Karney, 1990).

Even though there is a necessity to identify good numerical procedures, the interaction of a hydraulic model with an accurate water quality model is not easy to achieve over a wide range of hydraulic conditions. In this sense, Axworthy (1997) summarizes this concept by invoking a redirection of the numerical scheme through several key questions, such as the possibility of different transient states at the same time in different locations and how the transition between different states is considered.

Most so-called dynamic models are, in fact, sequential steady-state or quasi-steady hydraulic models (Islam and Chaudhry, 1998; Boulos et al. 1994, and Rossman et al., 1994). This approach simplifies the sequence of events in the continuum process. Specifically, the one-dimensional representation of flows in pipe systems, with both velocity profiles and friction losses, is assumed as the basis of equivalent steady-state values of velocity (Brunone et al., 1999). The inertial effects are not considered (compressibility and deformability assumed negligible), any change in mass storage is neglected and flow conditions can only change slowly; such assumptions are clearly unreasonable for rapidly changing conditions. As a matter of fact, these simplifications are likely to account for the difference revealed by Islam and Chaudhry (1998) when comparing their model with the EPANET results (Rossman, 1993).

Yet another consideration is related to the friction factor "f", usually calculated from steady-state considerations. It is known that shear stresses at the wall are not in phase with the mean velocity. As a consequence, the unsteady energy dissipation may greatly exceed that of the assumed steady energy dissipation. In this way, the inappropriate evaluation of the friction factor "f" can affect water quality predictions, especially those related to corrosion and biofilm formation.

6   AN INTEGRATED FRAMEWORK

Different approaches have been considered to model the various mechanisms of transport of a disinfectant/constituent in water distribution systems. All of the approaches predict the mass transport of a given constituent, for a specified hydraulic condition and usually assuming the initial disinfectant-residual concentration as that required by the water treatment rules. However, increasingly strict regulations not only require an understanding of the appropriate conditions for the treatment/distribution systems but also drive the continuing development of integrated management systems to guarantee the quality of both source and demand flows, as those referred by Thomann (1998) for example.

A comprehensive approach requires a progressive integration of the water treatment/distribution process considering the influence of mechanisms inducing environmental degradation in the watershed. This is a complex process and only can be assessed if the numerical tools give an appropriate indication of the main water quality parameters. As a minimum, this process requires:

- the development of a numerical water quality procedure that simulates the effects of integrated water quality parameters such as disinfectant-residual, DBP formation, microorganisms regrowth, corrosion, water residence time, and biofilm formation;
- analysis of water quality changes in the distribution system under different hydraulic conditions;
- evaluation of interaction between velocity profile and shear stresses under different laminar, turbulent, and transient conditions addressed by external effects;
- analysis of compressibility effects and their influence on water quality parameters in the distribution system;
- evaluation of the impact of external actions over the flexibility of water treatment plants to refine the treated water and, as a consequence, over the numerical boundaries of the water quality models;
- evaluation of the impact of distributed leaks over water quality conditions in the system;
- development of associated calibration and data collection techniques.

The definition of this alternative numerical procedure will permit an evaluation of the significance of the behavior of the overall system under various boundary conditions, especially those required for the solution of governing equations in water quality models.

7   FINAL COMMENTS
Information presented in the literature reveals the complexity of the modelling, analysis and design processes for water distribution systems. Issues include the representation of the physical/topological characteristics of the network and the effectiveness of the numerical solution to achieve a meaningful representation of the overall physical processes for a wide range of spatial and temporal scales. Nevertheless, a more comprehensive analysis is required to evaluate the impact of the actions in the watershed over the physical, chemical and biological characteristics of a water distribution system.

238

REFERENCES

Axworthy, D. (1997). *Water distribution network modeling: from steady state to waterhammer*, Ph.D. thesis, Dept. of Civil Engineering, University of Toronto, Toronto.

Axworthy, D., B. Karney. (1996). Modeling low velocity/high dispersion flow in water distribution systems. *Journal of Water Resources Planning and Management*, vol. 122, no.3, 218-221.

Boulos, P. F., T. Altman, P. A. Jarrige and F. Collevati. (1994). An event-driven method for modelling contaminant propagation in water networks. *Journal of Applied Mathematical Modelling*, vol. 18, 84-92.

Brunone, B., B. Karney and M. Ferrante. (1999). Velocity profiles, unsteady friction losses and transient modelling. *26th annual Water Resources Planning and Management Conference*, ASCE, Tempe, Arizona (in press).

Chaudhry, M. (1987). *Applied hydraulic transients*. Van Nostrand Reinhold, New York.

Elton, A., L. F. Brammer, and N. S. Tansley (1995). Water quality modeling in distribution networks. *Journal of American Water Works Association*, 87(7), 44-52.

Islam, M., M. Chaudhry. (1998). Modeling the constituent transport in unsteady flows in pipe networks. *Journal of Hydraulic Engineering*, vol. 124, no.11, 1115-1124

Karney, B., and B. Brunone (1999). Water hammer in pipe networks: two case studies. *Water Industry Systems,* Vol. 1, Research Studies Press Ltd, UK.

Karney, B. (1996). Understanding transients in pipeline systems: computer power and engineering insight. *Uni-Bell PVC pipe news*, 17(1),8-12.

Karney, B. (1990). Energy relations in transient closed-conduit flow. *Journal of Hydraulic Engineering*, ASCE, 116(10), 1180-1196.

LeChevallier, M. W. (1990). Coliform regrowth in drinking water: a review. *Journal of American Water Works Association*, 82(11), 74-86.

Lu, C., P. Biswas, R. M. Clark. (1995). Simultaneous transport of substrates, disinfectants and microorganisms in water pipes. *Water Research*, 29 (3), 881-894.

Romain, D. (1996). *Modeling impacts of ozone disinfection on distribution system water quality*. M.A.Sc. thesis, Dept. of Civil Engineering, University of Toronto, Toronto.

Rossman, L. A., P. F. Boulos (1996). Numerical methods for modeling water quality in distribution systems: a comparison. *Journal Water Resources Planning and Management*, ASCE, 122(2), 137-146.

Rossman L. A., R. M. Clark and W. M. Grayman (1994). Modeling chlorine residuals in drinking-water distribution systems. *Journal of Environmental Engineering*, ASCE, 120(4), 804-820.

Rossman, L. A. (1993). EPANET Users manual. Risk reduction engineering laboratory, U.S. Environmental Protection Agency, Cincinnati, Ohio.

Singer, P. C. (1994). Control of disinfection by-products in drinking water. *Journal of Environmental Engineering*, 120, 727-744.

Thomann, R. V. (1998). The future "golden age" of predictive models for surface water quality and ecosystem management. *Journal of Environmental Engineering*, ASCE, 124(2), 94-103.

Vasconcelos, J. J., W. M. Grayman, L. A. Rossman, R. M. Clark, P. F. Boulos, J. A. Goodrich and J. W. Clapp. (1994). Characterization and modeling chlorine decay in distribution systems. *AWWA Water Quality Technology Conference*, San Francisco, 391-406.

Wable, O., and P. A. Jarriage. (1992). Evolution of water quality and modeling of chlorine concentration in distribution network. *Proc. AWWA Computer Conference*, Nashville, 899-912.

Wylie, E. B., and V. L. Streeter. (1993). *Fluid transient in systems*. Prentice-Hall, Inc.,Englewood Cliffs, N. J.

Sag Curve and River Parameters Determination: Sebou River

M. Saffi *and* D. Ouazar

ABSTRACT

In this paper, the sag curve for rivers is dealt with by combining optimisation techniques and finite elements. Both the direct and inverse problems are treated. The direct treatment simulates the concentration of the dissolved oxygen along the river, taking into account the following parameters: i) reaeration from the atmosphere, ii) photosynthesis of aquatic plants and algae. The rate of the photosynthesis is supposed to be sinusoidal; however, the respiration rate is assumed to be constant as it does not depend on light radiation, iii) deoxygenation due to organic matter: the main sink of the oxygen is the process of biological oxidation of organic matter which is supposed to be a first order reaction, iv) aerobic decomposition of the sludge, taking place at the interface between the benthic layer and the flowing water, which is another source of oxygen depletion. The solver used in the first problem is based on LU (Lower Upper) decomposition. The matrix is stored via the conventional skyline technique. In the second part, parameter identification of river coefficients using both the steepest descent, Fletcher-Powell methods and Genetic Algorithms is developed. The predetermined coefficients are the first order reaction rate constant (1/day), the flow velocity, which is assumed to be constant along the river (kms / day), and the magnitude of the ultimate biological oxygen demand of the source term. The GAs seem to be more efficient as a global optimisation tool, however, the steepest gradient and Flect* her-Powell methods either do not progress or diverge to infinity in their search for the global minima. The model can simulate accurately one pollution event. The numerical and the close form solution have shown good agreement. The model was validated on a real case for the Sebou river in Morocco.

1   INTRODUCTION

Rivers are subject to much domestic and industrial pollution. Thanks to their ability to purify themselves, rivers can absorb pollution via bacterial action on organic matter and the settling process taking place at its bottom. However, their capacity of assimilation is limited. Since the amount of the discharged wastes becomes important, reaeration and photosynthesis are unable to supply the required entity of oxygen to stabilise the organic matter and that is when degradation of the river quality starts. Its direct harmful effect concerns life within the river as an ecosystem and the cost of producing potable water is no longer economically reasonable. For instance, the reagent cost at Kariat Ba M'hammad station had already reached 0.48 \$/m^3 which is 14 times more expensive than the cost registered in 1991 [8]. The knowledge of the oxygen distribution along the river, especially the critical dissolved-oxygen deficit and its location are among the most helpful parameters for engineers in practical applications.

2   GOVERNING EQUATION

Sag curves for rivers are described by equation (1) as reported in [12]; minor modifications were incorporated :

$$\frac{\partial c}{\partial t} = \varepsilon\frac{\partial^2 c}{\partial t^2} - U\frac{\partial c}{\partial x} + R_d + R_r \quad \text{on} \quad \Omega = [\,0\,,L\,], \quad (1)$$

where c is the concentration of the dissolved oxygen, ε is the coefficient of diffusion, U is the flow velocity in the x direction (assumed to be constant), R_d is the deoxygenation rate and R_r is the reoxygenation rate.

The main source of oxygen depletion is oxidation of organic matter. The amount of oxygen required for this process is measured by the biochemical oxygen demand (BOD). In the first approximation, the kinetics of the BOD reaction is formulated in accordance with first order reaction kinetics [5 , 6]:

$$\frac{dL}{dt} = K_1 L \qquad\qquad (2)$$

Thus $L = L_0 e^{-K_1 t}$

where Lo is the ultimate BOD at the point of discharge and L is the ultimate BOD at the position x.

We assume that the river and the waste are completely mixed at the outlet, then the ultimate BOD is expressed as a function of the ultimate BOD of the river before mixing, namely l_r, and the ultimate BOD of the wastewater L_w :

$$L_o = \frac{Q_r l_r + q_w L_w}{Q_r + q_r}, \qquad\qquad (3)$$

where Q_r is the river flowrate and q_w is the wastewater flowrate.

In the case of slow-moving rivers, if the amount of settlable solids is important, sludge deposits become a non-negligible sink for dissolved oxygen. The maximum daily benthic (river bed) oxygen demand is reported in [6].

The primary sources of oxygen replenishment in a river are reaeration from the atmosphere and photosynthesis. The rate of oxygen transferred R1 from the atmosphere to the water is defined by the following formula [11]:

$$R_1 = K_2.(\,c_s - c\,), \qquad\qquad (4)$$

where : K_2 is the reaeration constant, cs is the dissolved oxygen concentration at saturation and c is the dissolved oxygen concentration.

Photosynthesis of aquatic plants and algae increases the concentration of the dissolved oxygen in the river during daylight; the relevant equation proposed by Bigura et al is reported in [12] as follows :

$$P - R = Ce^{\frac{E_a}{T+273}} \left(\sin\frac{\pi t}{12} \right)^n , \tag{5}$$

where P is the rate of photosynthesis, R is the rate of respiration, T is the temperature, and Ea is the activation energy. To make the formula (5) simpler let $F = Ce^{\frac{E_a}{T-273}}$.

To solve a particular problem, the governing equation (1) has to be supplemented by appropriate initial and boundary conditions. Initial conditions include the distribution of the dissolved oxygen along the river before pollution takes place.

$$c(x, t) = c(x) , \tag{6}$$

On the other hand, boundary conditions are of the form :
 1. Given concentration :
$$c(0, t) = c_0 \quad \text{for } t >= 0, \tag{7}$$

 2. Flux prescribed :
$$\left. \frac{\partial c}{\partial x} \right|_{x=L} = 0, \tag{8}$$

3   FINITE ELEMENT FORMULATION(FEM)
The response equation of the river is obtained numerically by transforming equation (1) using FEM; a trial function for the dissolved oxygen concentration is defined as :

$$c = \sum_{i=1}^{BBG} c_i N_i , \tag{9}$$

where C_i is the concentration at the node i, N_i are linearly independent trial functions and NNG is the global nodes number, using the Galerkin principle :

$$\int R N_i \, dx = 0, \tag{10}$$

where
$$R = \frac{\partial c}{\partial t} - \varepsilon\frac{\partial^2 c}{\partial x^2} + U\frac{\partial c}{\partial x} - R_d - R_r,$$

Equation (10) can be expressed in matrix formulation as :

$$A\frac{dc}{dt} + Bc + G = 0, \qquad (11)$$

$$C(0) = c_0,$$

where :

$$A_{ij} = A_{ji} = \frac{\Delta x}{6}, \quad if \ j = i + 1.$$

$$A_{11} = A_n = \frac{\Delta x}{2},$$

$$A_{ii} = 2\frac{\Delta x}{3},$$

$$B_{ij} = -2\frac{\varepsilon}{\Delta x} + \frac{2}{3}K_2\Delta x \quad if \ i = j \ and \ i \neq 1 \ and \ i \neq n$$

$$B_{ij} = -\frac{\varepsilon}{\Delta x} - \frac{U}{2} + \frac{K_2\Delta x}{6} \quad if \ i = j - 1,$$

$$B_{ij} = -\frac{\varepsilon}{\Delta x} + \frac{U}{2} + \frac{K_2\Delta x}{6} \quad if \ i = j + 1,$$

$$B_{11} = \frac{\varepsilon}{\Delta x} - \frac{U}{2} + \frac{K_2\Delta x}{3},$$

$$B_{nn} = \frac{\varepsilon}{\Delta x} + \frac{U}{2} + \frac{K_2\Delta x}{3},$$

$$G_1 = K_1 L_0 \Delta x \left(\frac{1}{\Phi} - \frac{1}{\Phi^2} + \frac{e^{-\Phi}}{\Phi^2} - \frac{K_2 C_s \Delta x}{2} + OT_b \right)$$

$$G_i = 2K_1 L_0 \Delta x \left(-1 + ch(\Phi) \right) e^{(1-i)\Phi} - \frac{K_2 C_s \Delta x}{2} + OT_b$$

$$for \ 2 \leq i \leq n$$

$$G_n = K_1 L_0 \Delta x e^{(2-n)\Phi} \left[-\frac{e^{-\Phi}}{\Phi} - \frac{e^{-\Phi}}{\Phi^2} + \frac{1}{\Phi^2} \right] - \frac{K_2 C_s \Delta x}{2} + OT_i,$$

where n is the number of nodes, Δx is the space step, $\Phi = Kr.\Delta x / U$, and the other terms OTi concern both photosynthesis and bottom sludge:

$$O T_i = \sum_{g=1}^{n_g} [(F(\sin\frac{\pi X_g}{12.U})^n + Y_{mg}) N_i(X_g).\omega_g ,$$

X_g are integration points, ω_g are weighting factors, n_g is the number of Gauss points, and Y_{mg} is the maximum daily oxygen demand at the Gauss point X_g.

The problem is defined here as: find a set of optimal or near optimal partial cumulative solutions forming a planning set for a progressive rehabilitation, replacement and/or expansion of an existing water distribution system using a fixed amount of money.

4   NUMERICAL EXAMPLE

The code is tested for known closed form solution, assuming a steady state condition. The effects of photosynthesis and bottom sludge are neglected. Other parameters are as follows:

$$\varepsilon = 0,$$

$$\frac{K_1}{U} = \frac{K_2}{U} = 0.5\frac{1}{day}, \frac{K_r}{U} = 0.125\frac{1}{day},$$

$$C_0 = C_s = 6\frac{mg}{l},$$

$$\Delta x = 0.5\ Km,$$

$$L = 29.5\ Km.$$

The analytical solution is:

$$C = C_s - \frac{K_1 L_0}{K_2 - K_r}\left[e^{\frac{-K_r x}{U}} - e^{\frac{K_2 x}{U}}\right] - (C_s - C_0)e^{\frac{-K_2 x}{U}},$$

as depicted in figure 1, the numerical and the analytical solutions are in good agreement.

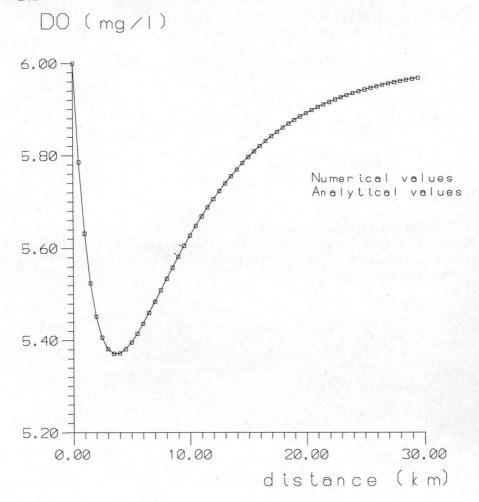

Fig 1: Analytical versus Numerical

5 INVERSE PROBLEM

The goal of the inverse problem is to determine the optimal values of the reaeration rate, the average velocity, the coefficient of diffusion and the initial load using experimental information. Let C_{obs} be the vector of observed values of the dissolved oxygen concentrations and C_{pre} the vector of the predicted concentrations at different observation sites, then the inverse problem is formulated as follows:

$$\text{Minimize} \qquad \|c_{pre}(\qquad K_1, K_2, K_r, U, \varepsilon, \qquad L_o)-c_{obs}\| \qquad (12)$$
$$(K_1, K_2, K_r, U, \varepsilon, L_o)$$

To solve equation (12) three techniques were compared, namely : two conventional methods i.e the steepest gradient method (SGM) and the Fletcher-Powell method (FPM), and Genetic Algorithms (G). The first two are well detailed in the literature [10]. Both the steepest gradient and the Fletcher-Powell methods were tested with explicit functions and from several starting points Xo; the code converges to the optimal solution after the required number of iterations:

$$f(x_1, x_2, x_3, x_4) = (x_1 - 0.5)(x_1 - 0.5)+(x_2 - 1)(x_2 - 1)+(x_3 - 9)(x_3 - 9)+$$
$$(x_4 - 43)(x_4 - 43),$$

starting point   optimal point

	FPM	SGM
(1200,100,-10,250)	(0.572,0.997,8.987,42.791)	(0.410,1.005,8.987,42.791)
(-1, 1 ,-1 , 1)	(0.507,0.994,9.004,42.992)	(0.507,0.993,9.004,42.992)

Several runs with different values of the starting vector Xo were made. The sequence (X_k) is either constant which means that the algorithm does not progress, in this case the correction to the solution from one iteration to another is negligible, or it diverges to infinity. To highlight this, the objective function surface was plotted in the following particular case : $\varepsilon = 0$, $K_r = 0.25$, $C_s = C_o = 6$. We know beforehand that the objective function reaches its minimum at (0.5 , 0.5). From figure 2, one can see that point (0.5 , 0.5) is almost a saddle point. (The upper diagram in figure 2 represents the objective function within a large domain [0.2, 2]x[0.2, 2], while the lower part is an enlargement over the optimum region.) This particular property is due to the sink term of the BOD in equation (1). This renders the inverse problem of the sag curve for rivers ill-posed.

248

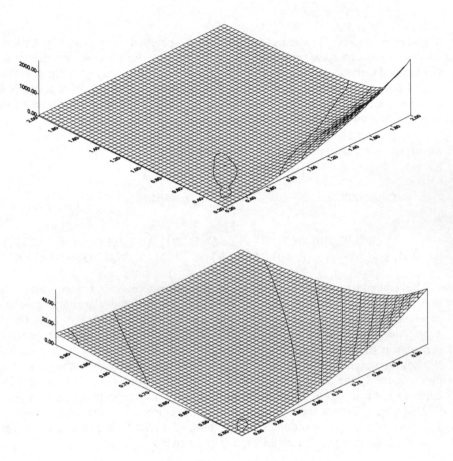

Fig 2: Objective function J(J1,J2)

6   GENETIC ALGORITHMS

Unlike traditional optimisation methods, genetic algorithms rely on a population of individuals in their search for the optimum. As generations progress, chromosomes undergo alterations. Each individual struggles to survive, the worst one (in the sense of the objective function) dies off, the best is selected to give more offspring. Simple genetic algorithms consist of three conventional transformations [2] i.e. selection, crossover, and mutation. GAs have been shown to be a powerful tool for solving hard optimisation problems such as a multiple-peak functions, noisy and discontinuous functions, deceptive problems [13]. GAs are now tested with a new type of function where the minimum is near to being a saddle point. The GAs code is written in the C++ language using object oriented concepts [1,10]. It consists of three classes: EP, SGA, SGA_RIVER. The class EP defines the main services common to GAs such as: initialisation (), evolution(), selection(), crossover(), mutation(), statistics(), objective_function(), display-informations().

All these actions are defined as virtual functions in order to make the exploitation of these classes simple and more general. SGA_RIVER class inherits all properties of the SGA class and adds the definition of the objective function which is computed using the direct code. On the other hand the class SGA is easily extensible for other GA implementations. It is well known that all optimisation techniques, even GAs, share the same limitation, which is premature convergence. However, in traditional optimisation methods this problem is inbuilt, whereas GAs offer the user the probability of mutation as a device to control their progress. In fact, when a gene mutates the information is lost, but this leakage is circumvented by the new templates added to the population. So one should manage to preserve the precarious equilibrium between a high rate of mutation, which makes the GAs operate as a random walk, and the risk of a premature convergence caused by a high selection pressure. Low rates of mutation (pm) are used in the literature. In the present study pm = 0.05, the other parameters are as follows: population size = 50, probability of the crossover = 1, chromosome size $= 30 \times 4 = 120$.

7   APPLICATION

A parameter estimation problem for the case of the Sebou river is considered. It consists of the predetermination of the following coefficients: ε, K_1, K_2, K_3, U, L_0. As we have assumed, the kinetics of assimilation of the organic matter is of first order reaction type, and so it is easy to find out values of L_0 and ω where $\omega = K_r / U$, from BOD measurements at observation points using a conventional least square method. Table 1 shows the measured value of the BOD along the studied section [8],

$$\Delta = \sum_{i=1}^{N_{obs}} (Y_{imes} - Y_{ipred})^2,$$ (13)

The error Δ describes the size of the misfit between the vector of the observed and the predicted BOD entities, therefore:

$$\Omega = \frac{(\sum_{i=1}^{N_{obs}} X_i \sum_{i=1}^{N_{obs}} Y_{imes} - N \sum_{i=1}^{N_{obs}} X_i Y_{imes})}{N \sum_{i=1}^{N_{obs}} X_i^2 - (\sum_{i=1}^{N_{obs}} X_i)^2},$$ (14)

$$L_o = Exp\left(\frac{\sum\limits_{i=1}^{N_{obs}} X_i^2 \sum\limits_{i=1}^{N_{obs}} Y_{i\,mes} - \sum\limits_{i=1}^{N_{obs}} X_i \sum\limits_{i=1}^{N_{obs}} X_i Y_{i\,mes}}{\sum\limits_{i=1}^{N_{obs}} X_i^2 - (\sum\limits_{i=1}^{N_{obs}} X_i)^2}\right), \qquad (15)$$

where summations in formulas (13, 14, 15) are extended on the observation points N_{obs}, X_i is the abscissa of the sample location and $Y_{i\,mes}$ is the logarithm of the measured BOD at this location. From the table 1, L_o = 134.29 mg / l, K_r = 0.025 * U (1 / day), U is expressed in Km / day.

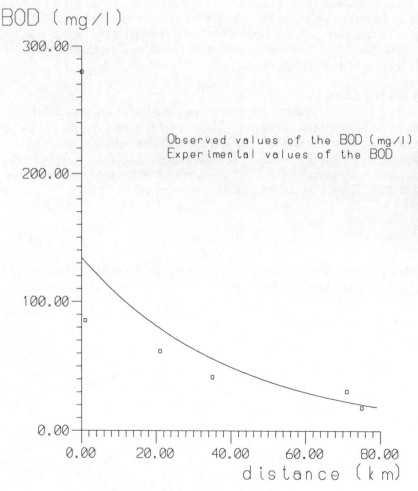

Fig 3: BOD (observed/constated)

Sample number	1	2	3	4	5	6
Distance from Oued Fes, KM	0	1	21	35	71	75
[O_2] in mg/l	10.10	4.30	0.77	2.14	6.94	6.88
BOD in mg/l	280	85.45	61.81	41.81	30.91	18.18

Figure 3 shows both predicted and measured data. Remaining parameters are determined using the inverse code based on GAs. In fact, it is natural to use directly the measured value of the average velocity, but since the corresponding measurement is not available, we assume that the velocity U is among unknown coefficients. From figure 4, one can notice that the fit between the observed and the modelled data is satisfactory. The determined coefficients are : K_1= 0.298 (1/day), K_2 = 1.979 (1/day), U = 38.67 (km/day), K_r = 0.966 (1/days), the coefficient of diffusion is assumed to be negligible. The simulated value of the critical dissolved-oxygen deficit is 10 mg/l, located 26 kms from Oued Fes.. GAs seem to be efficient for parameter estimation problems for rivers sag curves; however, to build a more flexible code, we suggest the following:

■ The direct code should be extended to several pollution sources and anaerobic conditions,
■ It is more realistic to solve the inverse problem in unsteady state conditions in order to exploit all the available information,
■ The kinetics of the BOD reaction should be extended.

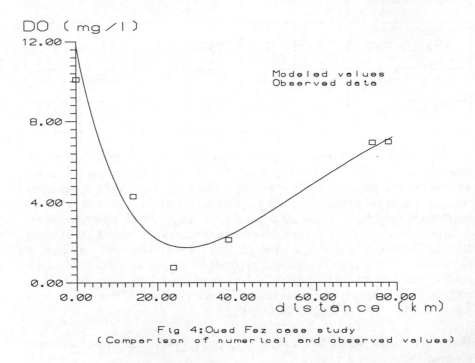

Fig 4:Oued Fez case study
(Comparison of numerical and observed values)

8 CONCLUDING REMARKS

In this study, both direct and inverse problems have been treated. The inverse problem is highly ill-posed. Both the steepest and the Fletcher-Powell methods fail in the search for the global optimum owing to the particular shape of the objective function whose global minima is almost a saddle point. It is natural that GAs sort out the correct solution since they do not rely on computing the gradient of the objective function to improve the quality of their potential solutions. Instead, they operate on a whole population and good templates are automatically preserved. In their search, GAs look for the better rather than for the best.

This tool is simple to implement, useful for practical applications and easily extendible since it is built in the object oriented concept. Maybe one of its most important applications that interest municipalities is in the determination of the optimal design and locations of wastewater treatment plants, that would satisfy the required quality standards on a minimum of dissolved oxygen, at minimum cost, in a regional treatment wastewater program.

REFERENCES
1. Claude Delannoy, Programmer en turbo C++, Eyrolles 1991.
2. D.E. Goldberg, Genetic Algorithms in search, optimisation and machine learning, Addison-Wiley Publishing company, Inc. 1989.
3. F. Edeline Trib. Cebedeau N 470 pp 37-45, 1983, Nouvelles recherches sur la courbe en sac, Deuxième partie.
4. F. Edeline, W. Binet, H. Ftticcioni, A Ichikawa, H. Peixoto trib. cebedeau N 329 pp 203-209 Avril 1971, Etude des eaux du canal Albert, Première partie.
5. F. Edeline Trib. cebedeau N 339, pp60-65, Février 1971, Les difficultés de l'établissement d'un modèle mathématique de rivière.
6. Metcalf & eddy, Wastewater engineering : treatment, disposal, reuse. McGraw-Hill, Inc. 1979.
7. P. Balland et S. Alsace trib. Cebedeau N498, pp 31-44, 1985, Modèle d'autoépuration en rivière en régime permanent.
8. Rapport Office National de l'Eau Potable (Maroc), Département Qualité des eaux, Division contrôle de pollution, Pollution de Oued Sebou en amont de Kariat Ba Mhamad, Mai 1989.
9. Rapport ONEP, Direction du Laboratoire de la Qualité des Eaux , pollution de l'Oued Sebou au niveau des stations de traitement de Kariat Ba Mhamad et de Mkansa, 1992.
10. Russel Winder, Developing C++ software, John Wiley & sons, 1995.
11. R.Willis, W.W-G. Yeh, Groundwater systems planning and management, prentice-Hall, Inc. 1987.
12. W.W. .Eckenfelder Gestion des eaux usées urbaines et industrielles. Technique et documentation. Lavoisier.
13. Zbigniew Michalewicz, Genetic Algorithms + Data structures = Evolution Programs, Springer-Verlag, 1992.

A Predictive Relational Fuzzy Logic Model for Activated Sludge Processes

E. Scheffer *and* I.M. MacLeod

ABSTRACT
Biological and chemical processes are generally non-linear, multi-variable and non-stationary. With the progression from linear systems, to describable non-linear systems, to non-describable, non-linear systems, the application of artificial intelligent techniques becomes more attractive. In this paper, the focus is on self-learning or relational fuzzy logic which was applied to a biological water purification process, namely activated sludge. Fuzzy logic was used to create a model of the characteristics of the in-flowing water, after which this model was extended to a predictive model. Using this method, the behaviour of the system was captured and the system's behaviour was predicted one step ahead in time. As is the case with most artificial intelligent techniques, less initial assumptions need to be made about the system. The accuracy of the model depends on the amount of data available and whether the fuzzy tuning was done accurately. This type of model does not only capture the system characteristics, but can also be formulated as a predictor for on-line control applications.

1   INTRODUCTION

Activated sludge is one of the most common biological methods used for water purification [1-4]. Activated sludge processes essentially involve an aeration phase where the water to be purified is brought into contact with a bacterial floc in the presence of oxygen, followed by a clarification phase in which the floc is separated. This process is an intensification of a phenomenon which occurs in the natural environment [1]. Phosphorus and nitrate are two of the most essential nutrients in the biological activity of the micro-organisms involved in activated sludge. Together with temperature, pH and dissolved oxygen, their concentrations have a strong influence on the activity and efficiency of the process. Therefore, a balance amongst the interdependent parameters, nitrate, phosphate, pH, temperature and chemical oxygen demand (COD) amongst others, is required at each stage of the biological process [4]. This has created a need for modelling tools that can assist in the design process of automatic control systems for activated sludge plants. In modern water purification plants, optimisation can be complicated because of many interacting processes. The International Associations of Water Quality (IAWQ) task group has recently produced a mathematical model for

255

activated sludge, Activated Sludge Model (ASM) No. 2, to simulate the activity within an activated sludge process [5]. According to this model there are nineteen processes involved in the system and nineteen components of importance. This indeed amounts to a multi-variable, interdependent system. ASM No. 2 has captured the behaviour of an extremely complex biological system, by way of mathematical equations. In order to apply this model successfully, a vast amount of information is required about the system, for example readily biodegradable substrate (g m^{-3} COD), fermentation products (g m^{-3} COD), bicarbonate alkalinity (mol m^{-3} HCO$_3$-), phosphorus accumulating organisms (g m^{-3} COD) and heterotrophic biomass (g m^{-3} COD). Process rate equations are determined from information gathered. These equations are not dynamic and when the initial conditions change drastically, the rate equations are insufficient to describe the system.

Another way of creating a model of such a complex system is to capture the characteristics of the system by observing the behaviour of the system over a long period of time. If observed data is used as an input to a learning algorithm, the system characteristics may be captured and the future behaviour may be predicted. One such modelling technique is self-adaptive fuzzy modelling, which has proven to be particularly well suited for systems with inherent, non-linear dynamics [6]. In this paper, self-adaptive fuzzy logic modelling was applied to in-flowing water at a purification plant in order to capture the characteristics of certain variables by creating a model of the system. After the model was captured, a predictor was generated, which could predict the behaviour of a certain variable one step ahead in time. By knowing the variable's characteristics one step ahead in time, the system non-linearities, such as dead time, can be accommodated. Being able to predict input behaviour has a special appeal to systems with large dead-time in the sense that the predicted values can be used in designing a feed-forward compensator.

2 FUZZY LOGIC AS A MODELLING TECHNIQUE

Fuzzy logic is a natural, continuous logic, patterned after the approximate reasoning of human beings. As a theoretical mathematical discipline, fuzzy logic reacts to constantly changing variables. It challenges traditional logic by not being restricted to the conventional binary values of zero and one. Instead, it allows for partial truths and multi-valued truths [7,8]. Fuzzy logic is especially suitable for problems which cannot easily be represented by mathematical modelling because data is either unavailable, incomplete or the process is too complex [6-14]. Fuzzy logic can be divided into two categories according to two different approaches: (a) rule-based fuzzy logic [15,6] and (b) self-adaptive or self-learning relational fuzzy logic.

2.1 Rule-based fuzzy logic

The rule-based fuzzy logic approach is based on the capturing of a human operator's behaviour in a set of 'if-then' rules, with a certain degree of flexibility [9,10]. The success of a rule based control technique depends on the accuracy with which the operator's behaviour was captured and the efficiency of the operator's behaviour. Unfortunately, the characteristic of the specific system is neglected when a generic rule-based fuzzy system is implemented, as can also be expected for most mathematical modelling techniques.

2.2 Self-learning relational fuzzy logic

The second category of fuzzy logic, self-learning relational fuzzy logic, has the advantage that the characteristics of the system are captured automatically by exposing the learning algorithm to a substantial amount of measured data regarding the variable in question. Relational self-learning fuzzy logic has two specific advantages in that this type of algorithm was proven to be highly immune to observation noise [16,17], and it is a self-learning system [11,14,18,19]. A self-learning fuzzy algorithm may either be used to build a model of the system, or it may be used to construct a predictor. When a self-learning fuzzy algorithm is used to construct a predictor, the process consists of two stages of operation: (a) the learning phase and (b) the estimation or prediction phase [11,14]. The learning phase of the algorithm is based on the approach of building a fuzzy relation [19] by means of adaptive clustering [6]. The clustering method used is based on probability theory, in particular, the weighted average of frequency of occurrence of the fuzzy rules that describe the system [18]. The fuzzy rules are encapsulated in the fuzzy relation, and contain the characteristic dynamics of the system under observation. When a model of the system is constructed using self-learning fuzzy logic, the fuzzy relation is taken as a function describing the relationship between the input (in the case of this project, the in-flowing water's characteristics), and the output (the out-flowing water's characteristics). Viewed in this way, the fuzzy relation can be seen as analogous to the transfer function of the system [6]. In the case of fuzzy prediction, the predictor is used in conjunction with the controller in order to improve the system's efficiency.

Another approach - time series prediction - uses two past points of the same variable to predict a future value. The input and the output is then of the same data series, i.e. the input is two historical data points, e.g. $u[k-3]$ and $u[k]$, and the output is the predicted value, $u[k+1]$. In this paper, time series prediction is applied to certain variables in the activated sludge process.

3   CHARACTERISTICS OF VARIABLES

The five most important variables in the activated sludge process are the concentrations of phosphate, nitrate, the chemical oxygen demand (COD), the pH and the temperature. All of these variables have an influence on the activity of the micro-organisms. Undoubtedly these are not the only variables which play important roles in the efficiency of the process, but these are used to illustrate the concept, and if any other variables' data are available, the same principles can be applied.

The interdependence of the process variables varies - COD shows interdependence with nitrate and phosphate respectively, but nitrate and phosphate does not show mutual interdependence. The pH and the temperature can be seen as conditions under which the process takes place, rather than individual variables. Although pH and temperature are not interdependent or dependent on nitrate, phosphate or COD, they do influence the activity of the micro-organisms directly.

258

The activity of the micro-organisms influences the concentrations of nitrate, phosphate and also the COD and therefore, the temperature and the pH influence the concentration of the phosphate and nitrate and the COD, indirectly [1]. Even with these few variables, it is already apparent that activated sludge is a complex process which is difficult to describe mathematically.

The in-flowing water's data were captured in the form of histograms, in terms of pH, COD, phosphate concentration and are shown in figures 1, 2 and 3 for a period of approximately four years. The temperature during the process was also measured and is shown in figure 4. The nitrate concentration at the in-flowing side is insignificant and therefore no further attention, in terms of time series prediction, will be given to the nitrate concentration. The data points were taken daily at an operating water purification plant. From these graphs it is clear that the values of the variables vary significantly over the set of possibilities. From the original time domain graphs (which were not included for the sake of brevity), it can be concluded that the parameters vary significantly from day to day and in some cases sudden increases or decreases in the values of the variables occur.

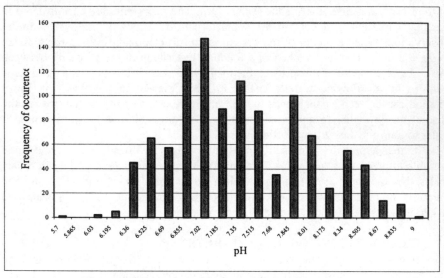

Figure 1: Histogram of measured in-flowing pH values

Sudden changes in the variable value and bursty behaviour resulting in large statistical variance make the system difficult to control with linear approximations [17,18]. Although these large peaks and valleys do not occur often, it is precisely these unexpected values that result in large errors that cause the model's accuracy and the control system's performance to decrease.

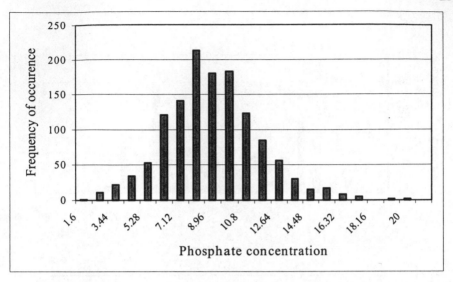

Figure 2: Histogram of measured in-flowing phosphate values

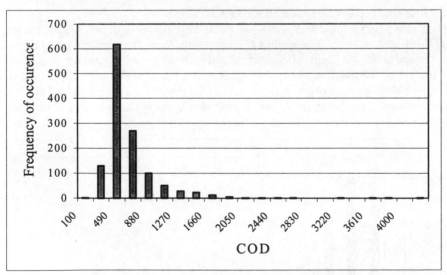

Figure 3: Histogram of measured in-flowing COD values

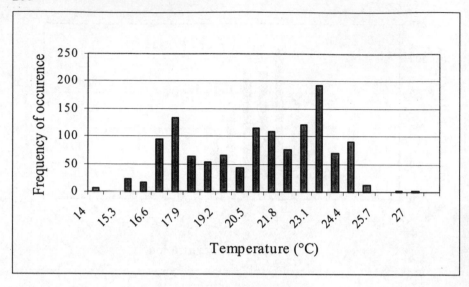

Figure 4: Histogram of measured temperature during the process

4 RESULTS OF TIME-SERIES PREDICTION

Self-learning fuzzy logic was used for modelling and time-series prediction of COD, pH, phosphate and temperature. The results from the prediction algorithm are presented in the form of histograms, this time showing the frequency of occurrence of the absolute error. From these histograms the size of the smallest error, as well as the size of the error for which the frequency is at its highest can be read. For a good predictor, the smallest error should occur the most frequently and the frequency of occurrence should decrease as the error increases.

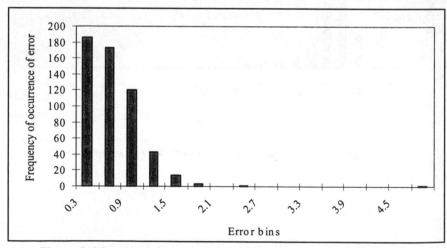

Figure 5: Histogram of error between predicted and measured pH values

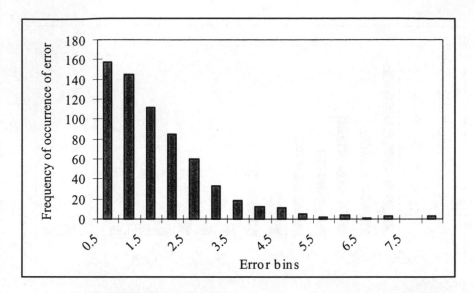

Figure 6: Histogram of error between predicted and measured phosphate values

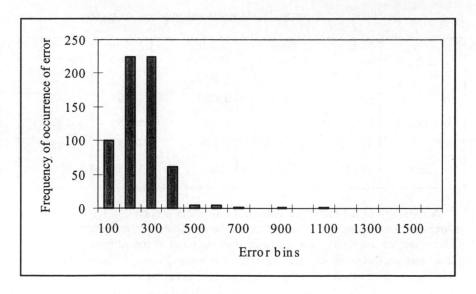

Figure 7: Histogram of error between predicted and measured COD values

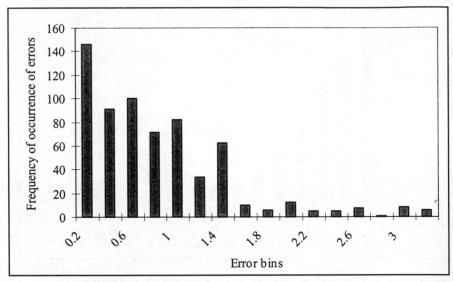

Figure 8: Histogram of error between predicted and measured temperature values

Table 1: Summary of results obtained in terms of the error

	Mean of measured values	Mean of Error	Standard Deviation of Error	Average % Error	Percentage errors within 10%
pH	7.61	0.476	0.3399	6.259	±34.57
Phosphate	7.91	1.407	1.234	17.79	±37.42
COD	423.92	199.19	106.35	46.98	±16.02
Temperature	20.45	0.723	0.645	3.535	± 36.68

A summary of the important statistical indicators is shown in table 1. The first column shows the average value (mean) of the variables that were measured at the in-flow side of the plant. This indicator was included in the summary in order to get an idea of the range of each variable. The second column shows the average error and the third column gives the standard deviation - an indication of how much the error deviates from the mean error. The fourth column, the average percentage error, is a term showing the relationship between the mean of the error and the mean of the measured in-flowing variable, as a percentage. Lastly, shown in the far right column, are the number of errors not larger than 10% of the maximum error - where the maximum error was taken as 100%. If the requirement for the performance of the model is based on an error less than 10%, these last values would give a good indication of the performance of the model.

From the graphs and table 1 it is seen that chemical oxygen demand is the variable which is the most difficult to predict. The performance of the prediction algorithm was the worst for this case and the percentage error was the largest. One of the most important advantages of fuzzy logic is its robustness and its ability to accommodate large variations in the input. From figure 3 it is established that COD is actually very predictable and that most values occur within the second bin. A reason for the low performance of the fuzzy prediction algorithm may be that the constraints were too 'soft' and too much variation is allowed with this technique. Most of the errors occur around 200 to 300, which causes the standard deviation and the average percentage error to be large. Another reason may be the occurrence of large COD values, which occur less frequently (otherwise known as burstiness) and are therefore very unexpected. Because these values are unexpected, they are not well predicted by the fuzzy prediction algorithm and because they are large, they cause large errors. Large errors in turn have a profound effect on the performance parameters.

Models are generally used to serve as a basis on which control systems are designed. The control system usually has to operate within certain boundaries of a variable value. In order to design a model-based controller, the model needs to adhere to the same performance specifications as the controller (or better). According to table 1, pH, phosphate and temperature were modelled with good accuracy. The average percentage error of temperature was the smallest (3.535%), pH followed at 6.259% and phosphate had an average percentage error of 17.79%. These performance parameters indicate acceptable modelling, as these errors are within the control specifications.

5   CONCLUSIONS

An adaptive fuzzy learning algorithm was applied to time series prediction of four important variables in the activated sludge process. The in-flowing variables were predicted one step ahead in time. This prediction can be implemented as a lead compensator as part of a complete control system in order to reduce the effect of dead zones, or it may just be used in order to capture the characteristics of the system. Three of the variables, pH, temperature and phosphate concentration, had good prediction performance results, but chemical oxygen demand (COD) was more difficult to predict. This can be ascribed either to the soft constraints of fuzzy reasoning or to the variable's bursty nature and unexpected large variations, which occur less frequently, but have a large impact on the predictor's performance.

ACKNOWLEDGEMENTS

I would like to thank the Greater Johannesburg Metropolitan Council for making the data freely available. The support of the Foundation for Research Development and the University of the Witwatersrand is gratefully acknowledged.

REFERENCES

[1]   Lyonnaise des eaux Dumez (1991): **"Water Treatment Handbook - Volume 1"**; Degremont - Water and the environment. (ISBN: 2.9503984.1.3).

[2]   Droste R.L. (1997): **"Theory and Practice of Water and Wastewater Treatment"**; John Wiley & Sons, Inc. (ISBN:0-471-12444-3).

[3]   Ramalho R.S. (1983): **"Introduction to Wastewater Treatment Processes - 2nd Edition"** ; Academic Press, New York; (ISBN: 0-12-576560-6).

[4]   Bolton R.L. & Klein L. (1961): **"Sewage Treatment - Basic Principles and Trends"** ; Butterworths, London.

[5]   International Association of Water Quality (1995): **"Activated Sludge Model No. 2"**; IAWQ Task Group on Mathematical Modelling for Design and Operation of Biological Wastewater Processes; Scientific Report No. 3; Bourne Press Ltd.

[6]   Kosko, B. (1992): **"Neural Networks and Fuzzy Systems - A Dynamical Systems Approach to Machine Intelligence"** Prentice-Hall International Editions, New Jersey.

[7]   Microchip Technology Inc. (1994): **"fuzzyTECH®-MP Fuzzy Logic User's Guide"** Microchip Technology Incorporated, USA.

[8]   Kosko, B. (1994): **"Fuzzy Thinking"** ; Flamingo Publishers, London.

[9]   Zadeh, L.A. (1972): **"A Rationale for Fuzzy Control"** ASME Transactions Journal of Dynamic Systems, Measurement and Control, Vol. 94, pp. 3-4.

[10]   Zadeh, L. (1968): **"Fuzzy Algorithm"** Information and Control, Vol. 12, pp. 94-102.

[11]   M.F. Scheffer & I.S. Shaw (1994): **"Application using VLSI Hardware Realisation of Self-learning Recursive Fuzzy Model"** Proc. SICICA 94, Budapest.

[12]   E.H. Mamdani (1977): **"Application of Fuzzy Logic to Approximate Reasoning using Linguistic Synthesis"** IEEE Trans. Computers, Vol. C26, No. 12, pp. 1182-1191.

[13]   C.C. Lee (1990): **"Fuzzy Logic in Control Systems: Fuzzy Logic Controller - Part 1"** IEEE Trans. Man. Cybern., Vol. 20, No. 2.

[14]   M.F. Scheffer & I.S. Shaw (1994): **"VLSI Hardware Realisation of Self-learning Recursive Fuzzy Model for Dynamic Systems"** Proc. IFAC World Congress, Sydney.

[15]   J. Nie, A.P. Loh & C.C. Hang (1996): **"Modelling pH neutralisation processes using fuzzy-neural approaches"** Fuzzy Sets and Systems, Vol. 78, pp. 5-22.

[16]   B. Postlethwaite (1991): **"Empirical Comparison of methods of fuzzy relational identification"** IEE Proc. -D, Vol. 138, No. 3.

[17]   B. Postlethwaite (1991): **"Probabilistic fuzzy model in a noisy environment"** Internl. Pub. Dep. of Chem. and Process Eng., Univ. Strathclyde, Glasgow.

[18]   J.N. Ridley, I.S. Shaw & J.J. Kruger: (1988): **"Probabilistic Fuzzy Model for Dynamic Systems"** Electr. Letters, Vol. 24, No. 14.

[19]   C. Xu & Y. Lu (1987): **"Fuzzy model identification and self-learning for dynamic systems"** IEEE Trans. Syst. Man Cybrn., Vol. SMC-17, No. 4.

Optimal Pollution Control Models for Interceptor Sewers and Overflow Chambers

N. S. Thomas, A. B. Templeman *and* R. Burrows

ABSTRACT
A robust method has been described for the optimal pollution control of interceptor sewer systems possessing overflow chambers. The hydraulics of the interceptor sewer has been idealised as a slug flow approach, which enables a computationally efficient solution of global control actions. Overflow chambers have been included in the control model allowing the determination of control strategies for realistic interceptor sewer systems.

The results from the application of the optimal pollution control model on the northern leg of the Liverpool Interceptor Sewer have shown considerable reductions in pollutant load spilled when compared to traditional fixed local control.

The formulation of the optimal pollution control model enabled the inclusion of non-linear equations that governed the continuation flow through the overflow chambers between time step control solutions, which did not significantly reduce the computational efficiency of the model. Therefore, the approach offers promise for practical implementation of optimal real time control (RTC).

1 INTRODUCTION

Interceptor sewer systems are designed to reduce the environmental impacts ofurban drainage systems (UDS) by diverting the flows from existing combined sewer outfalls to Wastewater Treatment Works (WwTW). Most combined sewer overflows (CSOs), at the junction between interceptor sewer and sewer system outfalls, are operated locally with static operating rules, commonly to 'Formula A' consents in the UK. These systems have, in general, considerably improved the quality of the receiving waters, especially aesthetically since the overflow structures are often designed to maximise gross solid retention. However, spills from CSOs during storm events can still be significant and now form the dominant source of pollution in many watercourses. In fact, it has been estimated that the CSOs contribute about one third of the pollutant load to urban streams (Andoh, 1994).

Many interceptor sewers have the facility for flexible active and remote control but these facilities are only used to regulate local control actions (where only the local measurements are used by the controller). This type of operating procedure has deficiencies since the loading of the entire interceptor sewer varies temporally and spatially. This is due to the heterogeneity of rainfall, the variations in response characteristics of the sub-catchments and the temporal and spatial variations in dry weather flow (DWF). Therefore by adopting only local control, it is likely that there will be needless overspills when storage is available elsewhere in the system. Global control, where measurements taken across the system are used to operate the flow regulators throughout the system, enhances the efficiency of the control actions. Here, the control procedure would reduce the frequency of spills by allowing overflow (spill) only when the entire interceptor sewer storage is fully utilised (if this is physically achievable). This has been an active area of research for the last decade (notably Schilling, 1989, 1994 and 1996).

An improved control procedure (for environmental improvements) of interceptor sewers, and UDS in general, is active pollution control. The development of these control models has been hindered by the complexities of synthesising the pollutant concentrations temporally within the sewer flow. However, this control procedure determines control actions (or control strategies) that not only maximise the sewer volume utilisation but also maximise the pollutant load retention, i.e. only the least polluting sewage would be spilled and this only when the sewer was completely full (if this was physically achievable). This has become a topical area of research in the last few years, especially with the increasing pressure of imposed targets of frequency of spills in regulations often arising from EC Directives. An example of this type of research is Weinreich et al (1997) who investigated pollution-based real time control of combined sewer systems and applied it to an interceptor sewer system. Their linear programming formulation differs from the model now discussed here.

An optimal pollution control model has recently been developed at the University of Liverpool for interceptor sewer systems where linear (LP) or dynamic programming (DP) can be used to maximise pollutant load retention within the sewer. The formulation of the control model for idealised interceptor sewers using a slug flow approach has been presented in Thomas et al (1998 and 1999b). The results illustrated the viability of using LP or DP and that significant environmental improvements could be achieved, in terms of pollutant load reduction to the receiving waters, when compared to fixed local control procedures. The model assigned a varying pollutant concentration factor to each inflow to synthesise pollution load from the flow hydrographs. The control model objective therefore minimised pollutant load in the spills from the sewer system. In this model, sub-catchment inflows over and above the controlled interceptor inflows were assumed to spill without retention in overflow chambers.

The control models have been verified and validated against hydraulic criteria (Thomas et al, 1999a) utilising the WALLRUS (HRS, 1991) flow simulation package. For this a post-processing hydraulic verification routine was included in the model where a quasi-steady approach was used to develop approximate interceptor sewer water profiles using the Manning equation. Comparison against WALLRUS solutions demonstrated the adequacy of the slug flow approach in the control models.

This paper presents the extension of the original control models to include the secondary storage effects of CSO volume. The formulation of this extension is described and applied to a simplified version of the northern leg of the Liverpool Interceptor Sewer System. Both fixed and variable local control procedures are included for the purpose of comparison.

2 OPTIMAL CONTROL MODEL EXTENSION TO STORAGE OVERFLOW CHAMBERS

The original control models determine optimum interceptor inflow rates based on incoming pollutant concentrations to maximise pollutant load retention within the interceptor sewer. The model is formulated using a slug flow approach where the 'slugs' are tracked through the interceptor and the control model determines the amount of sewage that should be added from the individual catchments based on the appropriate time delays and their respective pollution loadings. This optimisation problem is solved using two procedures, Linear Programming (LP) and Dynamic Programming (DP). A detailed description of the model formulation and validation is presented in Thomas et al (1999b). In these models, flows in excess of the interceptor inflows were assumed to be spilled.

The control models have now been extended to include storage chambers at the intercept points. A typical chamber arrangement is shown in Figure 1.

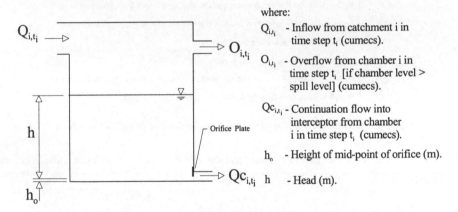

where:

Q_{i,t_i} - Inflow from catchment i in time step t_i (cumecs).

O_{i,t_i} - Overflow from chamber i in time step t_i [if chamber level > spill level] (cumecs).

Qc_{i,t_i} - Continuation flow into interceptor from chamber i in time step t_i (cumecs).

h_o - Height of mid-point of orifice (m).

h - Head (m).

Fig. 1   Typical chamber arrangement.

The continuation flow rate, Qc, into the interceptor is governed by the non-linear equation :

$$Qc = C_d a\sqrt{2gh} \tag{i}$$

where C_d is the coefficient of discharge of the orifice (dimensionless), a is the area of the orifice (m^2), g is the acceleration due to gravity (m/s^2) and h is the head (m). However, the optimal control model is formulated to determine the control strategies within discrete time steps and equation (i) can be solved between the time step solutions of the objective function.

The fundamentals of the extended optimal control model formulation are shown in Figure 2.

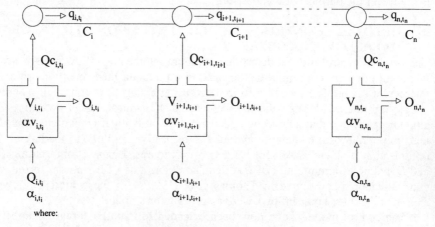

where:

q_{i,t_i} - Interceptor flow rate below chamber i in time step t_i (cumecs).

O_{i,t_i} - Overflow from chamber i in time step t_i [if chamber level > spill level] (cumecs).

$\alpha v_{i,t_i}$ - Chamber pollutant concentration factor in chamber i in time step t_i (dimensionless).

Q_{i,t_i} - Inflow from catchment i in time step t_i (cumecs).

V_{i,t_i} - Chamber volume in chamber i in time step t_i (m³).

α_{i,t_i} - Polluntant concentration factor of inflow from catchment i in time step t_i (dimensionless).

Qc_{i,t_i} - Throughflow into interceptor from chamber i in time step t_i (cumecs).

Fig. 2. Fundamentals of the extended optimal control model formulation.

The chamber pollutant concentration factor, $\alpha v_{i,t_i}$, is determined by the mixing model:

$$\alpha v_{i,t_i} = \frac{\alpha v_{i,t_i-1} V_{i,t_i-1} + \alpha_{i.t_i} Q_{i,t_i} \Delta t}{Q_{i,t_i} \Delta t + V_{i,t_i-1}}$$

(ii)

For computational convenience here the pollutant concentration factor is defined as a coefficient assigned to the inflow at each time step. For general illustration this coefficient can be considered to range from 0 to 1, i.e. 'absolutely clean' through to absolutely 'dirty' inflows, though more generally it might be the concentration (typically mg/l) of the chosen determinant.

Additionally, the control model maintains volumetric continuity within the chambers:

$$A \frac{\Delta h}{\Delta t} = Q_{i,t_j} - Qc_{i,t_j} \qquad \text{(iii)}$$

where A is the storage chamber area (m^2) and h is the chamber level (m). However, equation (iii) applies to conditions when the chamber level h is less than the spill level, which is normally the invert level of the overflow pipe. When the chamber level h is greater than this level then:

$$A \frac{\Delta h}{\Delta t} = Q_{i,t_j} - Qc_{i,t_j} - O_{i,t_j} \qquad \text{(iv)}$$

The overflow term also has to be included in the chamber pollutant concentration factor mixing model (ii) under these conditions to maintain continuity.

In application, the initial state of the interceptor sewer system is assumed to be known. During each time step, therefore, the control model adds the inflow volume $Q_{i,t_i} \Delta t$ to the known chamber volume $V_{i,t_{i-1}}$ to obtain the possible chamber retention volume $V \max_{i,t_i}$, assuming that the entire inflow volume is retained within the CSO chamber. The corresponding chamber level h for this volume is used in the calculation of (i) to determine the maximum possible inflow rate $Qc \max_{i,t_i}$ when the orifice is completely open, i.e. when a is at a maximum in (i). Additionally, the chamber pollutant concentration factor $\alpha v_{i,t_i}$ is calculated from (ii) for this volume $V \max_{i,t_i}$, again assuming that it is completely retained within the CSO chamber. These values are used within the optimisation routine where the objective function is solved within the appropriate capacity constraints to determine the optimum control strategy:

$$\text{Max} \sum_{i=1}^{n} \alpha v_{i,t_j} q_{i,t_j} \qquad \text{(v)}$$

subject to:

$$q_{i,t_i} \leq Qc \max_{i,t_i} \qquad \forall i \qquad \text{(vi)}$$

$$\sum_{j=1}^{i} q_{j,t_j} \leq C_i \qquad \forall i \qquad \text{(vii)}$$

where:-

n - number of intercept points;

t_i - time step position within the interceptor of intercept point i;

$\alpha v_{i,t_i}$ - pollutant concentration in chamber i in time step t_i;

q_{i,t_i} - interceptor sewer flow rate below intercept point i in time step t_i;

$Qc\max_{i,t_i}$ - maximum inflow into interceptor from chamber i in time step t_i;

C_I - interceptor sewer pipe full capacity below intercept point i.

The actual continuation flows Qc_{i,t_i} that satisfy the objective function (v) are then calculated. From these values the orifice area a is calculated implicitly from (i) to determine the control action of the flow regulator (orifice gate or penstock). The chamber pollutant concentrations $\alpha v_{i,t_i}$ and chamber volumes V_{i,t_i}, as a consequence of the control strategy, are determined from (ii) and (iii) respectively. These values now represent the state of the CSO before the next time step in the solution procedure.

In the next time step, the procedure again adds the subsequent inflow volume $Q_{i,t_{i+1}} \Delta t$ to the stored volume V_{i,t_i} to determine the maximum possible continuation flow $Qc\max_{i,t_{i+1}}$ in this time step. The respective pollutant concentration factor is mixed with the chamber pollutant concentration in (ii). These values are then used within the optimisation routine, for LP (v), (vi) and (vii). This continues until all discrete time steps solutions have been determined within the control time horizon.

The objective function (v) represents decisions to be made that maximise the pollutant load received by a slug of sewage travelling through the interceptor incrementing inflows from the CSOs with the highest pollutant concentrations. However, equation (v) corresponds to only one time step t_i and the control strategies throughout the control time horizon are determined by altering the time step position (i.e t_i+1, t_i+2, ...etc.). Since, in this approach, successive 'slugs' of water are assumed not to interact, then the sequence of optimal controls derived for each time step also represents the optimal control strategy for the entire event.

Overall, the optimisation is little changed from the original control model but the effects of storage (in the overflow chambers) on the pollutant concentration factors and inflow hydrographs are now accounted for. In this extended model formulation the control strategies are governed by the mixed pollutant concentrations in the storm chambers not the pollution concentration of the inflow hydrographs as in the original model.

3. TEST CASE – NORTHERN LEG OF LIVERPOOL INTERCEPTOR SEWER SYSTEM

The northern leg of the Liverpool Interceptor Sewer has been simplified and used as a test case for the extended optimal pollution control model. A longitudinal section of the sewer can be seen in Figure 3 and the input data for this sewer is shown in Table 1. The input data for the overflow chambers at the intercept points is shown in Table 2.

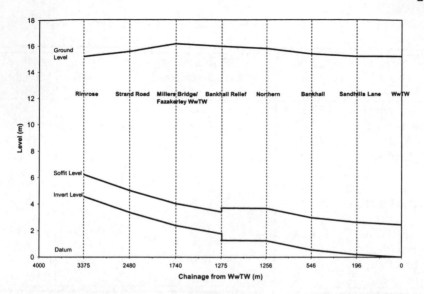

Fig. 3. Longitudinal section of the simplified northern leg of the Liverpool Interceptor Sewer (not to scale).

Table 1 Input data for test interceptor sewer.

Intercept Point (catchment)	Sewer Diameter (m)	Sewer Gradient	Sewer Capacity (cumecs)	D.W.F. (cumecs)	Fixed Inflow Setting (cumecs)
RIMROSE	1.66	1/750	3.26	0.30	1.24
STRAND RD	1.66	1/750	3.26	0.09	0.25
MILLERS BRIDGE/ FAZAKERLEY WWTW	1.66	1/750	3.26	0.04	0.97
BANKHALL RELIEF	2.44	1/1000	7.72	0.14	0.69
NORTHERN	2.44	1/1000	7.72	0.50	2.13
BANKHALL	2.44	1/1000	7.72	0.11	0.29
SANDHILLS LANE	2.44	1/1000	7.72	0.09	0.31

Table 2 Storm chambers input data for test interceptor sewer.

Intercept Point (catchment)	Chamber Area (m²)	Spill Level [above invert level] (m)	Orifice Width (m)	Orifice Height (m)
Rimrose	282.82	5.42	1.250	1.450
Strand RD	136.03	6.91	1.700	0.625
Millers Bridge/ Fazakerley WwTW	50.31	7.95	0.354(E)	0.354(E)
Bankhall Relief	169.78	8.04	2.075	0.625
Northern	328.24	8.18	2.650	1.450
Bankhall	167.06	8.47	1.800	0.625
Sandhills Lane	147.95	9.26	1.650	0.625

(E) – Equivalent dimensions.

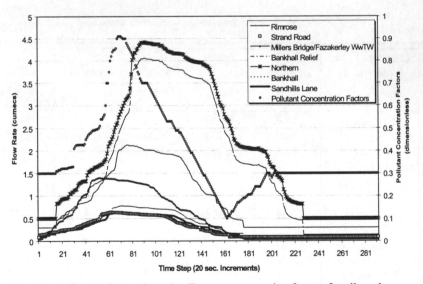

Fig. 4   Inflow hydrographs and pollutant concentration factors for all catchments.

The control model was run with hypothetical runoff hydrographs (Figure 4), which were loosely based on the catchment's response characteristics, and the respective pollutant concentration factors (Figure 4). For computational convenience the pollutant concentration factors were taken to be identical for each catchment. Two control procedures were considered in the test case illustrated here: fixed local control (FLC), where inflows up to the fixed inflow setting are passed forward and no account is taken of the interceptor sewer's pollutant concentrations; and optimal pollution control (OPC), where the optimal control model uses global information including pollutant concentrations within the interceptor and the storage chambers in making its decisions. Both control procedures use the slug flow approach within the interceptor sewer to convey the inflows along the interceptor and only the decision criteria differ.

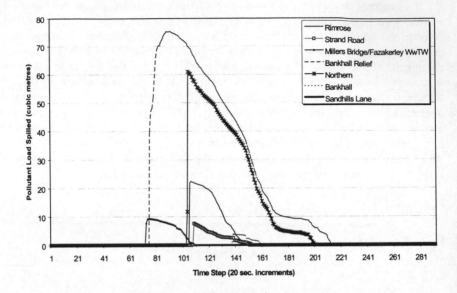

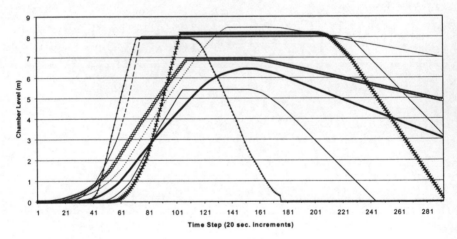

Fig. 5 Fixed local control (FLC) chamber levels (bottom) and pollutant load spilled (top) – common legend. (Pollutant load spilled = spill volume × pollutant concentration)

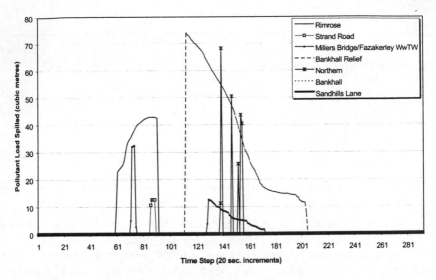

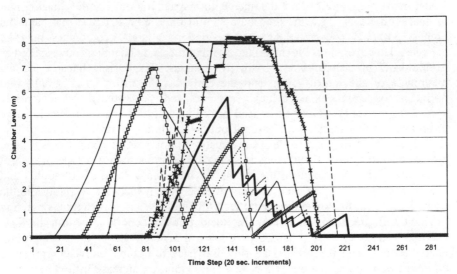

FIG. 6 Optimal pollution control (OPC) chamber levels (bottom) and pollutant load spilled (top) – common legend. (Pollutant load spilled = spill volume × pollutant concentration)

Table 3 Comparison of control procedures.

Pollutant Load to Receiving Waters (Spill Volume × Pollutant Concentration)			Improvement (OPC v FLC) %	Improvement (OPC v VLC) %
Fixed Local Control (FLC)	Variable Local Control (VLC)	Optimal Pollution Control (OPC)		
8860.06	6203.25	5189.48	41.43	19.54

Figures 5 and 6 show the variations of chamber levels and pollutant load spilled from each overflow chamber along the northern leg of the Liverpool Interceptor using the FLC procedure and the OPC procedure. They show that the OPC procedure reduced the pollutant load spilled compared to the FLC procedure. Under moderate rainfall where the interceptor capacity is only slightly insufficient to deal with flows, the OPC procedure should only spill at CSOs that have the highest pollutant concentrations so a reduction in spill occurrences may be expected. However, it is evident in this case that the same chambers have spilled as in the FLC procedure, but at reduced volumes. This arises because of the magnitude of the run-off event and the fact that the inflow hydrographs and pollutant concentrations were highly synchronised. A spatially and temporally varied storm would generate localised peaks in pollutant load that would more clearly show the potential benefits of the optimal pollution control model.

Figures 5and 6 illustrate the effect of using optimal active control in that there are considerably more fluctuations in storage conditions with this type of control procedure. This is evident in the results for both the pollutant load spilled and chamber levels. This is because the control procedure responds to variations in the operative state of the sewer system through time rather than using a fixed operating procedure. Figure 5 does not show the fluctuations in conditions within the CSO chambers and all the chambers follow similar 'filling' and 'emptying' characteristics. This is, perhaps, the only drawback of optimal active control since an increase in control actions will have a corresponding increase in flow regulator activity. It is likely that an increase in regulator activity would increase the frequency of operating problems. Therefore, a compromise is needed where constraints are used within the optimisation to reduce the activity of the flow regulators to within acceptable limits even though the resulting control solution would then be sub-optimal. The inclusion of such additional constraints is a topic of current research.

Figure 6 shows that the OPC procedure improves the chamber recovery times compared to the FLC procedure. In fact, Figure 5 shows that the control time horizon selected for the calculations was not of sufficient length in this case since some of the chambers had not fully recovered from the storm event. This shows an additional benefit of using optimal active control. The OPC procedure should be better able to control multiple-peak inflow hydrographs or pollutographs compared to the traditional FLC procedure.

Table 3 shows a comparison between the FLC and OPC procedures. In this application of a reasonably low intensity highly synchronised storm event (i.e. peak inflow to the system at approximately 2 times fixed inflow settings), the OPC procedure produced a 41% improvement in pollutant load spilled compared to the FLC procedure. An additional result has been included in Table 3, arising from adoption of variable local control VLC, where interceptor inflows are permitted up to the interceptor sewer's capacity locally but no account is taken of pollutant concentrations. In this case, the interceptor sewer volume is fully utilised, if physically possible. The OPC procedure produced a 20% improvement compared to VLC. The results from the VLC procedure (not presented) show that only the downstream CSO chambers spill in each of the two sewer sections. This is because the VLC procedure fills the interceptor as soon as possible in the upstream section of the sewer. Other results (which have not been presented) show that, generally, the improvements of using the OPC procedure reduce as the severity of the storm event increases. This is expected because spills are inevitable with larger inflows where the interceptor sewer and overflow chamber storage are quickly utilised. It is likely that most improvements would be encountered when the hydrographs and corresponding pollutant concentrations vary spatially and temporally in low to moderate rainfalls that occur more frequently.

The inclusion of overflow chambers has increased the computational demand of the optimal pollution control model. This is because equation (i) is solved implicitly between time step solutions to determine the actual control strategies on the penstocks. This is obviously sensitive to the numerical method used and an increase in computationally efficiency may be achievable with the use of other numerical methods; this is also a topic of ongoing research. The effect is not, however, considered to be too prejudicial since the model runs considerably faster than real time.

4   CONCLUSIONS

A robust method has been described for the optimal pollution control of interceptor sewer systems that include overflow chambers. The hydraulics of the interceptor sewer has been idealised by a slug flow approach, which allows for a computationally efficient solution of global control actions. Overflow chambers have now been included in the control model to allow control strategies to be determined for more realistic interceptor sewer systems.

The results from the application of the optimal pollution control model on the northern leg of the Liverpool Interceptor Sewer have shown considerable reductions in pollutant load spilled when compared to traditional fixed local control.

Overall, the formulation of the optimal pollution control model enabled the inclusion of non-linear equations that governed the continuation flow through the overflow chambers between time step control solutions, which did not significantly reduce the computational efficiency of the model. Therefore, the approach offers

278

promise for practical implementation of optimal real time control (RTC). As a final step towards practical implementation, efficient methodologies will later be used for the simulation of inflows from rainfall (Mehmood, 1996) and for specification of time-varying pollutant concentrations (Gupta, 1995).

ACKNOWLEDGEMENTS
This paper presents the findings from an ongoing PhD programme at the Department of Civil Engineering, University of Liverpool. The authors would like to thank North West Water Limited and Liverpool City Engineers Department for their assistance in the supply of data relating to the Liverpool Interceptor Sewer System.

REFERENCES
Andoh, R. Y. G. (1994): Urban Runoff: Nature, Characteristics and Control. *JIWEM*, **8**, (4), 371-378.

Gupta, K. (1995): *A Methodology to Predict the Pollutant Loads in Combined Sewer Flow*. PhD thesis, Department of Civil and Structural Engineering, University of Sheffield.

HRS (1991): WALLRUS User's Manual. Wallingford Procedure Software, 4th Edition, Hydraulics Research Station, Wallingford, UK.

Mehmood, K. (1996): *Studies on Sewer Flow Synthesis with Special Attention to Storm Overflows*. PhD thesis, Department of Civil Engineering, University of Liverpool.

Schilling, W. (1989): *Real-Time Control of Urban Drainage Systems. The State of the Art*. Scientific and Technical Reports No.2. IAWPRC.

Schilling, W. (1994): Smart Sewer Systems: Improved Performance by Real Time Control. *European Water Pollution Control*, **4**, 24-31.

Schilling, W., Andersson, B., Nyberg, U., Aspegren, H., Rauch, W. and Harremöes, P. (1996): Real Time Control of Wastewater Systems. *Journal of Hydraulic Research*, **34**, 785-797.

Thomas, N., Templeman, A. B. and Burrows, R. (1998): Optimal Control Models for Interceptor Sewer Systems. *Fourth International Conference on Developments in Urban Drainage Modelling*, London, September 1998, 691-698.

Thomas, N., Templeman, A. B. and Burrows, R. (1999a): Optimal Pollution Control of Large Interceptor Sewer Systems. *Submitted for Eighth International Conference on Urban Storm Drainage*, Sydney, Australia, August 1999.

Thomas, N., Templeman, A. B. and Burrows, R. (1999b): Pollutant Load Overspill Minimisation of Interceptor Sewer Systems. *Submitted to Engineering Opimization Journal*, 1999.

Weinreich, G., Schilling, W., Birkely, A. and Moland, T. (1997): Pollution Based Real Time Control Strategies for Combined Sewer Systems. *Water, Science and Technology*, **36**, *(8-9)*, 331-336.

Optimal Location and Scheduling of Disinfectant Additions in Water Distribution Networks

J. G. Uber

ABSTRACT

Three optimization model formulations are presented for determining the locations of booster disinfectant additions in water distribution networks, and for determining their time–dependent dosage rates. Two of the formulations seek to minimize the total disinfectant mass required in order to maintain disinfectant residuals over time at selected monitoring locations. The last formulation seeks to minimize the number of booster stations required to achieve adequate spatiotemporal residual "coverage", drawing on previous work on maximum set–covering models. This latter model has the advantage of being particularly efficient. All models essentially embed an extended period hydraulic and water quality model within the constraint set of a linear optimization model, using the principle of linear superposition. Linear superposition holds so long as the time varying hydraulics are assumed to be known, and disinfectant decay is first order. A brief set of results is presented for one of the models, to suggest that it is computationally feasible for real networks.

1. INTRODUCTION

Water utilities and regulatory agencies often desire a detectable disinfectant residual at all points of consumption in a water distribution network. Such a residual can be difficult to maintain when disinfectant is added at a single location such as a treatment plant, because the spatially distributed nature of water storage facilities and consumer demands leads naturally to a wide distribution in water travel times. Since all disinfectants that can maintain a residual will react with a variety of inorganic and organic compounds, they decay over time; thus any single disinfectant addition must be sufficient to maintain a detectable residual for the longest travel time. This requirement can lead to unacceptably high residuals close to the point of addition and to an uneven spatiotemporal residual distribution. Both these may be undesirable from the perspective of unpleasant tastes and odors or public health (e.g., unpleasant tastes of excessive chlorine or chlorinated organic compounds, or health impacts of chlorinated by–products [Bull and Kopfler, 1991]).

To address the difficulties of residual maintenance in distribution systems, one may decide to distribute injections of disinfectant to strategic locations in the distribution network. This practice of "booster disinfection" has been practiced by utilities for some time. It can be difficult, however, to determine: i) the best number and locations of disinfectant additions, and ii) the best way to schedule or control the dose at each location over time. These difficulties arise principally from complex dynamic hydraulics which affect disinfectant transport, coupled with decay of disinfectant due to bulk and wall reactions.

This chapter presents mathematical programming formulations (optimization models), and an associated design approach, for locating and scheduling disinfectant additions in distribution networks. Three formulations are presented. The first formulation attempts to identify optimal disinfectant dosage rates at booster stations — in the sense that the total disinfectant dosage is a minimum — by assuming that the number of booster stations and their locations are known. The resulting linear programming (LP) problem can be solved by the standard simplex algorithm. The second formulation combines the function of the first formulation with determining the best locations of the booster stations from among a set of potential locations. This formulation is a mixed–integer linear programming (MILP) problem with no particular structure, and is much more difficult to solve using standard techniques. The third formulation shows how a pure optimal location problem, which minimizes the number of booster stations such that the ratio of the minimum to maximum residual concentrations is controlled, can be expressed as a particular type of integer linear programming (ILP) problem called the maximum set–covering problem. It is a great advantage to do so, because such ILP problems can often be easily and efficiently solved using the simplex method or heuristics.

All the above formulations share the following characteristics: i) they assume known but time–varying network hydraulics, and first–order disinfectant decay kinetics, ii) they consider explicitly the dynamic hydraulic behavior in the network, for a goal of adequate long–term operation, iii) they embed the information and assumptions contained within standard extended–period network water quality models within linear optimal design models, and iv) with the possible exception of the MILP formulation, they can be solved in reasonable computer time for practical networks.

Following presentation of the model formulations, solution approaches are discussed along with a suggested design approach which uses the optimization model formulations in concert, such that the maximum set–covering model is used as a screening model to select locations for use by either the optimal location and scheduling (MILP) or optimal scheduling (LP) models. Available software to facilitate development and solution of these models is also discussed. Finally, computational results suggest that the methods can be used to locate booster stations in real networks.

2. BACKGROUND I – LINEAR SUPERPOSITION

In a water distribution system, the monitoring and booster locations will coincide with nodes of a dynamic hydraulic and water quality network model (see, e.g., Liou and Kroon, 1987, Grayman et. al., 1988, Rossman et. al., 1993, Boulos et. al., 1995, Rossman and Boulos, 1996). The disinfectant concentration at monitoring node j and time period m, denoted $c_j^m(u)$ [M/L^3], will, in general, depend on the disinfectant mass injection rate at booster node i and time period k, denoted u_i^k [M/T], for all booster locations i and time periods $k \leq m$ (u is the vector of all such mass injection rates). This dependence can be complicated, owing to significant time delays for transporting water (and thus disinfectant) between booster and monitoring nodes, and to multiple transport paths between any pair of booster and monitoring locations. Under certain assumptions that are common to network water quality simulation models, the principle of linear superposition can be applied to express the concentration $c_j^m(u)$ as a *linear function* of the mass injection rates u. This linear relationship between injection rates and concentration is a powerful tool for developing practical optimal design formulations, and it underlies each optimization model formulation discussed in this chapter; thus it is worthwhile to discuss briefly how linear superposition can be applied to modeling chlorine concentration dynamics resulting from multiple booster station dosages.

We assume: 1) the water distribution system hydraulics are known and time varying, 2) disinfectant decay kinetics are first–order with respect to disinfectant concentration, and 3) reaction rate coefficients are independent of the booster injections u. Boccelli et al. [1998] show that, under these conditions, disinfectant transport in water distribution networks with time–varying hydraulics and disinfectant doses can be described mathematically as a linear dynamic system; thus the principle of linear superposition is applicable (e.g. Luenberger, 1979, pp. 108–112). They also give a more intuitive explanation of linear superposition by way of a simple network example. Using a different approach, Zierolf et. al. (1997) derived a recursive expression for $c_j^m(u)$ by back–tracking through the distribution system over all transport paths to find the superimposed impact of all disinfectant sources. These works show that, for any arbitrarily complicated distribution system, the monitoring concentrations $c_j^m(u)$ can be defined as a linear summation of individual booster injection influences:

$$c_j^m(u) = \sum_{i=1}^{n_b} \sum_{k \leq m} a_{ij}^{km} \cdot u_i^k \qquad (1)$$

where $a_{ij}^{km} = \partial c_j^m / \partial u_i^k$ $((M/L^3)/(M/T))$ corresponds to the coefficients of the discretized impulse response function (Chow et. al., 1988, pp. 204–213; Oppenheim and Willsky, 1997, pp. 77–90), describing the effect of dose u_i^k on the concentration at monitoring node j and time m. The impulse response coefficients can be calculated via network water quality simulation software (see below); once computed, eq. (1) is an accurate mathematical statement of the effect of changes in u on residual concentrations and, because of its simple form, allows development of efficient mathematical models to optimize dose magnitudes and their locations.

3. BACKGROUND II – THE USE OF DYNAMIC NETWORK WATER QUALITY MODELS IN A PLANNING CONTEXT

To satisfy engineering objectives, the booster disinfection dosages should maintain adequate disinfectant residual for all times, much as the hydraulic infrastructure should be designed to always maintain minimum pressures. Nevertheless, in a planning context, assumptions about the *future variation* in: 1) dose injections, and 2) network hydraulics are needed to construct practical models for locating and scheduling booster dosages. Such assumptions are similar in concept to adopting peak demand plus fire flow as a basis for design of hydraulic networks; one assumes that networks designed on that basis will operate in an acceptable manner despite unknown future disturbances (i.e. the actual demands will not, in general, ever equal the demand scenario assumed for design, but the assumed scenario is nevertheless useful as it leads to acceptable designs). The essential problem is, when using dynamic models as a basis for design (as opposed to steady–state models), one must treat the initial conditions carefully, for although the dynamic water quality response to booster dosing must be reflected in the design, these dynamics can not be based on *any* arbitrary initial conditions. The idea is to make certain that the water quality dynamics are *periodic*, and then base the design on a representative "snapshot" of those periodic dynamics. These conceptual design issues are discussed more fully below, after presentation of the design assumptions and conditions.

We assume that discrete mass injection rates are periodic, so that the mass dosage variation at each booster station has a cycle time $T_s = n_s \Delta t$, where n_s is the number of injection rates contained in one scheduling cycle, and Δt is the duration of one discrete mass injection period (cycle time is analogous to wavelength). Such an assumption defines periodic mass dose rates v_i^k such that $v_i^k = u_i^{k-qn_s}$, $\forall q$, $k = 1, \ldots, n_s$. Thus eq. (1) becomes:

$$c_j^m(v) = \sum_{i=1}^{n_b} \sum_{k=0}^{n_s-1} \alpha_{ij}^{km} v_i^k, \tag{2}$$

where the composite impulse response coefficients α_{ij}^{km} quantify the response of concentration at a monitoring location to a unit periodic dose at a booster station.

If, in addition to injection rates, the α_{ij}^{km}, $\forall i, k$, are periodic with an impulse response cycle time $T_a = n_a \Delta t_m$ where n_a is the number of monitoring times contained in one impulse response cycle and Δt_m is the disinfectant residual monitoring time step, then the monitoring concentrations c_j^m are themselves periodic with a cycle time T_a (see eq. 2). Accordingly, the composite impulse response coefficients are assumed periodic with cycle time T_a, in which case it is sufficient for design purposes to consider only one cycle of these dynamics (this assumption, as well as the conditions under which α_{ij}^{km} is periodic, is discussed below). Such a periodicity assumption is one practical solution to the indeterminacy inherent in many long–range planning problems, where the underlying processes are dynamic and suggest a strong cyclical component.

If the optimal disinfectant concentrations are to be periodic, then other system characteristics must be consistent with such behavior. First, we have already discussed that optimal disinfectant doses are assumed periodic with a cycle time T_s. Intuitively, the monitoring node concentrations could not be periodic if a key driving force – the dose schedules – were not. Second, the hydraulic dynamics obviously play an important role in the resultant concentration dynamics. Specifically, we assume periodicity of network hydraulic dynamics, with cycle time T_h (i.e., the flow rates, tank elevations, and pressures are periodic with cycle time T_h). Periodic network hydraulics would, on a practical level, require periodic water demands at the network nodes. Of course, the actual demands will differ from those assumed for design, and thus the design, if implemented, will perform somewhat differently from model predictions. Again, there is no way completely around this difficulty; assumptions about the future are a routine and necessary part of every long–term design or planning problem.

Boccelli et al. [1998] show that the monitoring node concentrations, and the composite impulse response coefficients, have a cycle time $T_a = \eta T_s = \mu T_h$ for some integers $\eta, \mu > 0$. Thus long–term periodicity of monitoring node concentrations requires long–term periodicity of both dose schedules and assumed hydraulic dynamics, and the relationship between the independent cycle times T_s and T_h is sufficient to determine the cycle time T_a. To take a specific example, if hydraulic dynamics are periodic on a 24 hour cycle, and booster dosages are periodic on a 12 hour cycle, then the residual concentrations will become periodic on a 24 hour cycle ($\eta = 2; \mu = 1$).

By achieving periodic disinfectant concentrations, assuming periodic dose schedules and network hydraulics, important dynamic processes are allowed to influence the optimal schedule while maintaining reasonable data requirements. This point is emphasized by considering other plausible assumptions about the disinfectant concentration dynamics, and the hydraulic forcing that drives those dynamics. For example, one might assume steady–state concentrations forced by steady–state hydraulics. Arguably, the assumptions of dynamic concentrations forced by dynamic hydraulics – even with the pragmatic assumption of periodicity – lead to a more realistic representation of disinfectant transport and decay in water distribution systems, a point emphasized by some of the results discussed by Boccelli et al. [1998].

4. OPTIMAL SCHEDULING OF BOOSTER STATION DOSAGES AS A LINEAR PROGRAMMING (LP) PROBLEM

In this section we present a formulation for optimization of booster dose schedules, defined by the periodic dose rates v_i^k, $i = 1,\ldots,n_b$, $k = 1,\ldots,n_s$. The locations of n_b booster stations are assumed known. The design objective is to minimize the total disinfectant mass rate applied over one period of the concentration dynamics. This objective is intended as a surrogate for minimizing disinfection by–product (DBP) formation, chemical costs, and objectionable tastes and odors associated with chlorination of natural waters. The mass injections are required to satisfy lower and upper bound constraints on disinfectant residual at n_m monitoring locations and over all time. These constraints are consistent with environmental regulations calling for a detectable residual at all points of consumption, to serve as a barrier against microbiological contamination of the distribution system.

The model formulation is stated as the following linear programming problem:

$$\min_v \sum_{i=1}^{n_b} \sum_{k=1}^{n_s} v_i^k \tag{3}$$

subject to,

$$c^{\min} \le c_j^m(v) = \sum_{i=1}^{n_b} \sum_{k=1}^{n_s} a_{ij}^{km} \cdot v_i^k \le c^{\max} \tag{4}$$

$$j = 1,\ldots,n_m, \quad m = 1,\ldots,n_a$$

$$v_i^k \ge 0, \quad i = 1,\ldots,n_b, \; k = 1,\ldots,n_s \tag{5}$$

where $c^{\min}$ and $c^{\max}$ are the minimum and maximum concentration limits within the distribution system, and n_a is the number of monitoring time periods contained in one period of the concentration dynamics.

5. OPTIMAL LOCATION AND SCHEDULING OF BOOSTER STATION DOSAGES AS A MIXED–INTEGER LINEAR PROGRAMMING (MILP) PROBLEM

This formulation extends the above optimal scheduling model to consider also the optimal booster station locations. The extension of eqs. 3–5 yields a mixed integer linear programming (MILP) problem (Hillier and Lieberman, 1980). Once again the objective is to minimize the total mass dose rate during one period of the concentration dynamics:

$$\min_{v,\delta} \sum_{i=1}^{\pi_b} \sum_{k=1}^{n_s} v_i^k \tag{6}$$

subject again to restrictions on the concentrations at monitoring nodes, over all time:

$$c^{\min} \leq c_j^m(v) = \sum_{i=1}^{\pi_b} \sum_{k=1}^{n_s} a_{ij}^{km} \cdot v_i^k \leq c^{\max} \tag{7}$$

$$j = 1,\ldots,n_m, \quad m = 1,\ldots,n_a$$

Binary variables δ_i are introduced to determine whether a new booster station is (δ_i = 1) or is not ($\delta_i = 0$) to be built at location i. If there is no booster station at location i, then additional constraints must ensure that the total mass dose is zero at that location:

$$v_i^k \leq M_i \cdot \delta_i, \quad i = 1,\ldots,n_b, \quad k = 1,\ldots,n_s \tag{8}$$

where M_i is a positive constant equal to an upper bound on the mass dose at booster location i, equal to the maximum dose for which an entering concentration of $c^{\min}$ would exit at concentration $c^{\max}$, which is determined from the total flow rate exiting location i over time. Thus when $\delta_i=1$ the total dose at location i is unrestricted, while when $\delta_i=0$ the total dose is constrained to equal zero. This "big M" formulation is common in facility location problems or "fixed–charge" problems in general, and is used because it preserves the linearity of the formulation.

A restriction is placed on the total number of booster stations to be built, $\bar{n}_b < n_b$, which is a surrogate for the cost of installing booster stations:

$$\sum_{i=1}^{n_b} \delta_i \leq \bar{n}_b \tag{9}$$

Finally, the variables δ_i are restricted in value to zero or one, and again the dose rates must be positive:

$$\delta_i = \{0, 1\}, \quad i = 1,\ldots,n_b \tag{10}$$

$$v_i^k \geq 0, \quad i = 1,\ldots,\bar{n}_b, \quad k = 1,\ldots,n_s \tag{11}$$

The above MILP problem (eqs. 6–11) can be solved by the branch and bound technique with the simplex method (e.g., Hillier and Lieberman [1980]). A minor modification of the formulation allows consideration of booster stations that already exist, combined with new potential locations, so that the operation of pre–existing facilities can be optimized simultaneously with the location of new ones. This MILP model is significantly more difficult to solve, however, than the LP scheduling model (solution of the scheduling model is essentially incorporated as a sub–task of the branch–and–bound procedure for solution of the MILP location model). In practice it may be that the MILP model can only be applied for analysis of large networks if a relatively small set of potential locations has been determined by experience or other means (i.e., if n_b is small — much smaller than the total number of network nodes). This computational concern leads us to the following formulation of the location model as a maximum set covering model, which might be used as a method of locating booster stations in its own right, or as a screening tool to select potential locations for the MILP formulation.

6. OPTIMAL LOCATION OF BOOSTER STATIONS AS A MAXIMUM SET COVERING (MSC) PROBLEM

Here we show how the booster station location problem can be formulated as a classical integer linear programming (ILP) model called the maximum set covering model (MSC). Such a formulation brings with it significant computational advantages for medium to large networks; it may be possible to consider all nodes as potential booster locations, while maintaining a computationally tractable model. These computational advantages require that we give up optimization of dose schedules, and instead treat all booster stations as if they are operated in a typical fashion. So, unlike the MILP formulation, the MSC formulation does not build upon the optimal scheduling (LP) model directly, and thus does not combine the two functions of optimal scheduling and optimal location. We suggest the MSC model can be used in concert with either the optimal scheduling or optimal scheduling/location model formulations presented earlier.

The MSC formulation draws upon results from facility location models, and in particular the maximal covering location model [Church and Revelle, 1974], which has also provided inspiration for models that locate wells in groundwater monitoring networks [Meyer and Brill, 1988]. The basic MSC booster station location model is relatively simple to state mathematically; we do this first, leaving some of the details and motivation for later. Define a set of indices of potential booster locations, N_j^m, such that booster location i is included in N_j^m if, and only if, disinfectant dosing at i elicits a "significant response" at monitoring location j and time period m. Thus N_j^m is the set of indices of all potential booster locations that are assumed to "cover" monitoring location j and time m. We discuss the precise definitions of a "significant response" and the sets N_j^m below; for now, it can be assumed that if a booster location elicits a "significant response" at a monitoring location and time then, under reasonable conditions for operation of that booster station, an acceptable disinfectant residual can be maintained at that location and time by operation of only that one booster station.

Again define the binary variables δ_i to indicate if a booster station is ($\delta_i = 1$) or is not ($\delta_i = 0$) to be constructed at location i. We seek to minimize the number of booster stations:

$$\min_{\delta} \sum_{i=1}^{n_b} \delta_i \tag{12}$$

where n_b is again the number of potential booster station locations, but likely a much larger set than would be allowed by the MILP formulation. In minimizing the number of stations, we require that each monitoring location and time include a significant response from at least one selected booster location (or, be "covered" by a location):

$$\sum_{i \in N_j^m} \delta_i \geq 1, \quad j = 1, \ldots, n_m, \quad m = 1, \ldots, n_a \tag{13}$$

Where as with the above LP and MILP models, the constraints (13) are written for one cycle of the periodic residual concentration dynamics. Solution of (12) subject to (13) will yield the minimum number of booster locations that provide spatial and temporal residual coverage, in that the residual concentration at each monitoring location and time is influenced by at least one selected location.

6.1. Coverage Sets and Residual Coverage Parameter

The coverage set N_j^m is defined, more precisely, to include booster location i if, and only if, a *unit set point concentration* at location i elicits a residual concentration response exceeding a threshold, $0 < \eta \leq 1$, at monitoring location j and time period m. The reason for specifying the response in terms of a unit set point concentration is to decouple the decision about the optimal dose schedules from that about the optimal dose locations; if this decoupling is to be accomplished, then one must assume a logical scenario for operation of the booster stations. Here, we define such a scenario as that where each potential booster station is operated identically to produce a constant unit exit concentration from the booster station.

The coverage sets can now be defined in terms of the impulse response coefficients α_{ij}^m:

$$N_j^m = \{i \mid \alpha_{ij}^m \geq \eta\} \tag{14}$$

where the impulse response coefficients α_{ij}^m are defined in a similar fashion as in section 3, except they indicate the periodic concentration response at monitoring location j and time m due to a particular periodic mass dosage at booster location i *such that a unit set point concentration is maintained* at location i at all times.

The *residual coverage parameter, η,* is an important design parameter of the MSC model. Note that we could just as well define the coverage sets as follows:

$$N_j^m = \{i \mid c_j^m \geq c^{\min}\} \tag{15}$$

where c_j^m is the concentration that results from maintaining a set point concentration at booster station location i, and $c^{\min}$ is again the minimum disinfectant residual allowed or desired. Further, the maximum residual concentration in the network is the set point concentration maintained at the booster locations, because of disinfectant decay. Since all booster stations are assumed to be operating identically (at the same exit set point concentration), let us call this set point concentration $c^{\max}$. Linear superposition lets us define the concentration c_j^m in (15) in terms of the impulse response coefficient α_{ij}^m: $c_j^m = \alpha_{ij}^m c^{\max}$; substitution in (15) and rearranging yields:

$$N_j^m = \{i \mid \alpha_{ij}^m \geq c^{\min}/c^{\max}\} \tag{16}$$

Thus, comparing eqs. (14) and (16), the residual coverage parameter $\eta = c^{\min}/c^{\max}$ has a meaningful interpretation as the ratio of the minimum to maximum residual concentration in the network — arguably a useful measure of residual variability.

The residual coverage parameter will be a key determinant of solutions to the MSC booster location model. As a thought exercise, one may consider a network where every node (and thus every monitoring location) is a potential booster location. The optimal MSC solutions for the extreme values of $\eta = \epsilon$, where ϵ is an arbitrarily small strictly positive constant, and $\eta = 1$ are known immediately. When $\eta = \epsilon$ a minimal set of booster locations is comprised of those associated with network water sources (this is the conventional case where disinfectant is added only at treatment nodes). When $\eta = 1$ a minimal set of locations is every node in the network, which is the only way to have a perfectly uniform concentration at each monitoring node. Thus η will evidently have a marked effect on the number of booster stations required to achieve "coverage" of the monitoring locations and times, and a plot of the optimal number of booster stations (from solution of the MSC problem) versus the residual coverage parameter indicates the optimal tradeoff between variability of disinfectant residual and a measure of residual maintenance cost.

7. SOLUTION OF THE OPTIMIZATION MODELS

A critical step in solution of each optimization model described above is calculation of the appropriate impulse response coefficients. Further, these impulse response coefficients must convey the influence of a periodic dosage on the resulting periodic concentration dynamics. We stress that, as was discussed earlier, hydraulic and disinfectant dose periodicity is a necessary condition for long–term residual concentration periodicity, yet little can be said about the *time* required for the residuals to become periodic (say, in a network simulation). Briefly, the dynamic concentrations can be said to consist of two components: a non–periodic dynamics due to the initial conditions (initial concentrations in pipes and reservoirs, as specified in a simulation), and a periodic dynamics from the periodic dosages and hydraulics. According to the design approach advocated here, the initial conditions are considered to be arbitrary and thus their effect must be ignored in computation of the impulse response coefficients. Software is available which accomplishes this when calculating the response coefficients via perturbation methods (see the following section).

The LP scheduling model is relatively straight forward to solve using existing methods for general linear programming problems, and specifically the simplex method. The problem size for practical networks is within the capability of commercial LP software, as the number of decision variables depends on the selected number of scheduling periods times the number of booster stations, $n_s \times n_b$, and is otherwise independent of the pipe network size and system dynamics. The size of the constraint set will, however, depend on the pipe network through the number of monitoring locations n_m and the impact cycle time T_a. The total efficiency of solution is likely to hinge more on the problem setup than the actual linear programming solution; the calculation of composite impulse response coefficients a_{ij}^{km} will at least be a significant component of the total computational burden. These same computational issues apply to solution of the MILP scheduling/location model. Solution by branch–and–bound with the simplex method is, however, a great deal more computationally intensive, and no reliable methods exist for estimating the solution time *a priori*. It is also likely that algorithmic options or problem–specific details will greatly influence the computation time. In short, solution is not straightforward, but is possible for practical networks, provided that the number of potential booster locations is limited, and the effects of algorithmic options and tolerances on branch and bound performance is explored.

The MSC model formulation is a special category of integer linear programming (ILP) optimization models. Models of this type are usually very efficient to solve compared with IP models of the same dimension that lack any particular structure; it is not unusual that the relaxed solution of an MSC model (where the binary variables are allowed to vary continuously between zero and one, yielding a pure LP problem) is a natural integer solution, in which case the solution is very efficient to obtain via the simplex algorithm. Heuristics have also been invented, based on the special form of the constraints. Similar models (in other application contexts) have been solved using linear programming plus heuristics, linear programming plus branch and bound implicit enumeration, and simulated annealing [Church and Revelle 1974, Meyer and Brill 1988, Meyer et al. 1994].

8. AVAILABLE SOFTWARE

A computer code called BDDA (Booster Disinfection Design Algorithms) has been developed to interface with EPANET (Rossman, 1993), a distribution system water quality model, to automatically produce a specification of the above optimal booster scheduling and/or location problems in standard MPS format (Murtagh, 1981). This specification of the LP/MILP/MSC problem can then be solved using any available implementation of the simplex algorithm (LP) or branch–and–bound with the simplex method (MILP/MSC). Alternatively, the code interfaces directly with the CPLEX LP and MILP solution algorithms [CPLEX, 1998], for users of that commercial software. The standard EPANET data file is used, containing information about the physical and chemical characteristics of the distribution system. The BDDA application requires additional information relevant to the optimization model formulation, including potential and existing booster locations, booster schedule cycle time (T_s), number (n_s) and length of mass injection intervals for each booster location, monitoring node locations, and lower and upper residual bounds (c^{min} and c^{max}).

The BDDA proceeds by first simulating the network hydraulics for use during subsequent water quality simulations. The composite impulse response coefficients are then computed for each injection by a straightforward perturbation method. Each booster location i and periodic dose interval k is selected in turn and modeled as a periodic source of disinfectant (such that $u_i^{k+pn_s} = \hat{v}$, $\forall p \geq 0$). The simulated concentrations (c_j^m) are used to calculate the composite impulse response coefficients corresponding to the periodic mass injections ($a_{ij}^{km} = c_j^m/\hat{v}$, $m = M,\ldots,M + n_a - 1$). (The value of the time period M is an input parameter that is related to the selected simulation duration.) The BDDA records the column of the LP coefficient matrix corresponding to v_i^k. This iterative procedure continues until all booster locations and injection periods have been analyzed. The BDDA then adds the necessary constraint information to the MPS description of the LP problem for solution by a commercial optimization algorithm, or interfaces directly with the appropriate CPLEX modules.

9. EXAMPLE RESULTS

To illustrate briefly the type of results that can be obtained, the MSC model was used to locate candidate booster locations for a utility in the western United States. A network map of the distribution system is shown in Figure 2; the model includes 3323 junctions, 3829 pipes, 1 reservoir, 34 tanks, 61 pumps, and 2 valves. All water is treated at a single source in the north, and is pumped in stages to successively higher service elevations. This utility was experiencing difficulty in maintaining residual coverage in the southeast portion of the network, where average junction travel times can exceed three weeks.

The MSC model was applied by selecting 1200 potential booster locations at random from among the 3323 junctions, and adding to these the 35 tank and reservoir junctions. The model was then solved for different values of the residual coverage parameter. As shown in Fig. 1, as the ratio $\eta = c^{min}/c^{max}$ increases, the minimum number of booster stations increases to accommodate the stricter definition of residual coverage. For each solution an injection station at the source is selected (of necessity), and other booster stations are selected at locations adjacent to tanks or at key junctions which deliver signif-

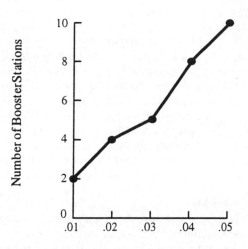

Fig. 1 —Number of booster stations versus coverage for example network.

icant flow to critical parts of the network. Among all five designs in Fig. 1, a total of 16 different booster locations are selected. These locations are shown in Fig. 2. The selected locations are geographically distributed throughout the network.

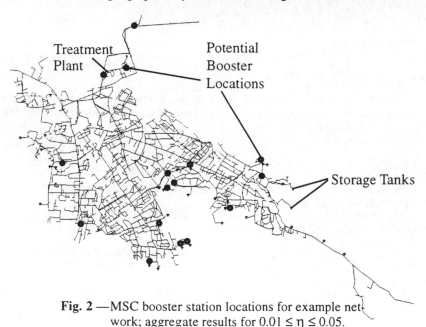

Fig. 2 —MSC booster station locations for example network; aggregate results for $0.01 \leq \eta \leq 0.05$.

The results required approximately 12 hours of CPU time for problem setup on a 300 MHz Pentium II personal computer (mainly for calculation of impulse response coefficients). This setup time was required once for all five designs, as the impulse response coefficients are not affected by η, and the BDDA code allows reading from a binary file of saved impulse response coefficients. The solution time for the MSC problem was under one minute using the CPLEX v6.5 MILP algorithm and employing its presolution algorithms to reduce model size. It is argued that these models are thus practicable for actual networks, and furthermore that the complexity of network dynamics would present a challenge for non–systematic, model–based design procedures. Given the relatively small subset of selected locations, however, further analysis may be conducted in a more focussed manner using simulation, or these locations may be input to the MILP or LP formulations for developing optimal dose schedules.

10. SUMMARY

Optimization model formulations were described for optimal location and scheduling of booster disinfection dosages throughout a distribution network and throughout time. Objectives include minimizing the total mass of disinfectant applied at multiple points in a distribution system and minimizing the total number of booster stations constructed. Constraints ensure the maintenance of adequate disinfection residuals at all monitoring nodes and at all times. The key to these formulations is knowing that linear superposition can be applied to disinfectant transport in distribution networks, assuming that the future dynamic hydraulics are known. By assuming periodicity of the hydraulics and requiring optimal dose schedules to be periodic, formulations were derived which logically incorporate the import influence of disinfectant concentration dynamics into the booster scheduling and location designs. Software that is available for solution of these models was briefly described, along with example results.

11. REFERENCES

1.   Boccelli, D., Tryby, M., Uber, J., Rossman, L., Zierolf, M., and M. Polycarpou, "Optimal Scheduling of Booster Disinfection in Water Distribution Systems," *J. of Water Res. Plan. Manag.*, A.S.C.E., 124(2), Mar/Apr, 1998.
2.   Boulos, P. F., Altman, T., Jarrige, P., and Collevati, F. (1995) "Discrete Simulation Approach for Network–Water–Quality Models." *Journal of Water Resources Planning & Management*, ASCE 121(1), 49–60.
3.   Bull, R. J., and Kopfler, R. C. (1991). *Health Effects of Disinfectants and Disinfection By–Products*, American Water Works Association Research Foundation, Denver CO.
4.   Chow, V. T., Maidment, D. R., and Mays, L. W. (1988) *Applied Hydrology*, McGraw–Hill, New York, 204–213.
5.   Church, R., and C. Revelle, "The Maximal Covering Location Problem," *Pap. Reg. Sci. Assoc.*, 32, 101–118, 1974.
6.   Clark, R. M., Grayman, W. M., Males, R. M., and Hess, A. F. (1993). "Modeling Contaminant Propagation in Drinking Water Distribution Systems." *Journal of Environmental Engineering*, ASCE, 119(2), 349–364.
7.   Grayman, W. M., Clark, R. M., and Males, R. M. (1988) "Modeling Distribution–system Water Quality: Dynamic Approach." *Journal of Water Resources Planning & Management*, ASCE. 114(3), 295–312.
8.   Hillier, F. S. and Lieberman. G. J. (1980) *Introduction to Operations Research*, Holden–Day, Inc., Oakland, CA, 68–91.
9.   Liou, C. P., and Kroon, J. R. (1987) "Modeling The Propagation Of Waterborne Substances In Distribution Networks." *Journal American Water Works Association.* 79(11), 54–58.

10. Meyer, P., and E. D. Brill, Jr., "A Method for Locating Wells in a Groundwater Monitoring Network Under Conditions of Uncertainty," *Water Resour. Res.*, 24(8), August, 1988.

11. Meyer, P., Valocchi, A., and J. W. Eheart, "Monitoring Network Design to Provide Initial Detection of Groundwater Contamination," *Water Resources Research*, 30(9), 1994.

12. Murtagh, B. (1981) *Advanced linear programming: computation and practice*, McGraw–Hill International, New York.

13. Oppenheim, A. V., and Willsky, A. S. (1997) *Signals and Systems*, 2nd Edition, Prentice Hall, Englewood Cliffs, NJ, 77–90.

14. Rao, H. S., and Bree , D. Jr. (1977) "Extended Period Simulation of Water Systems – Part A." *Journal of the Hydraulics Division*, ASCE, 103(HY2), 97–108.

15. Revelle, C. S., Loucks, D. P. and Lynn, W. R. (1967) "A Management Model for Water Quality Control." *Journal of the Water Pollution Control Federation,* 39(6).

16. Revelle, C. S., Loucks, D. P. and Lynn, W. R. (1968) "Linear Programming Applied to Water Quality Management." *Water Resources Research*, 4(1), 1–9.

17. Rossman, L. A., (1993) *EPANET Users Manual*. Risk Reduction Engineering Laboratory, U.S.E.P.A., Cincinnati, Ohio.

18. Rossman, L. A., Boulos, P. F., and Altman, T. (1993) "Discrete volume–element method for network water–quality models." *Journal of Water Resources Planning & Management*, ASCE. 119(5), 505–517.

19. Rossman, L. A., Clark, R. M., and Grayman, W. M. (1994) "Modeling chlorine residuals in drinking–water distribution systems." *Journal of Environmental Engineering,* 120(4), 803–820.

20. Rossman, L. A., and Boulos, P. F. (1996) "Numerical methods for modeling water quality in distribution systems: A comparison." *Journal of Water Resources Planning & Management*, ASCE. 122(2), 137–146.

21. Seborg, D. E., Edgar, T. F., and Mellichamp, D. A. (1989) *Process Dynamics and Control*, John Wiley & Sons, Inc., New York, 224–245.

22. Tryby, M. E., Boccelli, D. L., Koechling, M. T., Uber, J. G., Summers, R. S., and Rossman, L. A. (1997) "Booster Disinfection for Managing Disinfectant Residuals," *Journal American Water Works Association*, 91(1), January, 1999.

23. Wagner, H. M. (1969) *Principles of Operations Research with Applications to Managerial Decisions*, Prentice–Hall, Englewood Cliffs, NJ, 361–365.

24. Zierolf, M. L., Polycarpou, M. M., and Uber, J. G. (1997) "Development and Auto-Calibration of an Input–Output Model of Chlorine Transport in Drinking Water Distribution Systems," *IEEE Transactions on Control Systems Technology* (to appear).

PART VI

OPTIMISATION

Practical Experience in the Successful Implementation of Optimisation Systems for Water Supply Management*

M.R. Fowler, S.C. Cook *and* J.P. Lumbers

ABSTRACT

The paper describes the key features of a number of applications of optimisation methods for the design and management of water resources and supply systems.

Common to all the applications is a mass-balance modelling approach, permitting the use of linear or mixed integer programming tools, although non-linear solvers have also been used where appropriate. Principal optimisation issues concern handling multiple objectives using prioritisation and weighting methods, and long-term/short-term interaction where the "single-step" method can be used to ensure an overall optimum is achieved without the need for control rules or licence apportioning.

The first application described is an on-line pump scheduling system which operated continuously for eight years. An implementation strategy which involved operations staff in the design process and focused on robustness and reliability issues ensured its acceptance. The second area, that of water resources optimisation software, relates to an off-line tool which has been applied to problems such as what-if analyses, optimal design, derivation of control rules and derivation of medium-term operation. Part of its success is due to its ease of use, enabling advanced analysis without a high degree of expertise. The third area relates to optimal response to failure events to ensure a consistent and widely applicable approach to consequence modelling for quantified risk analysis.

Optimisation will undoubtedly play an increasingly prominent role in the future, potentially further strengthening the link between strategic planning and operations scheduling, exemplified by a major system currently being developed for one of the largest water and sewerage companies in the UK.

---

1 INTRODUCTION

There is growing regulatory pressure on the UK water industry to achieve greater efficiencies in both capital investment and operations. Capital investment must be targeted in the most cost-effective manner and operating costs must be reduced, for example through the exploitation of electricity tariffs.

The assessment and prioritisation of capital investment is often made on the basis of risk reduction. In order for comparisons to be made between schemes on a consistent basis it is necessary that the systems are always operated in an optimal manner under the various configurations considered. Making the case for the development of new sources similarly demands that optimal usage of existing resources be demonstrated. The implication of the forecast climate change scenarios is that new surface water sources may be needed particularly in central and southern England. The greater variability in climate will also require a higher degree of control and faster response rate facilitated by optimisation systems.

Achieving significant reductions in energy costs has been an element of most of the water company business plans. Whilst there have been a number of attempts in recent years to implement various forms of pump scheduling, there have been few successful applications that have been fully taken up by operations staff. However, optimal pump scheduling is proven technology as described below and can deliver energy cost savings reliably, as well as enhancing the overall management of the system.

The successful application of optimisation methods depends on a wide variety of factors, including the selection of an appropriate technique, problem formulation, data availability and validation, demand forecasting, the temporal and spatial resolution required, solution speed, reliability, user expertise and interface design, on-line/off-line operation, support arrangements, etc. Some of these considerations are discussed below.

2 OPTIMISATION CONSIDERATIONS

2.1 Modelling the System

In most of the applications considered in this paper, optimisations are constructed as mathematical programming problems incorporating fundamental behaviour of the water supply system in the form of a mass-balance model. This formulation uses a conventional node and element approach but does not take pressures in the supply system directly into account, considering principally conservation of mass at each node. Optimisation problems for relatively straightforward supply systems can be constructed in an entirely linear formulation which offers the advantages of:

- Configuration: models are easily configured by the user (via a schematic for example) with the associated optimisation problem derived directly. This allows the modelled supply system to be rapidly reconfigured and analysed.
- Performance: well-established solution methodologies are available which, in comparison with those for non-linear optimisation problems, offer high

performance in terms of both scale of problem that can be accommodated and speed of solution.

- Robustness: the solution is generally independent of initial conditions, problems of local optima are avoided and solutions are produced reliably with little attention.
- Accessibility: linear optimisation is a mature development area with a wide range of solver tools available.

Many water supply optimisation problems which require the introduction of non-linear components can be formulated as a mixed integer programming (MIP) problem. MIP still offers many of the advantages of a purely linear approach and allows the modelling of control rules, switching constraints, non-linear cost curves and so on. The majority of components can be adequately represented using a mass-balance model and MIP problem formulation covering the supply network from sources such as boreholes and rivers, to impounding reservoirs, works, service reservoirs and demand centres at DMA level or their aggregates.

2.2   Multiple Objectives

The optimisation problem to be examined often requires multiple objectives to be met whether these are explicitly or implicitly stated. These objectives often involve a primary need to meet demand if possible, followed by the need to keep reservoir levels high enough to minimise risk of supply interruptions and a desire to minimise costs. Aside from these principal objectives, there may be other objectives such as those associated with "flexible" constraints. These relate to targets or limits which are viewed as desirable but not essential and are handled as goals which are to be achieved if possible and approached closely if not: for example, a flexible upper limit on the frequency with which flow along a bi-directional main changes direction. These goals can be prioritised so that objectives are met to suit user-defined preferences: for example, meeting demands as a priority over maximising reservoir storage. They can also be aggregated with weights in a conventional approach for handling multiple objectives.

Although the priority of certain goals may be relatively predictable (demand generally being of highest importance for example) this is not always the case and depends upon the problem and application under consideration. A particular issue for the successful implementation of water supply optimisation therefore is provision of a flexible approach to multiple objectives that allows users to select for each run the priority of each goal.

2.3   Short-term/Long-term Interaction

A significant aspect of optimisation is the interaction between long-term, strategic planning and short-term, operational objectives and constraints. Long-term constraints include regulating the use of annual and five-yearly licences, controlling the use of impounding reservoirs, exploiting seasonal tariff variations and meeting seasonal and long-term forecast demand variations. Short-term objectives on the other hand might include maintaining service reservoir levels,

298

minimising short-term costs, meeting peak demands, keeping pump switches to a minimum and so on.

There are two issues which are of most importance here. Firstly how in each optimisation consideration is given to phenomena which effect the whole or a substantial part of the optimisation horizon, and secondly how long-term and short-term objectives can both be met.

The first issue can be dealt with by undertaking optimisation of the entire horizon in a single step rather than in a series of independent optimisations. The benefit of this technique can be illustrated by firstly considering the alternative approach of breaking the horizon down into a number of timesteps and optimising each timestep individually in turn. In order to take account of future conditions and long-term constraints (e.g. annual licences and reservoir volume usage), an estimate would be made of future operation and limits placed on the current timestep (e.g. pro-rata licence usage or reservoir control curves). This method, although quick to optimise, represents a de-coupling of temporal aspects of the problem which may require some significant off-line work and is not guaranteed to provide results which are optimal overall.

If, however, the whole horizon is optimised as a single formulation, the need for licence apportioning or control rule derivation is removed; future timesteps in the horizon are considered at the same time as current timesteps and an overall optimal solution is found. Although such "single-step" optimisation is likely to take longer to optimise than "multi-step" (comprising as it does one large optimisation rather than many small optimisations), this must be weighed against the additional effort required to derive the long-term limits for the multi-step approach.

The second issue of trying to meet both long-term and short-term objectives has been addressed using a hierarchy of optimisations in which the optimal solution from one (or more) long-term optimisations serves as constraints for another, shorter-term optimisation. In some ways this is similar to the multi-step optimisation described above, except that the estimates of future operation and limits are themselves derived from optimisations. In practice, the degree to which decomposition of an optimisation into a spatial and/or temporal hierarchy occurs is a matter of balancing the potential of sub-optimality against computational load and data manipulation. Models need only be specified at a level of detail which suits their intended use. Excessive detail leads to more data preparation, analysis of results and computational effort than is necessary.

3   EXAMPLE APPLICATIONS

3.1   Pump Scheduling

Much of the experience described here relates to the development and support of an on-line pump scheduling suite of software for Thames Water. The system was first installed in 1988 and was operated continuously for eight years.

The concept of exploiting electricity tariffs to minimise energy costs is not new and in many simple networks substantial savings can be achieved by strategies formulated by operations staff based on a combination of desk-top analysis and experience. However, where the supply system and tariff structures are more complex, the optimal means of fully exploiting the electricity tariff is not intuitive and more advanced methods are required.

The introduction of software, which in some cases supersedes existing practices, can be seen as a threat to the operations staff. For the successful adoption of pump scheduling software it is essential that the operations staff are supportive of its implementation and are closely involved in the development or customisation. Wherever possible the operations staff should have an input into the design of the interface to suit individual requirements. A sense of ownership is engendered through involvement and hence the system is not regarded as an imposition from 'head office' or the 'R&D department' for example.

The first on-line pump scheduling system to be installed for Thames Water was at the Bourne End control centre. The ready adoption of the software by a succession of control room staff over the years can be attributed to the simplicity of the user interface and also the confidence that existed in the system. An example of the information that was provided automatically each morning for the control room staff is shown in Figure 1.

```
┌──────────────────────────────────────────────────────────────────────────┐
│ Optimising @ 07:00 09-Mar        Comms disabled      12:54:17 08-Mar       │
│ Schedule of ALL PLANT Switchings for : ALL ZONES                           │
│                                                                            │
│ Starting on: 11:15 WED 25-MAY-1994      Ending on: 07:00 THU 26-MAY-1994   │
│                                                                  Exp.Flow  │
│ Time   Device Name            Action                    Status   (l/s)     │
│                                                                            │
│ 11:15  Latton-Glos.Rd."DUTY"  Switch all pumps          OFF        .0      │
│        Baunton-St.Park "DUTY" Switch all pumps          OFF        .0      │
│        Baunton-Rapsgate "DUTY" Switch Booster No.1      ON  ]     18.0      │
│                               Switch Booster No.2       OFF ]              │
│        Baunton BHPs "DUTY"    Switch BHP No.1           ON  ]    176.0      │
│                               Switch BHP No.3           ON  ]              │
│                               Switch all other pumps    OFF ]              │
│        Ashton Keynes "DUTY"   Switch BHP No.3           ON  ]     69.0      │
│                               Switch all other pumps    OFF ]              │
│        Marlborough "DUTY"     Switch BHP No.1           ON  ]     25.0      │
│                               Switch BHP No.2           OFF ]              │
│        Clatford "DUTY"        Switch all pumps          OFF        .0      │
│        Bibury "DUTY"          Switch BHP No.1           ON  ]     69.0      │
│                               Switch BHP No.2           OFF ]              │
│                                                                            │
│ Continue, Print or Quit ?  _                                               │
└──────────────────────────────────────────────────────────────────────────┘
```

Figure 1. Example output from pump scheduler

Confidence in decision-support systems such as on-line pump scheduling depends on a number of factors:

- model performance - the ability of the model to predict the behaviour of the system under the forecast demands and recommended optimal control settings.
- the initial introduction of the system to operate in parallel with existing manual methods, for example through the production of off-line pump schedules for trial.
- the robustness of the software to survive problems such as telemetry system failure and always to provide a solution.

- the robustness of the software to provide a solution even where infeasibilities have been identified, for example arising from the non-availability of plant.
- the provision of simple but informative performance monitoring to allow the user to check position against targets easily and on a regular basis.
- the facilities provided to assist operators in determining the most efficient return to target profiles where deviation has occurred, for example owing to unplanned outage.
- the ready availability of on-line support and guidance plus training for new staff.

The design of new systems must take account of the particular company approach to control which may range from centralised active control to decentralised local control. Similarly there are differences between companies regarding the type of data collected by telemetry systems and the facilities included.

Where telemetry systems have been designed to gather data on the basis of exceptions rather than regular polling, data in-filling techniques may be required to generate the necessary input data for optimisation and issues of data currency must be considered. Inevitably accuracy is compromised, as is the ability to monitor system performance against the optimal targets.

The facilities for data validation within modern telemetry systems are often extensive. However, the extent to which these facilities are exploited varies considerably. The successful implementation of any modelling and optimisation system depends critically on the quality of the data available. Where adequate data validation is not practised then this must be added to the functionality of the optimisation software, as was the case originally at the Bourne End control centre.

The benefits of on-line pump scheduling have been proven, and in some cases the pay-back period for investment in the technology has been less than 1 year. Additional benefits have been obtained through closer operation of the system that in turn allows greater exploitation of storage. However, on-line optimisation may not always be cost justified and for some areas off-line scheduling combined with local controls may achieve most of the potential energy cost savings.

3.2   Water Resource Optimisation Software

Whereas pump scheduling is concerned with optimisation at a sub-daily level, water resource optimisation relates to operation over a longer timeframe. A package has been developed (MISER) with the principal objectives of:

- modelling strategic impounding and pumped storage reservoirs with long-term cyclic behaviour.
- modelling long-term groundwater constraints, e.g. annual and five yearly licences, as well as more complex licences such as restricted usage (where a daily licence can be exceeded for a given number of days in each year).
- providing optimal operating strategies over the long-term using time-varying demands, river flows, reservoir catchment flows, costs, etc.

- providing operational targets and constraints for implementation at the short-term level.

The package is now in use in six major water utilities in the UK with more expected. Of key importance to its successful implementation has been its ease of use, in terms of schematic interface design (see Figure 2) and required level of user expertise, and flexibility, in terms of capability to model a wide range of network features and operating scenarios.

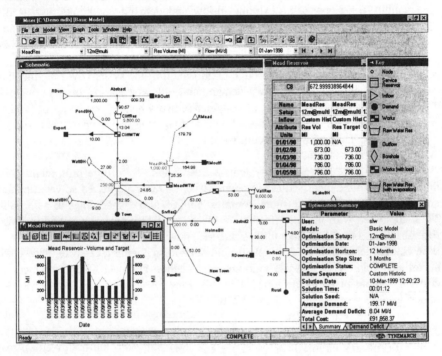

Figure 2. Example screen shot of the MISER user interface

Applications of the software (known to date) include:

1. *Analysis of historical performance:* results achieved through use of a set of control rules were compared with those of single step optimisation using *a priori* knowledge. This project indicated how the control rules required adjustment in order to accommodate changes to the relative costs and demands since the control rules had been produced, and how the control rules were inadequate in dealing with extremely low inflow conditions.

2. *What-if analysis:* the reliability of operation under alternative capital expenditure schemes (capacity improvements, inter-resource links, etc) was analysed by optimising against extreme historical data and probabilistic sequences. With the removal of the need for operational control rules (Section 2.3), changes were made to the system and optimal operation determined without the need for any changes in control rules or licence allocation.

3. *Optimal design:* MISER has been used to determine required capacities of new works directly. Rather than using an iterative method of testing the system against alternative outputs, in this analysis minimisation of the capacity of a works was specified as an objective and output obtained from a single optimisation. Similar techniques can be applied to reservoir sizing design.

4. *Derivation of medium-term operation:* in this current application, optimal operation in the medium-term is being derived by optimising over 18 months at monthly timesteps, using extreme inflow and demand conditions, and is then applied as a lower envelope for short-term 10-day optimisation. The envelope is in the form of target reservoir levels at the end of the short-term period and/or total borehole output over the period.

5. *Derivation of control rules:* through optimisation of the system over the medium-term against a variety of known demand and inflow scenarios, optimal output of reservoirs (potentially as part of conjunctive use schemes) are being derived for different reservoir level initial conditions and target levels, from which a series of reservoir control rules can be constructed.

Water resource optimisation systems are set to take on a more prominent role in the future. With the increased use of client-server systems, improved solver capability and hardware performance, more complex and advanced modelling is achievable. This, combined with the implementation of on-line pump scheduling and increased formalisation of the hierarchical structure, means that optimisation across all timescales is achievable so that strategic and operations planning can take place within a common framework and use a common modelling approach. Such advances are currently taking place in the form of the bespoke resources system WATERMAN under development for Severn Trent Water.

3.3 Risk Analysis

A significant application of optimisation is in the field of quantified risk analysis (QRA). In the context of this paper, QRA relates to the quantification of risks (where risk is defined as some combination of probability and severity) of interruptions to supply resulting from component failures such as treatment shutdowns, mains bursts, power cuts or pollution events. QRA is used for example to identify and quantify system risks, and to determine the effectiveness of remedial capital schemes.

The general methodology involves the identification of potential failure events, derivation of their associated occurrence and duration distributions, and quantified determination of the supply consequences of the failures. Alternative capital and operational schemes are assessed in terms of their risk-reduction effectiveness through implementation in the system model, additional consequence modelling and comparison of resultant risk levels.

Of key importance to the modelling of the consequences of failure is the identification of appropriate operational responses to each failure event, such as the level of increase of any works outputs and how water should be re-directed within or to/from the affected zone. The modelled response should match operational decisions as closely as possible, both in terms of the actions themselves and in terms of their timescales (e.g. such as time taken to identify the failure or time taken to change a valve setting or increase a works output). In addition, the chosen modelling method should permit analysis of the consequences of failure events for a range of different storage and demand scenarios so that a representative quantification of risk is achievable.

For analyses of non-complex systems, the operational response can often be manually derived, but for systems of reasonable size and complexity, the operational response may be far from obvious. In addition, where multiple analyses are required (such as for Monte Carlo simulation) and where the rules of operational response are dependent on the prevailing demand and storage scenario for example, a more automated approach is required. One such, successfully applied, technique is to use optimisation to provide a simulated "best" response to each failure. Here the whole operational problem is formulated as an optimisation problem, where the system constraints are formulated as optimisation constraints and the optimisation objectives match those of the water network operators. This approach offers several benefits including:

- Wide applicability: the optimisation approach is able to provide the optimal operational response to a wide range of complex systems.
- Automatic: when different failure events, demand or storage scenarios or network configurations are to be analysed, the operational responses will be modified accordingly by re-optimisation.
- Consistent: the optimisation approach always finds the "best" response to each failure and therefore each failure event is analysed in a consistent manner.
- Controllability and visibility of objectives: the optimisation objectives can be configured directly to represent the same objectives as required operationally, e.g. where the objectives are to delay the occurrence of any demand deficit, this can be framed directly in the optimisation objective.
- System modelling: the important characteristics of the system relevant to emergency responses such as maximum works outputs, works ramp rates, blending constraints and operational response times can be directly embedded as constraints in the optimisation problem. Where embedding of fixed rules in the operational response is required, e.g. to identify that a particular works is always used under certain circumstances, rules can be added as conditional constraints.

As with any modelling method, care is required to ensure that sensible results are obtained, which can be achieved by review of individual test cases. Similarly, actual operation of the system ought to make use of the same optimisation models on which the risk analysis is based to ensure that failure consequence analysis is representative. With the introduction of water resource optimisation systems

304

(Section 3.2), or the use of the consequence modelling results to generate contingency plans, this becomes less of an issue.

QRA for Investment Planning: High level

In this analysis, failure events of strategic importance are analysed, such as the loss of major works or transfer routes. Since such a failure can have widespread impact, a large-scale high level model of the system is required where strategic storage, major sources and significant transfers for all potentially affected areas are directly incorporated.

Non-strategic storage can have a considerable impact on the consequences of a failure but the separate modelling of all storage would be excessive. Simple aggregation of the local storage can give a distorted representation of the capability of a local area to distribute water between service reservoirs, and a suitable methodology has been developed to overcome the spatial resolution issue based on curves of the affected population manually derived, against duration.

The method chosen resulted in the problem being non-linear in nature and not open to linear optimisation. A GRG (general reduced gradient) non-linear solver was found to be suitable with the objective function:

$$\text{Min} \sum_{i=1}^{N} \frac{p_i}{t_i}$$

where:
   N is the total number of population/duration curve steps
   p_i is the population affected by a supply interruption in each step
   t_i is the time following failure occurrence after which p_i is affected

This method has been applied to the analysis of around £150m of capital schemes proposed for the Severn Trent Region (1). The risk-reduction effectiveness of each scheme was assessed and the most risk-reductive set of schemes at each level of investment identified to derive a curve as given in Figure 3. This curve provided substantial decision support to Severn Trent in determining a suitable level of investment, based upon reduced returns for increasing expenditure, and the appropriate allocation of that investment.

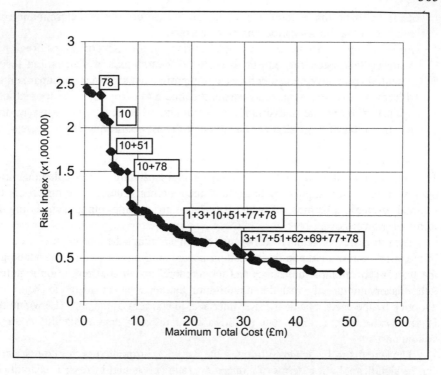

Figure 3. Variation in Resultant Risk Against Investment (with scheme identifiers)

Local level

In this analysis, failure events and schemes at a distribution level are assessed. Temporal and spatial resolution are increased, such that service reservoirs and demand centres are directly modelled within sub-daily timescales. Linear optimisation can be utilised (providing associated benefits as in Section 2.1), with the objective function:

$$\text{Min} \sum_{i=1}^{N} p_i (N - i + 1)$$

where: p_i is population affected in timestep i
 N is the total number of timesteps

Applications at this level have included:

- *Reservoir sizing/operation:* the level of security provided by a reservoir has traditionally been associated with the equivalent number of hours of average demand storage held by the reservoir. However, QRA of the failures whose consequences are influenced by buffer storage permits a more informed decision to be made about the sizing of new reservoirs or determination of suitable operating levels for current reservoirs.

- *Risk strategy development:* studies similar to the Severn Trent Regional analysis have been carried out to derive risk strategies at distribution level. Potential risks have been identified and quantified, permitting a comparison of risk levels on a universal scale across the area and company-wide (potentially against a corporate maximum or maximum risk level) to assist in the identification of key risk areas and the effectiveness of schemes.

4   CONCLUSION

The application of optimisation techniques and software to the planning and operation of water supply systems has been demonstrated, ranging from the optimal strategic planning of water resources to on-line pump scheduling to minimise power costs.

The successful implementation of such systems requires the resolution of a variety of technical issues, including the modelling approach, the optimisation methodology, the handling of multiple objectives and the temporal and spatial resolution required. Other factors not directly related to optimisation that must be considered to ensure full adoption by the users include the user interface, integration of the software within the IT and control room environment, performance monitoring, data availability and data validation.

The benefits to be achieved through the use of appropriate optimisation systems can be significant both in terms of reduced operating costs and increased reliability of water supply.

REFERENCES

1. Cook S., Fowler M., and Banyard J. (1999) "Quantitative Risk Analysis for Ranking Security of Supply Investment Schemes, *Computing and Control in the Water Industry*, Eds Powell and Hindi, Research Studies Press, 1999.

Optimization of Water Supply Using Successive Linear Programming

F. Guhl, B. Bremond *and* D. Gilbert

ABSTRACT

For several years we have been building models for the optimization of water networks. The optimization makes it possible to choose controls (opening of valves, start up of pumps) to be implemented over 24 hours.

The models we build are used on large water networks and are managed in real time. To enable short calculation times, linear programming has been chosen. Before optimization, linear relations between the network flows and water consumption are established.

In certain water networks, however, these linear relations do not match reality. In particular, when the influences of water levels in the tanks on costs and on the capacity of transit between tanks are taken into account, non-linear relations have to be integrated into the model.

To solve this sort of model and to maintain short calculation times, we propose the use of a successive linear programming method. A linearization of the program is performed at each stage. This method has been validated on a medium-sized drinking water supply network (20,000 inhabitants) located in the east of France and managed by the Compagnie Générale des Eaux. The previous method has been improved by the addition of a criterion enabling a choice between several solutions at identical cost. The method then converges in 4 iterations.

Finally, networks have been defined for which this method converges and those for which the field of the solutions is not convex. In this second case the solution obtained is not a single one. Only one approximate solution is adopted for which the error can be defined. The method is also implemented on the network presented previously and the solution is found in at most 1 second.

1   INTRODUCTION

For several years, we have been proposing models for the operational optimization of a drinking water network over a 24-hour period. The models built hitherto are based on linear programming, as is the case in [2] and [6]. The purpose of our models is to obtain controls to be sent over a network to minimize water production and transport costs.

As will be seen in section 3, the linearization of flows, on the one hand, and of transport and production costs on the other hand, is not always possible. To keep

307

calculation time as short as possible, we propose a successive linear programming method. As shown in [1], [3] and [5] the convergence of this type of method is not always obtained, irrespective of the degree of linearization adopted.

The linear programming method used hitherto is presented in section 2. Section 3 will explain the reasons for the non-linearity and present the new writing of our optimization problem. Section 4 will present the successive linear programming method proposed and the results obtained. Finally, section 5 will provide the opportunity to establish a typology of the networks for which the proposed method converges.

2   LINEAR OPTIMIZATION METHOD

The model that we customarily used can be presented as follows :

$$(P_1)\begin{cases} z_{\min} = \sum_{\theta=1}^{T}\sum_{i=1}^{I}\alpha_i(\theta)\pi_i(\theta) \\[2mm] (1)\forall\theta\in\langle 1;T-1\rangle, \forall r\in\langle 1;R\rangle, V_r(0)+\sum_{t=1}^{\theta}\sum_{i=1}^{I}\alpha_i(t)Q_{ri}(t)-\sum_{t=1}^{\theta}C_r(t)\le V_{r\max} \\[2mm] (2)\forall\theta\in\langle 1;T-1\rangle, \forall r\in\langle 1;R\rangle, V_r(0)+\sum_{t=1}^{\theta}\sum_{i=1}^{I}\alpha_i(t)Q_{ri}(t)-\sum_{t=1}^{\theta}C_r(t)\ge V_{r\min} \\[2mm] (3)\forall r\in\langle 1;R\rangle, V_r(0)+\sum_{t=1}^{T}\sum_{i=1}^{I}\alpha_i(t)Q_{ri}(t)-\sum_{t=1}^{T}C_r(t)= V_{r\max} \\[2mm] (4)\forall\theta\in\langle 1;T\rangle, \sum_{i=1}^{I}\alpha_i(\theta)=1 \\[2mm] (5)\forall\theta\in\langle 1;T\rangle, \forall i\in\langle 1;I\rangle, \alpha_i(\theta)\ge 0 \end{cases}$$

The implementation of this model pre-supposes the calculation before optimization of the flows $Q_{ri}(t)$ and the costs $\pi_i(\theta)$.

The flows $Q_{ri}(t)$ represent the volume of water entering the reservoir r over the time step <t-1;t> if the control i is applied, and the costs $\pi_i(\theta)$ represent the cost of the control i.

The unknowns of the optimization problem are the $\alpha_i(t)$. These determine the fraction of the time step <t-1;t> during which the control i is applied.

The constraints of type (1) and (2) translate the maximum and minimum limits in each of the reservoirs in the course of the optimization horizon. The constraint (3) makes it possible to set a final level which consists of filling the reservoirs. Finally, the constraint (4) results from the manner of constructing the variables $\alpha_i(t)$.

3   LIMIT OF LINEARITY

3.1   Hydraulic behavior

It is not always possible to calculate the $Q_{ri}(t)$ which may, in certain cases, depend on the water level in the reservoirs. Figure 1 shows the example of part of a drinking water network where the flow rate between two reservoirs strongly depends on the difference in level between the two reservoirs.

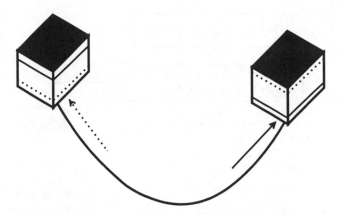

Figure 1 : Hydraulic configuration resulting in non-linearity

In the case of the two reservoirs in Figure 1 the flow depends on the water level in each reservoir. It can even change direction.

3.2 Writing of the model

If this dependency on the flow rates and costs with respect to the levels in the reservoirs is taken into account, the problem is written as follows :

$$
(P_2)\begin{cases}
z_{\min} = \sum_{\theta=1}^{T}\sum_{i=1}^{I}\alpha_i(\theta)\pi_i(V(\theta)) \\[2mm]
(1)\forall\theta\in\langle 1;T-1\rangle, \forall r\in\langle 1;R\rangle, V_r(0)+\sum_{t=1}^{\theta}\sum_{i=1}^{I}\alpha_i(t)Q_{ri}(V(t))-\sum_{t=1}^{\theta}C_r(t)\le V_{r\max} \\[2mm]
(2)\forall\theta\in\langle 1;T-1\rangle, \forall r\in\langle 1;R\rangle, V_r(0)+\sum_{t=1}^{\theta}\sum_{i=1}^{I}\alpha_i(t)Q_{ri}(V(t))-\sum_{t=1}^{\theta}C_r(t)\ge V_{r\min} \\[2mm]
(3)\forall r\in\langle 1;R\rangle, V_r(0)+\sum_{t=1}^{T}\sum_{i=1}^{I}\alpha_i(t)Q_{ri}(V(t))-\sum_{t=1}^{T}C_r(t)=V_{r\max} \\[2mm]
(4)\forall\theta\in\langle 1;T\rangle, \sum_{i=1}^{I}\alpha_i(\theta)=1 \\[2mm]
(5)\forall\theta\in\langle 1;T\rangle, \forall i\in\langle 1;I\rangle, \alpha_i(\theta)\ge 0
\end{cases}
$$

4   THE SUCCESSIVE LINEAR PROGRAMMING METHOD PROPOSED

4.1   First order successive linearization method

Initially, the method used consists of linearizing the flow and cost functions to the first order. Linearization is then written within the framework of a successive linear program. It comes in the form of the algorithm represented in Figure 2.

310

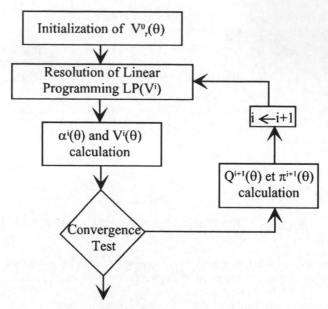

Figure 2 : Algorithm of Successive Linear Programming

The convergence test performed at each iteration i consists of calculating the sum $\sum_{\theta=1}^{T}\sum_{r=1}^{R}\left|V_r^i(\theta)-V_r^{i-1}(\theta)\right|$ and of comparing it to a set maximum value.

4.2 Network studied
We are working on a drinking water supply network located in the east of France on the frontier with Germany. This network serves approximately 20,000 inhabitants. It is illustrated in Figure 3.

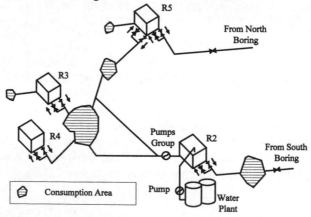

Figure 3: Diagram of an example network

In this network, the flow circulating between the reservoirs R2 and R4 typically depends on the consumption between the two reservoirs, but also on the level in the reservoir R4. The complete relation is written as follows:

$$Q_{inputR\,4}(\theta) = 40.58 + 0.43 \times Consum(\theta) - 0.026 \times V_{R4}(\theta)$$

4.3 Non convergence on the example network

When the method show in Figure 2 is applied directly, the convergence criterion behaves as shown in Figure 4.

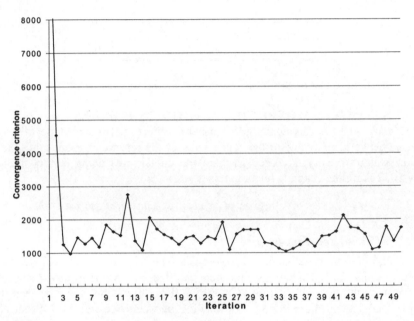

Figure 4: Evolution of the convergence criterion for the first method

The convergence criterion thus starts by decreasing over the first three iterations, after which the evolution of the criterion is no longer monotonous.

By examining the calculations more closely, it can be seen that there are several minimum costs for each iteration. The first algorithm will thus be transformed so as to make a choice from these solutions with an identical cost.

4.4 Choosing from several solutions with an identical cost

To enable a choice from among the solutions with the same cost, the linear program (P_1) is resolved first and the value of the objective function noted as $\tilde{z}^1$.

In a second stage, a new problem of linear optimization is posed, (P'_1), with the same constraints as the program (P_1) to which the constraint $\sum_{\theta=1}^{T}\sum_{i=1}^{I}\alpha_i(\theta)\pi_i(\theta) = \tilde{z}^1$

is added. The new objective function makes it possible to satisfy a second criterion, such as the safety of the network. This objective function may consist of :
- maximizing all the levels in all the reservoirs
- maximizing the levels in certain reservoirs
- maximizing the level of certain reservoirs at certain time steps

4.5   Convergence of the method

If at each iteration of the successive linear program the two linear programs (P_1) and (P'_1) are resolved, the criterion C then evolves as follows :

Iteration	1	2	3	4
C criterion	16052.6	1115.1	20.8	0.2

The method just presented converges on the example network.

5   NETWORKS WHERE THE METHOD CONVERGES

To verify whether this convergence can be generalized to any type of drinking water network, the value of the coefficients of the reservoirs influencing the flow rates was artificially increased, each of the coefficients being in fact multiplied by 10. Taking the example relation, this gives :

$$Q_{inputR4}(\theta) = 40.58 + 0.43 \times Consum(\theta) - 0.26 \times V_{R4}(\theta)$$

5.1   Case of non-convergence

The algorithm proposed no longer converges. The criterion C starts by decreasing over the first eleven iterations, and then stabilizes at a value of around 18 m³. On examining the level of the reservoirs during these iterations, it can be seen that as from the eleventh iteration there is a swing between two trajectories that can be examined in Figure 5 for the reservoir R2.

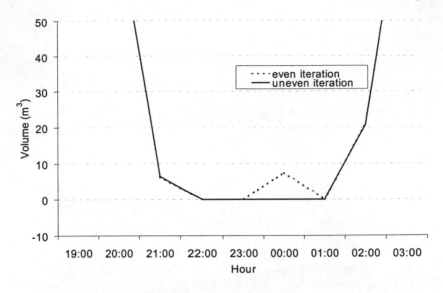

Figure 5 : Level in R2 between 19h00 and 3h00

This graph clearly demonstrates a phenomenon of convergence towards two distinct local minima. In this artificial case, there is no global minimum for the problem of optimization.

5.2 Characterization of the optimization problem

As the influence of the levels on costs is not taken into account, the program (P_2) that is solved is a program with a linear objective function and non-linear constraints; to be more precise, the field of the admissibles in a non-convex field.

The convergence towards two local minima can thus be explained by the fact that at each linearization, the optimum is found in a different part of the field of the admissibles. This was well presented by Minoux in [4] thanks to the example of Figure 6. In this example, the aim is to minimize a polynomial function in a convex field. By choosing to linearize the first order objective function at each iteration, starting from the point x^0 we obtain x^1 then x^2 then indefinitely x^1 and x^2.

314

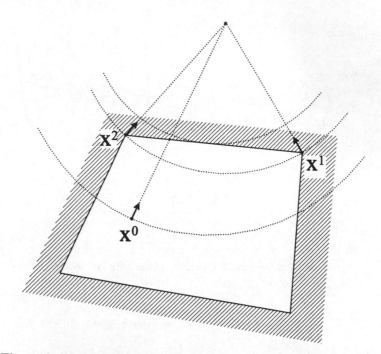

Figure 6: Convergence towards two local minima proposed by Minoux

5.3 Type of networks giving a single solution

The non convexity of the field of the admissibles previously evidenced is, from a mathematical viewpoint, due to two causes :

- in the program (P_l) maximum and minimum volume constraints systematically appear involving the same left member with different signs ;
- the functions may themselves be non convex if the levels of several reservoirs with different signs influence a flow rate.

The convergence towards two local minima only occurs in the second case. In our example, a check was made to see whether, if all the functions were made convex, the convergence of the method occurred again.

6 CONCLUSION

In cases where linear programming does not enable the modeling of the operational optimization of a drinking water network over 24 hours, a non-linear program appears. We sought to resolve this program by a successive linear programming method.

The method which converges is a first order linearization to which a method of choice taken from solutions with the same costs was added.

By transforming the data of the reference network artificially, it was possible to define the limits of the convergence. Thus in a drinking water network where the water levels in the reservoirs have a strong influence on the flow rates, and where that influence affects several reservoirs per flow, there may be convergence towards several local minima. In that case, the local minima found will nonetheless provide a good approximation of the solution.

REFERENCES

BAKER, Thomas E., LADSON, Leon S., **Successive Linear Programming at Exxon**. Management Science, N°3, Vol. 31, March 1985, pp. 264-274.

CRAWLEY, Philip D., DANDY, Graeme C., **Optimal Operation of Multiple-reservoir System**. Journal of Water Resources Planning and Management, N°1, Vol. 119, February 1993, pp. 1-17.

FERRIS, M.C., **Iterative Linear Programming Solution of Convex Programs**. Journal of Optimization theory and applications, N°1, VOL. 65, April 1990, pp. 53-65.

MINOUX, M., **Mathematical Programming. Theory and Algorithms**, Wiley, New York (1986)

PALACIOS-GOMEZ, F., LADSON, Leon S., ENQUIST, M. **Non-linear Optimization by Successive Linear Programming**. Management Science, N°28, Vol. 10, October 1982, pp. 1106-1120

ULANICKI, B., RANCE, J.P., DAVIS, D. et al. **Computer-aided Optimal Pump Selection for Water Distribution Networks**. Journal of Water Resources Planning and Management, N°5, Vol. 119, October 1993, pp. 542-562.

Random Search Methods in Model Calibration and Pipe Network Design

D.P. Solomatine

ABSTRACT
Many water-related problems are identified as optimisation problems and traditionally such problems were posed as single-extremum problems. Recently, increasing attention has been given to the multi-extremum (global) optimisation techniques that allow the identification of better solutions. This paper provides an overview of global optimisation (GO) methods – controlled random search, genetic algorithm, adaptive cluster covering and other methods implemented in PC-based GLOBE tool. Two examples of GO practical use are given; hydrological model calibration and pipe network optimal design. The application of GO methods in water distribution problems allows solving complex optimisation problems without the necessity of analytical formulation of an objective function. Within this paper the performance of different GO algorithms are compared and the most efficient algorithms are identified.

1   INTRODUCTION

A number of water-related problems require optimisation. These include reservoir optimisation, optimal allocation of resources and planning, calibration of models and design problems. Traditionally, optimisation problems were solved using linear and non-linear optimisation techniques which normally assume that the minimised function (objective function) is known in analytical form and that it has a single minimum (without a loss of generality we will assume that the optimisation problem is a minimisation problem).

In practice, however, there are many problems that cannot be described analytically and many objective functions have multiple extrema. In these cases it is necessary to pose multi-extremum (global) optimisation problems (GOP) where the traditional optimisation methods are not applicable and other solutions must be investigated. One of these typical GOPs is that of automatic model calibration, or parameter identification. The objective function is then the discrepancy between the model output and the observed data, i.e. the model error, measured normally as the weighted RMSE. One of the approaches to solve GOPs that has become increasingly popular during the recent years is the use of genetic algorithms (GAs) (*Goldberg 1989, Michalewicz 1996*). A considerable number of publications related to water-resources are devoted to their use (*Wang, 1991; Cieniawski, 1995;*

Savic & Walters, 1997; Franchini & Galeati, 1997). Evolutionary algorithms (EA) are variations of the same idea used in GAs, but were developed by different research groups. It is possible to say that EAs include GAs as a particular case.

Other GO algorithms are used for solving calibration problems (*Duan et al., 1993; Kuczera, 1997; Solomatine, 1995; 1998; 1999*), but GAs seem to be preferred. Our experience shows however, that many practitioners are unaware of the existence of other GO algorithms that are more efficient and effective than GAs. This serves as a motivation for writing this paper, which has the following main objectives:

- classify and briefly describe GO algorithms,
- demonstrate the applicability of several GO algorithms, including GAs, on two problems – automatic model calibration and pipe network design.

2   GLOBAL (MULTI-EXTREMUM) OPTIMIZATION

2.1   Posing the problem

A global minimisation problem with box constraints is considered: find an optimiser x^* which generates a minimum of the objective function $f(x)$ where $x \ O X$ and $f(x)$ are defined in the finite interval (box) region of the n-dimensional Euclidean space: $X = \{x \ O \ R^n: \ a \# x \# b\}$ (component wise). This constrained optimisation problem can be transformed to an unconstrained optimisation problem by introducing a penalty function with a high value outside the specified constraints. In cases when the exact value of an optimiser cannot be found, we deal with its estimate and correspondingly, the minimum of the estimate.

Approaches to solving this problem depend on the properties of $f(x)$:

1. $f(x)$ is a single-extremum function expressed analytically. If its derivatives can be computed, then gradient-based methods may be used: conjugate gradient methods; quasi-Newton or variable metric methods, such as DFP and BFGS methods (*Jacobs, 1977; Press et al., 1991*). In certain cases, e.g. in the calibration of complex hydrodynamic models, if some assumptions are made about the model structure and/or the model error formulation, then there are several techniques available (such as inverse modelling) that allow the speed up of the solution.

Many engineering applications use minimisation techniques for single-extremum functions, but often without investigating whether the functions are indeed single-extremum (unimodal). They do recognise however, the problem of the "good" initial starting point for the search of the minimum. This can be partially attributed to the lack of wide awareness within the engineering community of the developments in the area of global optimisation.

2. $f(x)$ is a single-extremum function which is not analytically expressed. The derivatives cannot be computed, and direct search methods can be used such as the *Nelder & Mead (1965)* method.

3. No assumptions are made about the properties of $f(x)$, so it is a multi-extremum function which is not expressed analytically, and we have to consider multi-extremum or global optimisation.

Most calibration problems belong to the third category of GO problems. At certain stages the GO techniques may use the single-extremum methods from category 2 as well.

2.2 Main approaches to global optimisation

Törn and Zilinskas (1989) and *Pintér (1995)* provide extensive coverage of various methods. It is possible to distinguish the following groups:
- set (space) covering techniques;
- random search methods;
- evolutionary and genetic algorithms (can be classified as random search methods);
- methods based on multiple local searches (multistart) using clustering;
- other methods (simulated annealing, trajectory techniques, tunnelling approach, analysis methods based on a stochastic model of the objective function).

Several representatives of these groups are covered below:
Set (space) covering methods. In these the parameter space X is covered by N subsets $X_1,...,X_N$, such that their union covers the whole of X. Then the objective function is evaluated in N representative points $\{x_1, ..., x_N\}$, each one representing a subset, and a point with the smallest function value is taken as an approximation of the global value. If all previously chosen points $\{x_1, ..., x_k\}$ and function values $\{f(x_1), ..., f(x_k)\}$ are used when choosing the next point x_k+1, then the algorithm is called a *sequential (active) covering* algorithm (and passive if there is no such dependency). These algorithms were found to be inefficient.

The following algorithms belong to the group of *random search methods*:
Pure direct random search (uniform sampling). N points are drawn from a uniform distribution in X and f is evaluated in these points; the smallest function value is the minimum f^* assessment. If f is continuous then there is an asymptotic guarantee of convergence, but the number of function evaluations grows exponentially with n. An improvement is to generate evaluation points in a sequential manner taking into account already known function values when the next point is chosen, producing thus an adaptive random search (*Pronzato et al., 1984*).

Evolutionary strategies and genetic algorithms. The family of *evolutionary algorithms* is based on the idea of modelling the search process of natural evolution, though these models are crude simplifications of biological reality. Evolutionary algorithms (EA) are variants of randomised search and use the terminology from biology and genetics. For example, given a random sample at each iteration, pairs of parent individuals (points), selected on the basis of their 'fit' (function value), recombine and generate new 'offspring'. The best of these are selected for the next generation. Offspring may also 'mutate', i.e., randomly change their position in space. The premise is that fit parents are likely to produce even fitter children. In fact, any random search may be interpreted in terms of biological evolution; generating a random point is analogous to a mutation and the step made towards the minimum after a successful trial may be treated as a selection.

Historically, evolution algorithms have been developed using three techniques - evolution strategies (ES), evolutionary programming (EP), and genetic algorithms (GA). *Back & Schwefel (1993)* provided an overview of these approaches, which differ mainly in the type of mutation, recombination and selection operators. There are various versions of GA varying in the way crossover, selection and construction of the new population are performed. In *evolutionary strategies (ES)*, mutation of coordinates is performed with respect to corresponding variances of a certain n-dimensional normal distribution and various versions of recombination are introduced. More on GA applications can be found in *Wang (1991), Cieniawski (1995), Savic & Walters (1997), Franchini & Galeati (1997)* and *Schaetzen et al. (1998)*.

Multistart and clustering. The basic idea of the family of *multistart* methods is to apply a search procedure several times and then choose an assessment of the global optimiser. One of the popular versions of multistart used in global optimisation is based on clustering, i.e., creating groups of mutually close points that hopefully correspond to relevant regions of attraction of potential starting points (*Törn & Zilinskas, 1989*). The *region (area) of attraction* of a local minimum x^* is the set of points in X starting from which a given local search procedure P converges to x^*. For the global optimisation using GLOBE, the tool used in the present study, two multistart algorithms were developed – *Multis* and *M-Simplex*. They are both constructed according to the following pattern:

1. Generate a set of N random points and evaluate f at these points.
2. Reduce the initial set by choosing p best points (with the lowest f_i).
3. Launch local search procedures starting from each of p points. The best point reached is the minimiser assessment.

In *Multis*, at step 3 the Powell-Brent local search (see *Brent, 1973; Press et al., 1991*) is started. In *M-Simplex* the downhill simplex descent of *Nelder & Mead (1965)* is used.

The ACCO strategy developed by the author also uses clustering as the first step, but it is followed by the global randomised search, rather than local search.

Adaptive cluster covering (ACCO) (first introduced by *Solomatine, 1995*, and later enhanced in *Solomatine, 1998, 1999*) is a workable combination of generally accepted ideas of reduction, clustering and covering (Fig.1).

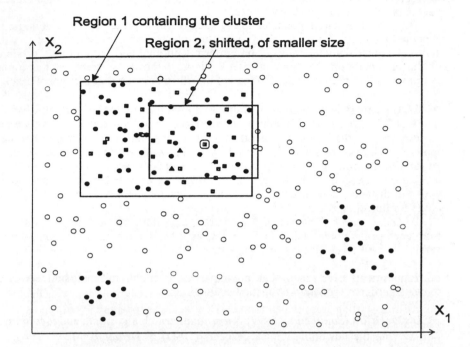

Region 1 containing the cluster

Region 2, shifted, of smaller size

○   initial population

●   'good' points grouped into three clusters

▫   points generated at regional iteration 1

◙   'best' point in region 1, around which region 2 is formed

    points generated at regional iteration 2

Fig. 1. Iterations in ACCO algorithm in 2-dimensional case

1. Clustering. Clustering (identification of groups of mutually close points in search space) is used to identify the most promising subdomains in which to continue the global search by active space covering.

2. Covering shrinking subdomains. Each subdomain is covered randomly. The values of the objective function are then assessed at the points drawn from the uniform or some other distribution. Covering is repeated multiple times and each time the subdomain is progressively reduced in size.

3. Adaptation. Adaptive algorithms update their algorithmic behaviour depending on the new information revealed about the problem. In ACCO, adaptation is in *shifting* the subregion of search, *shrinking* it, and changing the density (number of points) of each covering - depending on the previous assessments of the global minimiser.

4. Periodic randomisation. Due to the probabilistic character of point generation, any strategy of randomised search may simply miss a promising region for search. In order to reduce this danger, the initial population is re-randomised, i.e. the problem is solved several times.

Depending on the implementation of each of these principles, it is possible to generate a family of various algorithms, suitable for certain situations, e.g. with non-rectangular domains (hulls), non-uniform sampling and with various versions of cluster generation and stopping criteria. Figure 1 shows the example of an initial sampling, and iterations 1 and 2 for one of the clusters in a two dimensional case.

ACCOL strategy is the combination of ACCO with the multiple *local searches*:

1. *ACCO phase.* ACCO strategy is used to find several regions of attraction, represented by the promising points that are close (such points we will call 'potent'). The potent set P_1 is formed by taking one best point found for each cluster during progress of ACCO. After ACCO stops, the set P_1 is reduced to P_2 by leaving only several m (1...4) best points which are also distant from each other, with the distance at each dimension being larger than, for example, 10% of the range for this dimension;

2. *Local search (LS) phase.* An accurate algorithm of local search is started from each of the potent points of P_2 (multistart); to find accurately the minimum a version of the Powell-Brent search is used.

Experiments have shown that in comparison to traditional multistart, ACCOL brings significant economy in function evaluations.

ACD algorithm (*Solomatine, 1999*) is a random search algorithm, and it combines ACCO with the downhill simplex descents (DSD) of *Nelder & Mead (1965)*. It identifies the area around the possible local optimiser by using clustering and then applies covering and DSD in this area. The main steps of ACD are:
- sample points (e.g., uniformly), and reduce the sample to contain only the best points;
- cluster points, and reduce clusters to contain only the best points;
- in each cluster, apply the limited number of steps of DSD to each point, thus moving them closer to an optimiser;
- if the cluster is potentially 'good' i.e. that it contains points with low function values, cover the proximity of several best points by sampling more points, e.g. from uniform or beta distribution;
- apply local search (e.g., DSD, or some other algorithm of direct optimisation) starting from the best point in 'good' clusters. In order to limit the number of steps, the fractional tolerance is set to be, say, 10 times greater than the final tolerance (that is, the accuracy achieved is somewhat average);
- apply the final accurate local search (again, DSD) starting from the very best point reached so far; the resulting point is the assessment of the global optimiser.

ACDL algorithm, combining ACD with the multiple local searches, has been built and tested as well.

3   GLOBAL OPTIMIZATION TOOL *GLOBE*
A PC-based system GLOBE incorporating nine GO algorithms was built. Currently, GLOBE includes the following nine algorithms (described above):
-    *CRS2 (controlled random search, by Price, 1983)*;

- CRS4 (modification of the *controlled random search* by *Ali & Storey, 1994*);
- GA with a one-point crossover, and with a choice between the real-valued or binary coding (15 bits were used in our experiments); with the standard random bit mutation; between the tournament and fitness rank selection; and between elitist and non-elitist versions.
- *Multis* - multistart algorithm;
- *M-Simplex* - multistart algorithm;
- *adaptive cluster covering (ACCO)*;
- *adaptive cluster covering with local search (ACCOL)*.
- *adaptive cluster descent (ACD)*;
- *adaptive cluster descent with local search (ACDL)*.

GO algorithms implemented in GLOBE were used in solving various problems of calibration and optimisation. One of the important items in research was analysis of algorithm performance. Normally, three main *performance indicators* were investigated:

1. *effectiveness* (how close the algorithm gets to the global minimum);
2. *efficiency* (running time) of an algorithm measured by the number of function evaluations needed (the running time of the algorithm itself is negligible compared with the function evaluation time);
3. *reliability* (robustness) of the algorithms can be measured by the number of successes in finding the global minimum, or at least approaching it sufficiently closely.

Traditional benchmark functions used in GO with known global optima (*Törn & Zilinskas, 1989; Duan et al., 1993, Solomatine, 1999*) were used for performance analysis. Detailed analysis of algorithms' performance is made in *Solomatine (1998* and *1999)*. The general conclusion is that the most efficient, effective and reliable algorithms are ACCO and CRS4 (and not, for example GA).

4   USING GO ALGORITHMS IN MODEL CALIBRATION

The objective of *calibration (parameter optimisation)* of any model of a physical system is to identify the values of some parameters in a model which are not known a priori. This is achieved by inputting data into the model and subsequent comparison of the computed values of output variables and the values as measured in the physical system. In the trial-and-error calibration, the model parameters are adjusted according to a modeller judgement, in order to provide a 'better' match. Among the widely used evaluation criteria for the *discrepancy (error)* between the model results and observations the following criterion may be mentioned: given the vector OBS_t of observed output variables values, and the corresponding vector MOD_t of the modelled values at time moments $t = 1...T$,:

$$E = \frac{1}{T}\sum_{t=1}^{T} w_t (OBS_t - MOD_t)^{\gamma}$$

where γ taken between 1 and 2, and w_t is the weight associated with particular moments of time. For $\gamma = 2$ and $w_t = 1$, E is the mean squared error (MSE). *Duan et al. (1993)* and *Pintér (1995)* report the high degree of non-linearity, and often non-smoothness of such functions. This justifies the use of GO methods for minimising E.

Our experience of using GO algorithms implemented in GLOBE for model calibration includes:
- calibration of a lumped hydrological model (*Solomatine, 1995*);
- calibration of an electrostatic mirror model (*Vdovine et al., 1995*);
- calibration of a 2D free-surface hydrodynamic model (*Constantinescu, 1996*);
- calibration of an ecological model of plant growth;
- calibration of a distributed groundwater model (*Solomatine et al., 1999*);

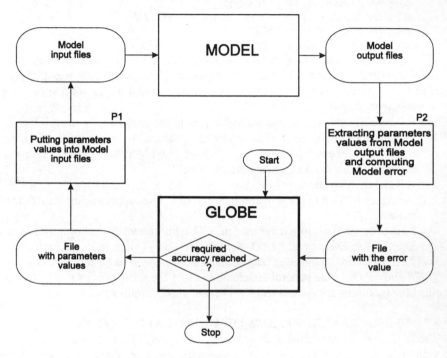

Fig. 2 Using GLOBE for model calibration

GLOBE can be configured to use an external program as a supplier of the objective function values. The number of independent variables and the constraints imposed on them are supplied by the user in the form of a simple text file. Figure 2 shows how GLOBE is used in the problems of automatic calibration. *Model* must be an executable module (program) which does not require any user input, and the user has to supply two transfer programs *P1* and *P2*. These three programs (*Model, P1, P2*) are activated from GLOBE in a loop. GLOBE runs in DOS protected mode (DPMI) providing enough memory to load the program modules. A Windows version is being developed. The user interface includes several graphical windows displaying the progress of minimisation in different coordinate plane projections. The parameters of the algorithms can be easily changed by the user.

As an example, Figure 3 shows a conceptual lumped rainfall-runoff model SIRT. It has 8 unknown parameters that have to be determined through calibration.

Before calibration, the ranges for all 8 parameters have been set. Figure 4 shows the performance of several algorithms used.

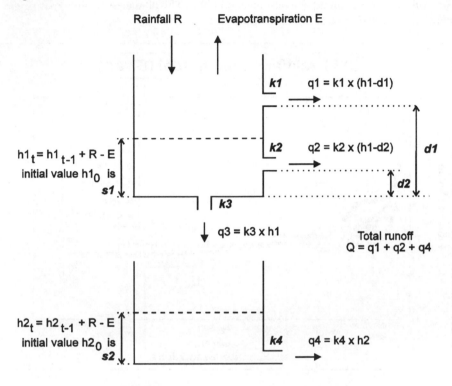

Parameters to be identified : *d1, d2, k1, k2, k3, k4, s1, s2*

Fig. 3 SIRT conceptual rainfall-runoff model with 8 parameters to calibrate

5   USING GO ALGORITHMS IN PIPE NETWORK OPTIMISATION
GO techniques have been applied to other optimisation problems:
- dynamic programming for reservoir optimisation (*Lee 1997*).
- optimisation of pipe networks (*Abebe & Solomatine, 1998*).

5.1   Approaches to pipe network optimisation
A water distribution network is a system containing pipes, reservoirs, pumps and valves of different types, which are all connected to provide water to consumers. It is a vital component of the urban infrastructure and requires significant investment. The optimal design of water distribution networks has various aspects to be considered such as hydraulics, reliability, material availability, water quality, infrastructure and demand patterns. Due to the complexity of these components they are often considered separately.

One of the important problems is determining the optimal diameters of pipes in a network with a predetermined layout. This includes providing the pressure and quantity of water required at every demand node. The diameters of the pipes in the network are considered as *decision variables*.

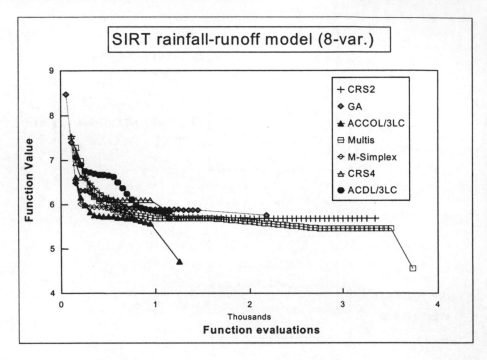

Fig. 4 Performance of GLOBE algorithms in model calibration

The *objective function* to be minimised is normally the cost of the network, however, there are a number of *constraints*. If the actual cost of the network is the sole objective function, then the search will end up with the minimum possible diameter allocated to each of the pipes in the network. Hence, the minimum head requirement at the demand nodes is taken as a constraint for the choice of pipe diameter. Additional *market constraints* dictate the use of commercially available (discrete) pipe diameters. Most of the constraints can be transferred to the *penalty* function added to the cost function; other constraints (such as ranges of pipe diameters) are handled as box-constraints. This allows the problem to be posed as a combinatorial problem.

Throughout the recent decades various researchers have addressed this problem in a number of different ways. Although enumeration techniques (explicit and implicit) are reliable global search methods (*Yates et al., 1984, Gessler, 1985*) their application to practical size networks is limited due to the extraordinarily wide search space and consequent enormous computational time.

Kessler & Shamir (1989) used the linear programming gradient (LPG) method as an extension of the method proposed by *Alperovits & Shamir (1977)*. It consists of two stages: an LP problem is solved for a given flow distribution and then a search is conducted in the space of flow variables. Later *Fujiwara & Khang (1990)* used a two-phase decomposition method extending that of *Alperovits & Shamir (1977)* to non-linear modelling. Also *Eiger et al. (1994)* used the same formulation as *Kessler & Shamir (1989)*, which leads to the determination of lengths of one or more segments in each link with discrete diameters.

Even though split pipe solutions obtained in the above cases are cheaper, some of the results obtained were not practical and others were not feasible. In addition to this, some of the methods impose a restriction on the type of hydraulic components in the network. For instance, the presence of pumps in the network increases the non-linearity of the problem and as a result, networks with pumps can not be solved by some of the methods.

Dissatisfaction with the results obtained forced researchers to pose the considered problem as a constrained discrete multi-extremum (global) optimisation problem. In recent years GAs (*Goldberg, 1989; Michalewicz, 1996*) have been applied to the problem of pipe network optimisation. *Simpson & Goldberg (1994), Dandy et al. (1993), Murphy et al. (1994)* and *Savic & Walters (1997)* applied both simple genetic algorithm (SGA) and improved GAs, with various enhancements based on the nature of the problem and reported promising solutions for problems from literature.

Problems associated with GAs are the uncertainty about the termination of the search and, as in all random search methods, the absence of a guarantee for the global optimum. To alleviate these problems we found it was appropriate to apply other GO methods.

5.2 Experiments with GLOBE

For network optimisation, a scheme presented on Figure 5 was realised – an appropriate interface was created between GLOBE and a network simulation model. EPANET (*Rossman, 1993*) was used for simulation – it can handle steady as well as dynamic loading conditions.

328

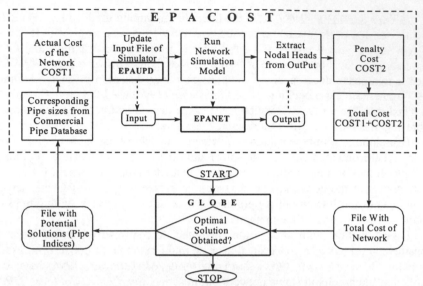

Fig. 5 GLOBE in solving network optimisation problem

The actual cost of the network C_a is calculated based on the cost per unit length associated with the diameter and the length of the pipe:

$$C_a = \sum_{i=1}^{n} c(D_i)L_i$$

where n is the number of pipes in the network and $c(D_i)$ is the cost per unit length of the ith pipe with diameter D_i and length L_i. The penalty cost is superimposed on top of the actual cost of the network in such a way that it will discourage the search in an infeasible direction. It is defined on the basis of the difference between the required minimum head (H_{min}) at the demand nodes and the lowest nodal head obtained after simulation.

Several test problems were run. Some of the details can be found in *Abebe and Solomatine, 1998* (and on *www.ihe.nl/hi/sol/global.htm*). For one of the test problems, the Hanoi network, data obtained from literature (*Fujiwara & Khang, 1990*) was used. The network contains 34 pipes, 31 demand nodes and a reservoir. The cost of the network was minimised by choosing a diameter for each of the 34 pipes out of a set of 14 possible values. The cost per unit length is calculated based on the analytical cost function $1.1HD^{1.5}$ used by *Fujiwara & Khang (1990)*. It must be noted that our approach does not require having an analytical function relating to the cost per unit length of the pipes to the diameter. Table 1 and Figure 6 give an indication of the performance of various algorithms for the Hanoi network.

Table 1. Comparison of the various solutions obtained for the Hanoi network

	Fujiwara & Khang (1990)	Fujiwara & Khang (1990)	Eiger et al. (1994)	Savic & Walters (1997)		Abebe & Solomatine (1998)	
				GA-1	GA-2	GA	ACCOL
Continuous solution*	Yes	No	No	Yes	Yes	Yes	Yes
Feasible**	No	No	No	Yes	Yes	Yes	Yes
No. of nodes with head deficit	18	18	6				
Cost of network (millions)	5.354	5.562	6.027	6.073	6.195	7.006	7.836

*   "No" implies split pipe solution.
**   Only in terms of nodal head violation.

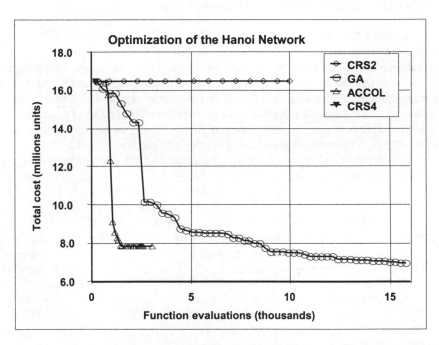

Fig. 6 Performance of GLOBE algorithms in minimising cost of the Hanoi pipe network

It is observed that the algorithm CRS2 terminated without any improvement in the network obtained by the first iteration. It stopped after exhausting the total number of function evaluations allowed. On the other hand, CRS4 in almost all of the experiments stopped after exhausting the initial iteration. One possible reason for the failure of these algorithms is that they were designed for continuous variables.

GA and ACCOL moved the search towards the global minimum. For GA it took relatively more function evaluations (about 17000, taking 75 minutes on Pentium 100MHz PC) and ended up with a better least cost solution. ACCOL on the other hand converged 5 times faster (about 3000 function evaluations, 15 minutes on the same PC) and reported a solution that was slightly more expensive (10%) than that of GA. The results obtained from all algorithms are feasible from the nodal head point of view.

For large networks it is sensible to use a suite of algorithms in order to have the choice between fast-running algorithms and algorithms oriented towards more exhaustive but longer searches. The fact that the problems considered do not have pumping facilities is simply because these problems were taken from literature. However, it is possible to optimise networks with any kind of hydraulic facilities as long as the network simulator has the handling capabilities. Since global optimisation methods work with any objective (cost) functions, they can also be efficiently used to solve other optimisation problems.

6 CONCLUSION

Advances in computer technology during the last decade have made it possible for researchers and practitioners to use ordinary PC's to apply computational techniques that have previously only been performed on expensive mainframe computers. One group of such methods is comprised of non-exhaustive empirical search techniques that do not require the knowledge of analytical expression for an objective function – they allow the search of the global (as opposed to local) extremum (a well-known representative of such GO methods is a GA).

This paper demonstrates how GO methods can be applied in solving two important problems arising in the water industry – model calibration and pipe network optimisation. For a calibration problem, GO methods allow for the search of model parameters, providing a better model fit, without requiring a tedious trial-and-error process. In pipe network optimisation these methods often allow the identification of solutions that are impossible to find using more traditional optimisation methods. Our current research is aimed at testing various parameter coding schemes allowing for better handling of pipe optimisation networks' discrete character.

GO methods allow the use of any objective (cost) functions, not necessarily analytically expressed but also represented as an algorithm or computer code of the third party. The proposed optimisation setup can handle any type of loading condition and makes no restriction on the type of hydraulic components in the network nor does it need analytical cost functions for the pipes. These methods can be efficiently used to optimise design, operation, maintenance and other aspects of water distribution where objective function may include a complex combination of various factors.

It is worth mentioning that the group of GO methods includes a whole variety of methods. The popularity of GAs does not mean however, that these are the only GO methods available. Our research shows that in almost all problems where GA is used, many other GO methods (CRS4 or ACCO), that are often more efficient, can be used as well.

Details on GLOBE tool can be found on *www.ihe.nl/hi/sol/global.htm*.

REFERENCES

Abebe A.J. & Solomatine D.P. 1998. Application of global optimization to the design of pipe networks. *Proc. Int. Conf. Hydroinformatics-98*, pp. 989-995.

Ali, M.M & Storey, C. 1994. Modified controlled random search algorithms. *Intern. J. Computer Math.*, 53, pp. 229-235.

Alperovits, E. & U. Shamir 1977. Design of optimal water distribution systems. *Water Res. Research*, 13(6), 885-900.

Back, T. & Schwefel, H.-P. 1993. An overview of evolutionary algorithms for parameter optimization. *Evolutionary Computation*, 1, No. 1, pp. 1-23.

Brent, R.P. 1973. Algorithms for minimization without derivatives. Prentice-Hall, Englewood-Cliffs, N.J., 195p.

Cieniawski, S.E, Eheart, J.W. & Ranjithan, S. 1995. Using genetic algorithms to solve a multiobjective groundwater monitoring problem. *Water Resour. Res.*, 31 (2), 399-409.

Constantinescu A. 1996. *Calibration of hydrodynamic numerical models using global optimization techniques*. M.Sc. thesis No. HH262, IHE, Delft, 85p.

Duan, Q., Gupta, V., Sorooshian, S. 1993. Shuffled complex evolution approach for effective and efficient global minimization. *J. of Optimiz. Theory Appl.*, 76 (3), pp. 501-521.

Eiger, G.,U. Shamir & A. Ben-Tal 1994. Optimal design of water distribution networks. *Water Res. Research*, 30(9), 26637-2646.

Franchini, M. & Galeati, G. 1997. Comparing several genetic algorithm schemes for the calibration of conceptual rainfall-runoff models. *Hydrol. Sci. J.*, 42 (3), 357 - 379.

Fujiwara, O. & D.B. Khang 1990. A two-phase decomposition method for optimal design of looped water distribution networks. *Water Res. Research*, 26(4), 559-5549.

Gessler, J. 1985. Pipe network optimization by enumeration. *Proc., Spec. Conf. on Comp. Applications/Water Res. Research*, 23(7), 977-982.

Goldberg, D.E. 1989. Genetic Algorithms in Search, optimization & Machine Learning. Addison-Wesley Publishing Co., Reading.

Jacobs, D.A.H. 1977. The state of the art in numerical analysis, Academic Press, London.

Kessler, A. & U. Shamir 1989. Analysis of the linear programming gradient method for optimal design of water supply networks. *Water Res. Research*, 25(7), 1469-1480.

Kuczera, G. 1997. Efficient subspace probabilistic parameter optimization for catchment models. Water Resour. Res, 33 (1), 177-185, January.

Lee H. 1997. *Optimal reservoir operation for flood control using a hybrid modelling approach*. M.Sc. thesis No. HH319, IHE, Delft.

Michalewicz, Z. 1996. Genetic Algorithms + Data Structures = Evolution Programs (3ed), Springer-Verlag, New York, N.Y.

Nelder, J.A & Mead, R. 1998. A simplex method for function minimization. *Computer J.*, vol. 7, No. 4 p. 308-313, 1965.

Neumaier. WWW page on global optimization: *solon.cma.univie.ac.at/ ~neum/glopt.html*

Pintér, J. 1995. Global optimization in action. Kluwer, Amsterdam.

332

Press, W.H., Flannery, B.P., Teukolsky, S.A., Vetterling, W.T. 1990. Numerical recipes in Pascal. The art of scientific computing. Cambridge University Press, Cambridge, 759p..

Price, W.L. 1983. Global optimization by controlled random search. *J. of Optimization Theory & Applications*, (40), 333-348.

Pronzato L., Walter E., Venot A., Lebruchec J.-F. 1984. A general purpose global optimizer: implementation and applications, in *Mathematics and Computers in Simulation*, 26, 412-422.

Quindry, G.E., E.D. Brill & J.C. Liebman 1981. Optimization of looped water distribution systems. *J. of Env. Engg.*, ASCE, 107(4), 665-679.

Rossman, L.A. 1993. EPANET, Users Manual. Risk Reduction Engg. Laboratory, Office of Research & Devt., U. S. Env. Protection Agency, Cincinnati, Ohio.

Savic, D.A. & G.A. Walters 1997. Genetic algorithms for least-cost design of water distribution networks. *Water Res. Planning & Management*, 123(2), 67-77.

Schaetzen, de, W.B.F., Savic, D.A. and Walters, G.A. 1998. A genetic algorithm approach to pump scheduling in water supply systems. *Proc. Int. Conf. Hydroinformatics-98*, pp. 897 - 899.

Simpson, A.R. & D.E. Goldberg 1993. Pipeline optimization using an improved genetic algorithm. Department of Civil & Env. Engg., *University of Adelaide, Research Report* No. R109.

Simpson, A.R., G.C. Dandy & L.J. Murphy 1994. Genetic algorithms compared to other techniques for pipe optimization. *Water Res. Planning & Management*, ASCE, 120(4), 423-443.

Solomatine, D.P. 1995. The use of global random search methods for models calibration. *Proc. XXVIth congress of the IAHR*, London, Sept. 1995.

Solomatine, D.P. 1998. Genetic and other global optimization algorithms-comparison and use in calibration problems. *Proc. Int. Conf. Hydroinformatics-98*, pp. 1021-1027.

Solomatine, D.P. 1999. Two strategies of adaptive cluster covering with descent and their comparison to other algorithms. *J. Global Optimiz. No.1.*

Törn, A. and Zilinskas, 1989. A. *Global optimization*. Springer-Verlag, Berlin, 255pp.

Yates, D.F., A.B. Templeman & T.B. Boffey 1984. The computational complexity of the problem of determining least capital cost designs for water supply networks. *Engg. Optimization*, 7(2), 142-155.

Vdovine, G., Middelhoek, S., Bartek, M., Sarro, P.M., & Solomatine, D.P. 1995. Technology, characterization and applications of adaptive mirrors fabricated with IC-compatible micromachining. *Proc. SPIE's Int. Symposium on Optical Science, Engineering and Instrumentation*, vol. 2534/13, San-Diego, USA, July 10-14.

On the Quality of a Simulated Annealing Algorithm for Water Network Optimization Problems

J. Sousa *and* M. da Conceição Cunha

ABSTRACT
A great deal of research has been conducted on water network design optimization. The literature reports many studies using classical optimization methods to solve this problem. The solutions obtained in those studies are of two types: continuous diameter solutions or split-pipe solutions. As many researchers have pointed out, they can not be considered realistic solutions.

Recently, the use of modern heuristics, such as genetic and simulated annealing algorithms, is helping to overcome the drawbacks of classical methods by achieving discrete diameter solutions. The implementation of this kind of heuristic technique has been presented in the literature using three well-known test problems (Alperovits and Shamir (1977), Hanoi and New York networks). However, there is still much work to do to evaluate these heuristics from the optimization standpoint. The heuristics have worked well for the standard test problems, but three problems are not enough for general conclusions to be drawn. This paper presents other case studies from the literature, with the aim of beginning a more detailed evaluation of simulated annealing algorithms.

1   INTRODUCTION

The least-cost network design can be stated as follows. The combination of diameters that will give the least-cost network needed to supply a set of demand nodes, within a prescribed range of pressures, has to be found, using a discrete commercial diameter set. This problem is difficult to solve, even for simple networks. The hydraulic behaviour of a looped network is expressed using a set of nonlinear equations. Although it is now relatively straightforward to solve them, their inclusion in optimization problem formulations is nevertheless quite complex. Because the problems are discrete and nonlinear objective functions are used to represent network costs, a non-linear mixed-integer problem results.

A great deal of research has been done on this subject. The literature contains many studies that use classical optimization methods to solve this problem. The solutions obtained with these methods are of two types: continuous diameter solutions or split-pipe solutions. As many researchers have pointed out, they cannot be considered realistic solutions. Comprehensive reviews of the optimization of water distribution systems are given by Rowell (1979), Walski (1985), and Simpson *et al.* (1994).

Recently, the use of modern heuristics, such as genetic and simulated annealing algorithms, is allowing the drawbacks to be overcome by achieving discrete diameter solutions. The implementation of such heuristics has been presented in the literature through three well-known test problems (Alperovits and Shamir (1977), Hanoi and New York networks). But much work still remains to be done to evaluate these heuristics from the optimization standpoint. The heuristics worked well for the standard test problems, but three problems are not enough for general conclusions to be drawn. This paper adds some more case studies to evaluate simulated annealing algorithms. The networks to be solved are taken from the literature, and their solutions from the application of other optimization methods are known.

2 EXAMPLES

Modern heuristics are particularly appropriate for dealing with the combinatorial nature of least-cost network design optimization problems. The authors have already presented some work in the use of simulated annealing to solve them. The guidelines for implementing this algorithm, and the parameter set used are discussed in Cunha and Sousa (1997, 1999).

The simulated annealing algorithm is based on an analogy with the physical process of the cooling and annealing of solids. The process involves increasing the temperature so that the molecules become mobile, and then decreasing the temperature so that the molecules arrange themselves at random. Once a low energy state is reached, a crystalline structure is achieved. If the temperature drops rapidly, a final crystalline structure does not occur because the low energy states are reached too quickly. Metropolis *et al.* (1953) described a simple algorithm to expresses these ideas. A parallel can be established between the states of the physical process and the values of the objective function of the optimization problem. The temperature will play the role of a parameter for controlling the search for the optimal solution.

The examples given below use data that can be found in the literature references. In all of them, the solutions obtained through simulated annealing respect the constraints considered by their authors. Head loss has been determined using the Hazen-Williams equation:

$$\Delta H = w \frac{L}{C^\alpha D^\beta} Q^\alpha$$

ΔH: head loss
L: pipe length
D: pipe diameter
w: numerical conversion constant, dependent on units
C: roughness coefficient

α, β: regression coefficients.

The values $w = 10.6792$, $\alpha = 1.85$ e $\beta = 4.87$ were used.

2.1   Example 1

The first example (Figure 1) is taken from Gessler (1981), and concerns a network comprising 17 nodes and 25 pipes defining nine loops. The network is supplied by a reservoir. All pipes but one are new. The existing pipe has a known diameter. There is one load condition and there are minimum pressure requirements to be observed.

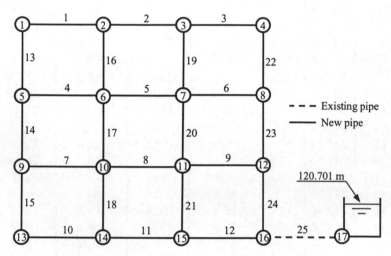

Fig. 1 - Example 1 - network

Gessler (1981) presents two solutions corresponding to different minimum diameter (D_{min}) requirements ($D_{min}=2$" and $D_{min}=4$"). Table 1 shows the solution design and cost determined, considering the two diameter constraints, using the simulated annealing algorithm and the Gessler method. Gessler (1981) uses a solution procedure that assumes that groups of pipes have the same diameter. This is done so that the problem size is decreased to enable an enumeration method to be applied. Simulated annealing can be implemented without using this kind of constraint, and is thus more flexible, producing better results (results are presented in Table 1).

Table 1 - Solution of Example 1

Pipe	Length	H. W.	Diameter (mm)			
Number	(m)	Coef.	Gessler (2")	S. A. (2")	Gessler (4")	S. A. (4")
1	365.76	120	152.4	50.8	101.6	101.6
2	365.76	120	152.4	50.8	101.6	101.6
3	365.76	120	152.4	101.6	101.6	101.6
4	365.76	120	50.8	203.2	101.6	101.6
5	365.76	120	50.8	50.8	101.6	101.6
6	365.76	120	50.8	50.8	101.6	101.6
7	365.76	120	50.8	50.8	101.6	152.4
8	365.76	120	50.8	203.2	101.6	152.4
9	365.76	120	50.8	203.2	101.6	101.6
10	365.76	120	203.2	50.8	203.2	101.6
11	365.76	120·	203.2	101.6	203.2	152.4
12	365.76	120	203.2	101.6	203.2	203.2
13	243.84	120	152.4	203.2	152.4	203.2
14	243.84	120	152.4	50.8	152.4	152.4
15	243.84	120	152.4	50.8	152.4	101.6
16	243.84	120	50.8	50.8	101.6	101.6
17	243.84	120	50.8	203.2	101.6	101.6
18	243.84	120	50.8	50.8	101.6	101.6
19	243.84	120	50.8	50.8	101.6	101.6
20	243.84	120	50.8	50.8	101.6	101.6
21	243.84	120	50.8	101.6	101.6	152.4
22	243.84	120	152.4	101.6	152.4	101.6
23	243.84	120	152.4	101.6	152.4	101.6
24	243.84	120	152.4	203.2	152.4	152.4
25	914.40	120	254.0	254.0	254.0	254.0
Solution cost			834,000	810,000	858,000	830,000

2.2 Example 2

The second example (Figure 2) presented by Walski *et al.* (1990), concerns increasing the capacity of the network. This can be achieved by building new pipes and reinforcing the existing pipes by using ones in parallel. There are three load conditions (one for the supply and the two other to satisfy fire demands) and the corresponding minimum pressure requirements. The existing network is supplied by two reservoirs and has nine nodes and nine pipes defining one loop. For the expansion, five new nodes and five new pipes are added. Three of the existing pipes can be reinforced. The extended network will have three loops.

The method consisting of assuming that groups of pipes have the same diameter is also used in Walski *et al.* (1990). Results are displayed in Table 2, and the effect of that method can once again be observed.

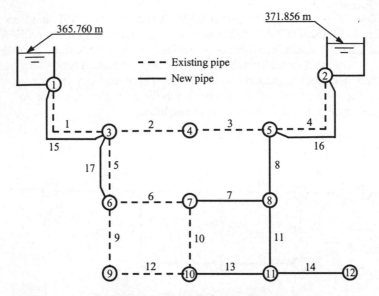

Fig. 2 - Example 2 - network

Pipe	Length	H. W.	Diameter (mm)	
Number	(m)	Coef.	Walski	S. A.
1	4,828.032	75	355.6	355.6
2	1,609.344	80	254.0	254.0
3	1,609.344	80	254.0	254.0
4	6,437.376	80	254.0	254.0
5	1,609.344	80	254.0	254.0
6	1,609.344	100	203.2	203.2
7	1,609.344	120	203.2	203.2
8	1,609.344	120	304.8	304.8
9	1,609.344	80	254.0	254.0
10	1,609.344	100	101.6	101.6
11	1,609.344	120	304.8	203.2
12	1,609.344	100	203.2	203.2
13	1,609.344	120	203.2	152.4
14	1,609.344	120	203.2	254.0
15	4,828.032	120	0.0	0.0
16	6,437.376	120	355.6	355.6
17	1,609.344	120	0.0	0.0
Solution cost			1,884,432	1,750,320

Table 2 - Solution of Example 2

2.3 Example 3

Example 3 (Figure 3) was first presented by Walski *et al.* (1990), and later solved by Taher and Labadie (1997). It is a new network supplied by a reservoir. A pump was installed to ensure supplies at all nodes, in good condition, even in those located at a great distance from the reservoir and on high ground. There are nine nodes and nine pipes organized into two loops. Three load conditions were studied. Two of them corresponded to different supply situations (serving to calculate energy costs) and the other related to fire conditions.

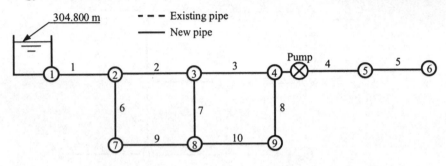

Fig. 3 - Example 3 - network

The pump operating point must be determined for each load scenario (the pump head is constrained by a maximum of 304.8 m). Taher and Labadie (1997) applied linear programming to solve this problem using a split-pipe approach for pipes and a linearized function to compute energy costs. In spite of this, smaller costs are obtained through simulated annealing (results are shown in Tables 3 and 4).

Table 3 - Solution of Example 3 (diameters and costs)

Pipe	Length	H. W.	Diameter (mm)		
Number	(m)	Coef.	Walski	Taher	S. A.
1	3,218.600	100	406.4	406.4	406.4
2	1,609.400	100	304.8	304.8	355.6
3	1,609.400	100	304.8	254.0/304.8	304.8
4	1,609.400	100	203.2	203.2	203.2
5	1,609.400	100	203.2	203.2	203.2
6	1,609.400	100	254.0	254.0	203.2
7	1,609.400	100	152.4	203.2	203.2
8	1,609.400	100	254.0	203.2	254.0
9	1,609.400	100	254.0	203.2/254.0	152.4
10	1,609.400	100	254.0	254.0	203.2
Energy cost			151,537	151,755	148,012
Pipes cost			1,948,848	1,890,239	1,858,058
Solution cost			2,100,385	2,041,994	2,006,070

Table 4 - Solution of Example 3 (pump head)

Pump operating points:		Walski	Taher	S. A.
Load 1:	Discharge (l/s) -	50.47	50.50	50.47
	Head (m) -	?	83.50	82.20
Load 2:	Discharge (l/s) -	35.33	35.30	35.33
	Head (m) -	?	55.10	52.00
Load 3:	Discharge (l/s) -	50.47	50.50	50.47
	Head (m) -	89.90	90.80	90.30

2.4 Example 4

In example 4 (Taher and Labadie (1997)) the network is supplied by a clear well, using a pump, the cost of which is not included in the optimization procedure. In this case the layout and the design of the network needed to satisfy demands and minimum pressure requirement had to be found. The pump head is constrained by a maximum of 79.25 m. There are 17 nodes than can be linked by 34 pipes (Figure 4).

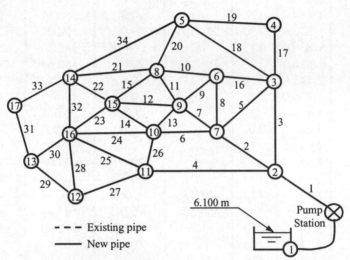

Fig. 4 - Examples 4 and 5 - network

There are five load conditions:
Load 1: Instantaneous peak (in this case energy costs are ignored)
Load 2: 0.67 Average demand (6h)
Load 3: 1.28 Average demand (6h)
Load 4: 1.08 Average demand (6h)
Load 5: 0.97 Average demand (6h).

Table 5 - Solution of Examples 4 and 5 (diameters and costs)

Pipe Number	Length (m)	H. W. Coef.	Diameter (mm)			
			Example 4		Example 5	
			Taher	S. A.	Lansey	S. A.
1	30.5	120	?	609.6	?	558.80
2	3,660.0	120	356/305	558.8	579.12	558.80
3	3,660.0	120	305	0.0	388.62	0.00
4	3,660.0	120	610	558.8	589.28	609.60
5	2,745.0	120	0	0.0	0.00	0.00
6	1,830.0	120	0	254.0	368.30	254.00
7	1,830.0	120	203	304.8	279.40	304.80
8	1,830.0	120	203	406.4	0.00	406.40
9	1,830.0	120	0	0.0	0.00	0.00
10	1,830.0	120	0	304.8	256.54	304.80
11	1,830.0	120	0	0.0	0.00	0.00
12	1,830.0	120	152	0.0	0.00	0.00
13	1,830.0	120	254/203	0.0	0.00	0.00
14	1,830.0	120	406	0.0	256.54	0.00
15	1,830.0	120	254/203	0.0	0.00	0.00
16	1,830.0	120	152	254.0	340.36	254.00
17	1,830.0	120	203/152	203.2	162.56	203.20
18	2,745.0	120	203/152	0.0	0.00	0.00
19	1,830.0	120	0	0.0	0.00	0.00
20	1,830.0	120	0	203.2	0.00	203.20
21	1,830.0	120	0	0.0	0.00	0.00
22	1,830.0	120	0	0.0	0.00	0.00
23	1,830.0	120	0	203.2	0.00	203.20
24	1,830.0	120	0	0.0	0.00	0.00
25	1,830.0	120	406/356	457.2	370.84	508.00
26	1,830.0	120	559/508	0.0	0.00	0.00
27	1,830.0	120	356	203.2	342.90	203.20
28	2,745.0	120	0	0.0	0.00	0.00
29	1,830.0	120	356/305	0.0	304.80	0.00
30	1,830.0	120	0	304.8	0.00	304.80
31	1,830.0	120	254	254.0	251.46	254.00
32	1,830.0	120	203/152	203.2	243.84	152.40
33	3,660.0	120	0	0.0	0.00	0.00
34	3,660.0	120	0	0.0	200.66	0.00
Energy cost			3,420,823	2,561,060	3,810,000	2,491,627
Pipes cost			5,575,609	4,987,696	5,370,000	5,103,380
Pump cost			0	0	3,120,000	3,219,103
Solution cost			8,996,432	7,548,756	12,290,000	10,814,112

Since energy costs are not considered for the first load, the pump head is at its upper limit. For the other loads, since energy costs are taken into consideration, the pump head in the final solution is at 61.75 m, 68.75m, 66.25m, and 64.75m respectively (results are in Tables 5 and 6).

2.5   Example 5

Example 5 (Lansey and Mays (1989)) uses example 4, but the pump cost is a function of maximum flow and the corresponding head. Therefore the final cost of the network consists of the cost of the pipes, energy (present value) and the pump. Lansey and Mays (1989) applied non-linear programming, using a continuous diameter approach to solve this network. Although the global optimal solution is not known, it can be asserted that the solution (Tables 5 and 6) found by those authors is a local optimum, because simulated annealing gives a better one.

Table 6 - Solution of Examples 4 and 5 (pump head)

| | Example 4 | | Example 5 | |
Pump operating points	Taher	S. A.	Lansey	S. A.
Load 1: Discharge (l/s) -	?	726.91	?	726.91
Head (m) -	?	79.25	?	79.25
Load2: Discharge (l/s) -	?	270.57	?	270.57
Head (m) -	?	61.25	?	61.75
Load 3: Discharge (l/s) -	?	516.92	?	516.92
Head (m) -	?	66.25	?	68.75
Load 4: Discharge (l/s) -	?	436.15	?	436.15
Head (m) -	?	64.25	?	66.25
Load 5: Discharge (l/s) -	?	391.73	?	391.73
Head (m) -	?	63.25	?	64.75

2.6   Example 6

Example 6 (Lansey and Mays (1989)) uses the network of example 5, but assumes that there are five pipes already installed (Figure 5). The final layout will comprise, besides these five pipes, the pipes chosen from the remaining 29 that will produce the least-cost solution. Energy costs are ignored and new load conditions are considered:

Load 1: Instantaneous peak
Load 2: Daily peak + Fire flow at node 9
Load 3: Daily peak + Fire flow at nodes 12 and 13
Load 4: Daily peak + Fire flow at node 9.

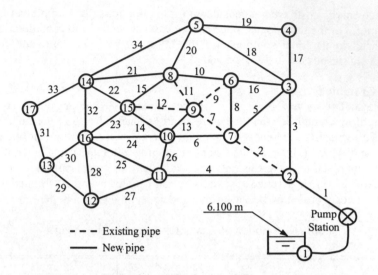

Fig. 5 - Examples 6 – network

Once again, the solution provided by Lansey and Mays (1989) is a continuous diameter solution, thus not realistic (results are in Tables 7 and 8).

Table 7 - Solution of Example 6 (pump head)

Pump operating points		Lansey	S. A.
Load 1:	Discharge (l/s) -	?	726.91
	Head (m) -	?	75.25
Load2:	Discharge (l/s) -	?	682.74
	Head (m) -	?	78.75
Load 3:	Discharge (l/s) -	?	651.19
	Head (m) -	?	78.75
Load 4:	Discharge (l/s) -	?	682.74
	Head (m) -	?	79.25

Table 8 - Solution of Example 6 (diameters and costs)

Pipe	Length	H. W.	Diameter (mm)	
Number	(m)	Coef.	Lansey	S. A.
1	30.5	120	?	558.8
2	3,660.0	120	406.40	406.4
3	3,660.0	120	439.42	355.6
4	3,660.0	120	513.08	609.6
5	2,745.0	120	0.00	0.0
6	1,830.0	120	287.02	0.0
7	1,830.0	120	355.60	355.6
8	1,830.0	120	0.00	0.0
9	1,830.0	120	304.80	304.8
10	1,830.0	120	0.00	0.0
11	1,830.0	120	254.00	254.0
12	1,830.0	120	254.00	254.0
13	1,830.0	120	0.00	0.0
14	1,830.0	120	0.00	0.0
15	1,830.0	120	0.00	0.0
16	1,830.0	120	347.98	304.8
17	1,830.0	120	132.08	254.0
18	2,745.0	120	149.86	0.0
19	1,830.0	120	0.00	203.2
20	1,830.0	120	0.00	0.0
21	1,830.0	120	0.00	0.0
22	1,830.0	120	157.48	0.0
23	1,830.0	120	322.58	355.6
24	1,830.0	120	0.00	0.0
25	1,830.0	120	360.68	558.8
26	1,830.0	120	0.00	254.0
27	1,830.0	120	401.32	304.8
28	2,745.0	120	0.00	0.0
29	1,830.0	120	378.46	0.0
30	1,830.0	120	0.00	355.6
31	1,830.0	120	210.82	254.0
32	1,830.0	120	0.00	203.2
33	3,660.0	120	0.00	0.0
34	3,660.0	120	0.00	0.0
Pipes cost				4,289,168
Pump cost				3,153,101
Solution cost			7,600,000	7,442,270

344

In all the examples described above, regarding a variety of hydraulic case studies, simulated annealing always obtains an improvement vis-à-vis the solutions obtained through the use of other optimization methods reported in the literature. The best enhancement of the solutions seems to be achieved in the examples where the pump head needs to be optimized, as in examples 4 and 5. In most of the examples there are different diameter solutions with the same cost.

The work started with this paper will have to be continued in order to compile a significant sample of case studies. A better understanding of the relationship between problem characteristics and optimum solutions will be achieved by analysing the corresponding results.

3   CONCLUSIONS

The authors have shown in previous works that the simulated annealing algorithm works quite well vis-à-vis the other methods reported in the literature. Nevertheless, it has been evaluated by resolving the three classic test problems (Alperovits and Shamir (1977), Hanoi and New York networks). In this paper six more networks are solved. The resolution of these networks confirms that simulated annealing is capable of coping with the combinatorial nature of least-cost design problems, always producing improvements on the solutions reported in the literature.

Modern heuristics could make a great contribution to solving realistically-sized network problems. It is important to evaluate the different heuristics by means of a wide variety of case studies so that general conclusions relative to optimization can be drawn (adequacy and accuracy of alternative methods). The creation of a network library for the use of all the researchers interested in this subject would greatly help in the accomplishment of this task.

REFERENCES

Alperovits, E., and Shamir, U., "Design of optimal water distribution systems." *Water Resour. Res.*, 13(6), 885-900, 1977.

Cunha, M.C.M.O., and Sousa, J.O., "Simulated Annealing Algorithms for Water Distribution Systems Optimization." *EURO XV-INFORMS XXXIV Joint International Meeting*, Barcelona, 1997.

Cunha, M.C.M.O., and Sousa, J.O., "Water Distribution Network Design Optimization: Simulated Annealing Approach". *J. of Water Resources Planning and Management, ASCE* 1999 (forthcoming).

Gessler, J., "Analysis of pipe networks" in *Closed-conduit flow*, Water Resources Publications, M. H. Chaudhry and V. Yevjevich Ed., 61-69, 1981.

Lansey, K.E., and Mays, L.W., "Optimization model for water distribution systems design.", *J. of Hydraul. Eng.*, ASCE, 115(10), 1401-1418, 1989 .

Metropolis, N., Rosenbluth, M., Rosenbluth, A., Teller, A., and Teller, E., "Equation of State Calculations by Fast Computing Machines." *Journal of Chemical Physics* , 21, 1087-1092, 1953.

Rowell W.F., "A methodology of optimal design of water distribution systems." *PhD thesis*, University of Texas, Austin, Tex., 1979.

Simpson, A.R., Dandy, G.C., and, Murphy, L.J., "Genetic algorithms compared to other techniques for pipe optimization." *J. Water Resour. Plng. and Mgmt.*, ASCE, 120(4), 423-443, 1994.

Taher, S.A., and Labadie, J.W., "Optimal design of water-distribution networks with GIS." *J. Water Resour. Plng. and Mgmt.*, ASCE, 122(4), 301-311, 1997.

Walski, T.M., "State-of-the-art: pipe network optimization." *Computer applications in water resources*, Buffalo, N.Y., 1985.

Walski, T.M., Sjostrom, J.W., and Gessler. J, *Water distribution systems: simulation and sizing* , Lewis publishers, Inc., 1990.

An Explicit Solution to the Colebrook-White Equation Through Simulated Annealing

J. Sousa, M. da Conceição Cunha *and* A. Sá Marques

ABSTRACT
Even though it is sixty years old, the Colebrook-White formula is still the equation that is most commonly used to calculate flow resistance through commercial pipes. It is considered quite accurate, giving errors of no more than 5 percent. Nevertheless, this equation has a less attractive feature: it is not possible to express the unknown variable (the friction factor) as a function of other variables. This makes its incorporation into the constraints of water network optimization problems impossible. The Hazen-Williams equation has therefore been used in these problems. Recently, however, this practice has been much criticized in the literature. In fact, its use may induce errors of up to 40 percent.

A great deal of research has been conducted to find explicit mathematical approximations of the Colebrook-White equation. This paper presents a new explicit equation obtained by using simulated annealing. The tests performed with this new equation yield results of high accuracy vis-à-vis the implicit Colebrook-White equation.

1 INTRODUCTION

Even though it is 60 years old, the Colebrook-White (C-W) equation (White and Colebrook (1937), and Colebrook (1939))

$$\frac{1}{\sqrt{\lambda}} = -2\log\left(\frac{k}{3.7\,D} + \frac{2.51}{R_e\,\sqrt{\lambda}}\right) \qquad (1)$$

still is the equation that is most commonly used to calculate flow resistance through commercial pipes. The C-W equation combines the theoretically based equations of Prandtl (1952)

$$\frac{1}{\sqrt{\lambda}} = -2\log\left(\frac{2.51}{R_e\,\sqrt{\lambda}}\right) \qquad (2)$$

appropriate for the turbulent regime in hydraulically smooth pipes, and of von Karman (1934),

$$\frac{1}{\sqrt{\lambda}} = -2\log\left(\frac{k}{3.7\,D}\right) \qquad (3)$$

347

appropriate for the turbulent regime in hydraulically rough pipes, to present a universal formula representing every kind of turbulent flow regime. It can be shown that the formula works quite well, giving errors of no more than 5 percent. However, this equation has a less attractive feature: it is not possible to express the unknown variable (the friction factor) as a function of other variables. The solution of pipe flow problems is not usually straightforward; many trials are needed. The C-W graphical representation through Moody's diagram facilitates this task. But it cannot be used for the computerised solving of pipe flow problems. This characteristic also makes its incorporation into the constraints of water network optimization problems impossible. For this reason the Hazen-Williams equation has been used in these problems. Recently, however, this practice has been much criticized in the literature. In fact, its use may induce errors of up to 40 percent.

A great deal of research has been conducted to find explicit mathematical approximations of the C-W equation. This paper first gives a brief literature review, and then presents a new approach to find an explicit version of the C-W, using a simulated annealing algorithm. The tests performed with this new explicit equation yield results of high accuracy vis-à-vis the implicit C-W equation.

2   LITERATURE REVIEW

The need to overcome the drawbacks of the C-W equation have given rise to a large number of works presenting a variety of explicit equations. Analyzing these works, it can be noted that two different ways of finding an explicit version of C-W formula have been used. One uses a method (Method 1) consisting of replacing the smooth turbulent term of C-W formula by an explicit equation. The formula is then calibrated to reduce its error vis-à-vis the C-W formula.

The other method (Method 2) consists of replacing the friction factor appearing in right-hand side of C-W formula by an approximation (usually an approximation of the smooth turbulent regime or an explicit equation obtained by Method 1). As such, the equation obtained can be regarded as a step in an iterative resolution of the C-W formula. This equation is then calibrated to obtain an accurate approximation of the C-W formula.

Table 1 - Synthesis of literature: explicit formulations (Method 1)

Author	Equation	Error
Altshul in Nekrasov (1968)	$$\frac{1}{\sqrt{\lambda}} = -1.8 \log\left(\frac{k}{10\,D} + \frac{7}{R_e}\right)$$	18.53%
Barr (1972)	$$\frac{1}{\sqrt{\lambda}} = -2 \log\left(\frac{k}{3.7\,D} + \frac{5.15}{R_e^{0.892}}\right)$$	2.77%
Churchill (1973)	$$\frac{1}{\sqrt{\lambda}} = -2 \log\left[\frac{k}{3.7\,D} + \left(\frac{7}{R_e}\right)^{0.9}\right]$$	3.42%
Barr (1975)	$$\frac{1}{\sqrt{\lambda}} = -2 \log\left(\frac{k}{3.7\,D} + \frac{5.1286}{R_e^{0.89}}\right)$$	2.93%
Swamee and Jain (1976)	$$\frac{1}{\sqrt{\lambda}} = -2 \log\left(\frac{k}{3.7\,D} + \frac{5.74}{R_e^{0.9}}\right)$$	3.35%
Round (1980)	$$\frac{1}{\sqrt{\lambda}} = -1.8 \log\left(0.135\frac{k}{D} + \frac{6.5}{R_e}\right)$$	10.18%
Haaland (1983)	$$\frac{1}{\sqrt{\lambda}} = -1.8 \log\left[\left(\frac{k}{3.7\,D}\right)^{1.11} + \frac{6.9}{R_e}\right]$$	1.38%
J. J. J. Chen (1984)	$$\lambda = 0.3164\left(0.11\frac{k}{D} + \frac{1}{R_e^{0.83}}\right)^{0.3}$$	8.00%
J. J. J. Chen (1985)	$$\frac{1}{\sqrt{\lambda}} = -2 \log\left(\frac{k}{3.7\,D} + \frac{4.52}{R_e}\log\left(\frac{R_e}{7}\right)\right)$$	2.62%
Nackab (1988)	$$\frac{1}{\sqrt{\lambda}} = -2 \log\left(\frac{k}{3.7\,D} + \frac{2.51}{R_e\sqrt{0.4\,R_e^{-0.3} + 0.0053}}\right)$$	2.95%

Table 2 - Synthesis of literature: explicit formulations (Method 2)

Author	Equation	Error
N. H. Chen (1979)	$\dfrac{1}{\sqrt{\lambda}} = -2\log\left[\dfrac{k}{3.7065\,D} - \dfrac{5.0452}{R_e}\log\left(\dfrac{1}{2.8257}\left(\dfrac{k}{D}\right)^{1.1098} + \dfrac{5.8506}{R_e^{0.8981}}\right)\right]$	0.324%
Barr (1980)	$\dfrac{1}{\sqrt{\lambda}} = -2\log\left(\dfrac{k}{3.7\,D} + \dfrac{5.02\log(R_e/4.518\log(R_e/7))}{R_e\left(1 + R_e^{0.52}(k/D)^{0.7}/29\right)}\right)$	0.403%
Malafaya-Baptista (1980)	$\dfrac{1}{\sqrt{\lambda}} = -2\log\left[\dfrac{k}{3.7\,D} - \dfrac{5.02}{R_e}\log\left(\dfrac{k}{3.7\,D} + \dfrac{2.51}{R_e}\dfrac{1}{0.49\,R_e^{-0.11} + 0.18\,R_e^{0.1}(k/D)^{0.6}}\right)\right]$	0.159%
Barr (1981)	$\dfrac{1}{\sqrt{\lambda}} = -2\log\left(\dfrac{k}{3.7\,D} + \dfrac{4.518\log(R_e/7)}{R_e\left(1 + R_e^{0.52}(k/D)^{0.7}/29\right)}\right)$	0.531%
Zigrang and Sylvester (1982)	$\dfrac{1}{\sqrt{\lambda}} = -2\log\left[\dfrac{k}{3.7\,D} - \dfrac{5.02}{R_e}\log\left(\dfrac{k}{3.7\,D} - \dfrac{5.02}{R_e}\log\left(\dfrac{k}{3.7\,D} + \dfrac{13}{R_e}\right)\right)\right]$	0.114%

The two methods mainly employed two types of smooth turbulent regime approximations:

$$\frac{1}{\sqrt{\lambda}} = -1.8 \log\left(\frac{7}{R_e}\right) = -2 \log\left[\left(\frac{7}{R_e}\right)^{0.9}\right] \cong -2 \log\left(\frac{5.76}{R_e^{0.9}}\right) \tag{4}$$

$$\lambda = a\,R_e^b + c \tag{5}$$

Tables 1 and 2 present a synthesis of the equations (and their authors) reported in the literature, obtained by the two methods, as well as the corresponding accuracy vis-à-vis the C-W formula.

It can be noticed that Method 1 gives rise to higher errors.

3   A NEW APPROACH

The approach proposed to find a new explicit version of the C-W formula consists of minimizing the function:

$$\text{Min} \quad \frac{\Phi(R_e,\frac{k}{D}) - \Psi(R_e,\frac{k}{D}, a, b, c, ...)}{\Phi(R_e,\frac{k}{D})}$$

where $\Phi(R_e,\frac{k}{D})$ is the C-W equation and $\Psi(R_e,\frac{k}{D}, a, b, c, ...)$ is the explicit formulation that is being looked for (R_e: Reynolds number; k/D: relative roughness). The parameters a, b, c, ... will play the role of decision variables in the minimization problem.

This optimization problem is solved using a simulated annealing algorithm. In the attempt to find a globally optimal solution by surmounting the considerable drawbacks inherent in the classical optimization methods, simulated annealing has been useful. This heuristic approach makes a series of moves to search the solution space. The classical optimization methods will only accept improving moves, but annealing algorithms search in such a way that worsening moves are also accepted, in keeping with Metropolis criteria (Metropolis *et al.* (1953)), so as to escape local optima. Temperature is the parameter used to manage the probability of accepting worsening moves, and it is reduced as the algorithm proceeds. High temperature values accept most uphill moves, but as the temperature drops, uphill moves are increasingly excluded. If the temperature drops very quickly, we can observe the situation that occurs when the traditional optimization methods are used, that is, there is a premature convergence to the local optima. The advantage of the new technique is that every part of the solution space can be searched.

There are two main features to implement this algorithm: a perturbation mechanism and a cooling schedule.

The perturbation mechanism is described in Lin (1965), Kirkpatrick *et al.*(1983) and Aarts and van Laarhoven (1985) and defines the random changes made to a solution so that a candidate solution could be obtained, for evaluation using the Metropolis criterion. With the perturbation mechanism employed here, a decision variable is chosen at random, and its value is changed at random. It will be decreased in 50% of the cases and increased in the other 50%. This is an iteration of the search procedure and is performed taking the range of variation employed in Moody's diagram for R_e and k/D.

The cooling schedule is devised to control the temperature parameter. A starting temperature is proposed, after which the algorithm has to define how and when this temperature is decreased during implementation. There must also be a stopping criterion.

Aarts and Van Laarhoven (1985) have shown, if certain conditions are satisfied then Markov chains can be applied to the analysis of the annealing algorithm, and so general convergence results can be found.

The way to establish the annealing parameters is described in Cunha and Sousa (1997, 1999).

A great many runs were needed, using various seed numbers for the pseudo-random generator, because the simulated annealing method is essentially a random search technique.

The explicit formulations obtained from the simulated annealing are shown in Table 3.

Table 3 - New explicit formulations

Equation		Error
Method 1: $$\frac{1}{\sqrt{\lambda}} = -2\log\left(\frac{k}{3.7D} + \frac{4.2}{R_e\sqrt{R_e^{-0.28} + (k/D)^{0.74}} + 0.012}\right)$$		0.498%
Method 2: $$\frac{1}{\sqrt{\lambda}} = -2\log\left[\frac{k}{3.7D} - \frac{5.16}{R_e}\log\left(\frac{k}{3.7D} + \frac{5.09}{R_e^{0.87}}\right)\right]$$		0.123%

Figure 1 and Figure 2 depict the Moody's diagram containing the values obtained using the explicit formulations given in Table 3. The values fit Moody's diagram quite well.

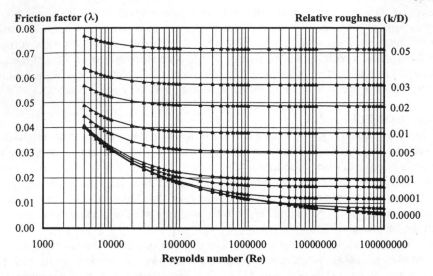

Fig. 1 - Moody's diagram and predicted values by Method 1

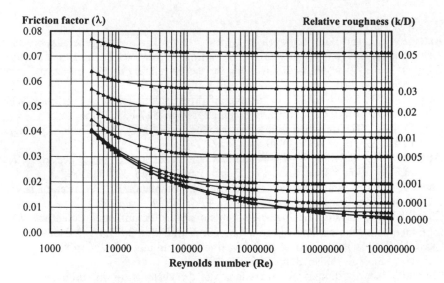

Fig. 2 - Moody's diagram and predicted values by Method 2

As other works in the literature report, Method 2 clearly works better than Method 1. In fact, it can be seen that Method 2 produces a more accurate explicit formulation. The small magnitude of the error encountered should be emphasized.

In the literature report regarding Method 2 presented before, only in one case (Zigrang and Sylvester (1982)) is the error slightly smaller than the error for the

354

explicit formula displayed in Table 3 for the same method. It can be said that they are errors of the same order. But it is necessary to point out that if a third iteration were performed (as was done by Zigrang and Sylvester (1982)), the new formula would produce errors of no more than 0.02%. Nevertheless, the version of the new explicit formulation shown in Table 3, using method 2, was chosen, because it is a very accurate approximation of the C-W formula and very easy to handle.

4 CONCLUSIONS

The C-W equation is the more accurate formula to use to determine the friction factor necessary to calculate the head losses in pipe flow problems. But its implicit formulation with regard to this friction factor makes it difficult to use. An explicit formulation of the C-W equation would, however, be more convenient to handle, especially if computers are to be used. Many authors have addressed this problem and various explicit formulations, with differing degrees of accuracy, are reported in the literature.

In this work, an attempt has been made to construct a new explicit formulation, using a simulated annealing algorithm. It worked well, and two approximations were obtained. The error reported using the approximation achieved with Method 2 shows an improvement vis-à-vis the works found in the literature. Thus this new explicit formulation should be considered when solving pipe flow problems in future.

REFERENCES

Aarts, E. H. L., and Van Laarhoven, P. J. M., "Statistical cooling: a general approach to combinatorial optimization problems." *Philips J. Res.* , 40(4), pp. 193-226, 1985.

Barr, D. I. H., "New forms of equations for the correlation of pipe resistance data", *Proc. Instn. Civ. Engrs.*, Part 2, 53, pp. 383-390, 1972.

Barr, D. I. H., "Two additional methods of direct solution of the Colebrook-White function", *Proc. Instn. Civ. Engrs.*, Part 2, 59, pp. 827-835, 1975.

Barr, D. I. H., "The transition from laminar to turbulent flow", *Proc. Instn. Civ. Engrs.*, Part 2, 69, pp. 555-562, 1980.

Barr, D. I. H., "Solutions of the Colebrook-White function for resistance to uniform turbulent flow", *Proc. Instn. Civ. Engrs.*, Part 2, 71, pp. 529-535, 1981.

Chen, J. J. J., "A simple explicit formula for the estimation of pipe friction factor", *Proc. Instn. Civ. Engrs.*, Part 2, 77, pp. 49-55, 1984.

Chen, J. J. J., "Systematic explicit solutions of the Prandtl and Colebrook-White equations for pipe flow", *Proc. Instn. Civ. Engrs.,* Part 2, 79, pp. 383-389, 1985.

Chen, N. H., "An explicit equation for friction factor in pipe", *Ind. Eng. Chem. Fund.*, A. C. S., Vol. 18, No. 3, pp. 296-297, 1979.

Churchill, S. W., "Empirical expressions for the shear stress in turbulent flow in commercial pipe", *Am. Inst. Ch. Engrs. J.,* Vol. 19, No. 2, pp. 375-376, 1973.

Colebrook, C. F., "Turbulent flow in pipes, with particular reference to the transition region between the smooth and rough pipe laws", *J. Instn. Civ. Engrs.*, 11, pp. 133-156, 1939.

Cunha, M. C. M. O., and Sousa, J. J. O., "Simulated Annealing Algorithms for Water Distribution Systems Optimization." *EURO XV-INFORMS XXXIV Joint International Meeting*, Barcelona, 1997.

Cunha, M. C. M. O., and Sousa, J. J. O., "Water Distribution Network Design Optimization: Simulated Annealing Approach". *J. of Water Resources Planning and Management, ASCE* 1999 (forthcoming).

Haaland, S. E., "Simple and explicit formulas for the friction factor in turbulent pipe flow", *J. Fluids Eng. Trans.,* ASME, 105, pp. 89-90, 1983.

Kirkpatrick, S., Gelatt, C., and Vecchi, M., "Optimization by Simulated Annealing." *Science,* 220(4598), pp. 671-680, 1983.

Lin, S., "Computer solutions of the traveling-salesman problem." *Bell Syst. Tech. J.,* 44, pp. 2245-2269, 1965.

Malafaya-Baptista, M., "Critérios de explicitação da expressão de Colebrook-White. Novas perspectivas", Laboratório de Hidráulica, FEUP, 1980.

Metropolis, N., Rosenbluth, M., Rosenbluth, A., Teller, A., and Teller, E., "Equation of State Calculations by Fast Computing Machines." *Journal of Chemical Physics,* 21, pp. 1087-1092, 1953.

Nackab, J., "Calcul direct, sans iteration, de la perte de charge en conduite par la formule de Colebrook", *La Houille Blanche,* N° 1, pp. 61, 1988.

Nekrasov, B., *Hydraulics,* Peace Publishers, Moscow, pp. 95-101, 1968.

Prandtl, L., "Essentials of fluid mechanics", Hafner Publishing, New York, 1952.

Round, G. F., "An explicit approximation for the friction factor-Reynolds number relation for rough and smooth pipes", *The Canadian Journal of Chemical Engineering,* Vol. 58, pp. 122-123, 1980.

Swamee, P. K. and Jain, A. K., "Explicit equations for pipe-flow problems", *Journal of the Hydraulics Division,* A.S.C.E., Vol. 102, No. HY5, pp. 657-664, 1976.

Von Karman, Th., "Turbulence and skin friction", *J. Aeronaut. Sci.,* 1 (1), 1934.

White, C. M. and Colebrook, C. F., "Fluid friction in roughened pipes", *Proc. Roy. Soc. A.,* 161, pp. 367-381, 1937.

Zigrang, D. J. and Sylvester, N. D., "Explicit approximations to the solution of Colebrook's friction factor equation", *Am. Inst. Ch. Engrs. J.,* Vol. 28, No. 3, pp. 514-515, 1982.

PART VII

OPERATIONAL
OPTIMISATION

Tariff Optimised Pump Control (TOP-C)

R. A. Bowmer

ABSTRACT

By the application of TOP-C it is possible to optimise pump run periods in pump fed reservoir systems to make substantial savings in electricity costs. TOP-C is a methodology involving site specific algorithms built into local controllers to manage pump run times to maximise use of off peak cheap rate electricity.

In the water industry the filling control of pump served reservoirs and storage tanks is predominantly level driven. The pumps are switched by upper and lower set control levels. This type of control often cycles only a small proportion of the available storage and leads to an inefficient use of available electricity tariffs.

Where the recharge period for normal daily reservoir throughput is well within 24 hours there is the opportunity to take advantage of less expensive electricity tariffs by moving the period of pumping and making greater use of the available storage capacity.

By using a site specific control algorithm it is possible to maintain operation security through automatic incremental adjustments to the pumping periods even when quite substantial changes in demand occur. The system does not require demand prediction or forecasting.

The control principal can be applied to many pumping configurations. The effectiveness is increased by the knock-on effect to borehole pumps when it is applied to in-line booster pumps. It can be applied to reservoir inlet control valves on pressure driven supply systems.

With relatively small design and implementation costs, worthwhile savings can be generated from small pumping stations that might be overlooked in multi-station optimisation schemes. It can be implemented, pumping system by pumping system, as part of Telemetry or SCADA systems providing regional optimisation.

1   INTRODUCTION

Electricity usage has a diurnal pattern just as the water supply industry. It relates to the activity of people and their demands. However, the substances being supplied are quite different, they have their own requirements in terms of demand and delivery.

Electricity moves at or near the speed of light. It has to be supplied as and when it occurs, as there is no way of storing large quantities. Generators have to be kept idling on the off chance of being required. This leads to inefficiency as well as unnecessary cost. The Electricity Companies seek to change the diurnal demand by offering less expensive electricity in the periods of low usage.

Water on the other hand moves more slowly, so supply has to be anticipated; water has to be moved to a position close to its use, prior to the demand. These positions are Service Reservoirs and store enough to supply about 24 hours of demand. Service Reservoirs are filled by pumps which are mostly electric. It would be better for pumps to run at the time of less expensive electricity tariff by making use of the available storage.

The control process described phase shifts the demand, optimising the tariffs used, whilst minimising the risks to supply. It uses a feed back process which operates automatically and takes account of demand variation without the need for any day to day intervention.

2   ELECTRICITY TARIFF STRUCTURE

Each Electricity Supply Company has its own tariff structure, which no doubt reflects the demand patterns it experiences. However, they seem to follow a general shape which is similar and can be described in the same way, albeit that the actual costs and times may vary between suppliers.

The normal tariff applied to an industrial user is based on a measurement of demand maximum. This type of tariff is known as a Maximum Demand Tariff and is relatively punitive. These tariffs are highest in the winter months and less during the summer months. There may be several bands, where the tariff reduces the more units that are used. This type of tariff is aimed at the user who cannot control the time at which it is used. In general it is to be avoided, as the only way to reduce the cost is to use less electricity. There are usually other tariffs, which have similar seasonal variation, but also vary through the day. The tariff is high during the day, less in the evening and lowest at night. The day units in the winter can be 5 - 6 times more expensive than the night units. Peak rate, between 16:00 and 19:00, can be as much as 15 times the night rate. The night units do not normally change the year round and probably represent the 'at cost' electricity.

A method of differentially choosing the cheapest electricity first, the more expensive next, avoiding the most expensive if possible, will provide the least cost option.

3   STRATEGIC VIEW

Both electricity and water usage have a diurnal pattern as previously stated, but the same cannot be said about the seasonal demand, the two are quite different. Light nights and warm days in summer mean that electricity is used less than in winter, whereas watering lawns and to some extent washing cars is rather pointless in winter. This difference in demand, besides inherent available storage, provides an opportunity for using the storage in Service Reservoirs to phase shift the re-charge pumping to night time and so reduce costs. Because the winter demand for water is less, the effective storage will be greater, allowing more scope to phase shift demand. In the summer when the demand is greater the electricity day tariffs are less and costs incurred will be less significant for day pumping.

The aim is to have the reservoir full at the end of the less expensive night rate tariff. There are two reasons for this. The first is to maximise the water available to supply demand during the day, reducing the need to use the pump during the expensive electricity tariff period. The second, by implication, being that the whole of the night period has been used to re-charge the reservoir using cheaper electricity. If the whole of this period has been used for pumping, that portion of the daily demand has been stored at the cheapest cost.

(Level controlled systems tend to spend the night period pumping very little or not at all, as most of the pumping is done during the day on demand and it is topped up to full during the evening.)

The evening electricity units are the next cheapest; any short fall should be selected from these units after the night units. Day units should only be selected if absolutely necessary. This is a perfect strategy in theory the only problem being that events occur in the reverse order.

4   METHODOLOGY

4.1   Evening & Night Cut-in Profile

The solution to the above problem is simply explained by considering a level profile, which instead of being a continuous level, (as in level control), is a steadily rising ramp to the reservoir full point at the end of the night period, the beginning of the ramp, let us say, being at zero level at the beginning of the day or the end of the previous night period.

Consider the basic mechanism: the demand occurring from the start of the day with a full reservoir will require no filling. The level will fall until the level coincides with the ramp profile. At this point the re-charge pump is switched on and the reservoir starts to fill, and will fill till full.

A variation in demand will alter the start time of the pump. If during a day, demand is increased, the level of the reservoir will drop more quickly and will reach the ramp at a lower level, sooner (the re-charge pump starts sooner). On the other hand, if during the day, demand is decreased, the level of the reservoir will drop more slowly and will reach the ramp at a higher point, later (the re-charge pump starts later).

As the pump is intended to stop at the end of the night period, each days variation in demand is accommodated by the variation in start time. This provides a mechanism that has the proper direction sense to produce a self-correcting control action which will restore the reservoir's capacity.

If the profile of this ramp can be arranged so that its position in level and time will cause this restoration of the reservoir by the end of the night period, it will provide the control function that is required.

This ramp profile would only be an appropriate function if there was no demand and the slope of such a ramp represented the rising re-charge rate of the pump. In practice a usable profile can be generated by taking account of the outflow distribution as well as the re-charge using an average demand distribution for the later part of the day, evening and night.

4.2   Morning & Day Cut-in Profile

The simplest way to look at this part of the profile at the beginning of the day is to consider an example. If the pump recharges the reservoir in 12 hours, and the reservoir has 24 hours storage at the average demand, then the pump will run for 12 hours, to the beginning of the day, re-charging the reservoir. The pump will be off for the first 12 hours of the day. If the demand were twice that of the average, the pump would be required to start at the beginning of the day, and run all day.

The control profile can start to be generated by subtracting twice the normal average outflow quantity from the reservoir full point. However, in the middle of the day, where the pumping would start normally with average demand, the control profile needs to have become the same as the average outflow distribution. The algorithm therefore, for the early part of the day, takes the normal average outflow distribution multiplied by two at the day start and linearly derates it to the normal pump start time, where it will join with the cut in profile for the later part of the day.

This process can be followed for different configurations, e.g. the storage might be 18 hours and again 12 hours pump recharge. In 24 hours the throughput is 1.333 times the volume of the reservoir of which it will take the pump 16 hours to deliver. The off period at average demand will be 8 hours, a period over which the pump could deliver a further 0.666 of the volume of the reservoir. This means the multiplier for the outflow distribution is 1.5 (e.g. 1.333+0.666/1.333 = 1.5), which will be derated to the average in 8 hours. By this means a cut in profile can be generated for any configuration.

4.3   Cut-out Profile

It was stated previously that pumping should cease at the start of the day period when the electricity is expensive. This could be achieved by constructing a profile which would consist of a spike coming from the 100% level position and extending downward. Although this would work, it is not the best strategy. In periods of high demand, when day units are going to be selected in any event, the better approach would be to select them at the beginning of the day rather than later. This would maintain the level of the reservoir higher than it otherwise would be. A better approach can be obtained by taking the cut in profile and adding a few percent to its level (leaving a gap between profiles to prevent rapid pump switching). This will then allow a roll over of the pumping at the beginning of the day if the rising re-charge level curve is somewhat below where it should be.

These curves can be seen in figure 1 & 2: Profile Control (24 Hours Storage). Figure 1 shows the average demand situation with the control profiles with modification, pump operation and flow distribution. Figure 2 shows a family of level excursions with 20% increments above and below the average demand to give a better

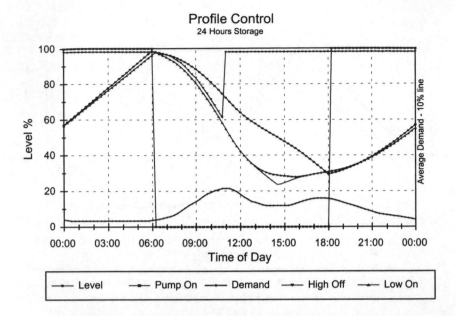

Figure 1

indication of the way the control process operates. As the demand increases, the pump start time moves to an earlier time and towards the more expensive tariff. Whereas, the pump stop times stay substantially the same until the start time has advanced into the day tariff period. A reduction in demand reverses the process, moving the pumping to the cheaper tariff periods. The tariff time parameters are 06:30 for the end of night period, 18:30 for the end of day and start of evening and 00:30 for the start of the night period, though the last parameter has no real bearing, as it is a continuous process.

The junction of the early day discharge algorithm and the latter part of the day recharge curve in figure 1 is not parallel and does not meet as a continuous contour (shown as a minima at 14:30). The curve (Low On) has been modified by smoothing and elevating to produce a more continuous function, to raise the minimum level to which the reservoir will fall. Elevating the profile increases the level at which the reservoir re-charges, which will have a knock-on effect for the following day. If this causes the level to peak too soon, it will spoil the optimisation. This elevating of the profile is only undertaken for the family of re-charge curves which are down on the maximum level at the beginning of the day, (usually at periods of higher demand, see figure 2 +20% & +40%).

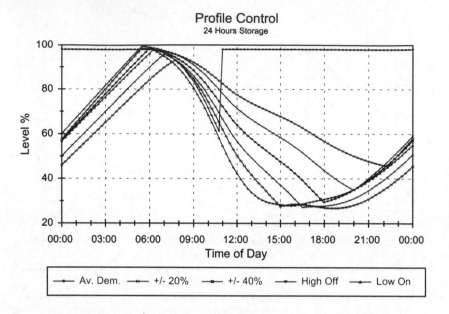

Figure 2

It should be pointed out that there is not a unique solution to these profiles. There are many possible profiles which will provide an optimised solution. The shapes of these curves are derived from the volume of the reservoir, the pump characteristics and the demand flow distribution; as such they are unique to each particular system. Within a particular system there may be many constraints, which will require adaption and modification to the curves to make them a practical solution.

The basic curves developed from the algorithms, in this way, represent the natural optimisation curves and will provide the maximum savings. System constraints may override the full implementation of this control process, requiring modification of the profile curves to accommodate the constraints. The reservoir level in this instance will fall to around 24% at the minimum, even with the elevated smoothing that has been added to the profile. Operational requirements may preclude the system being drained down to this level. In this event ramps and spikes may be added to the profile to alleviate the low levels. This can result in day-time pumping which reduces the optimisation, producing a trade off between lower cost and higher level.

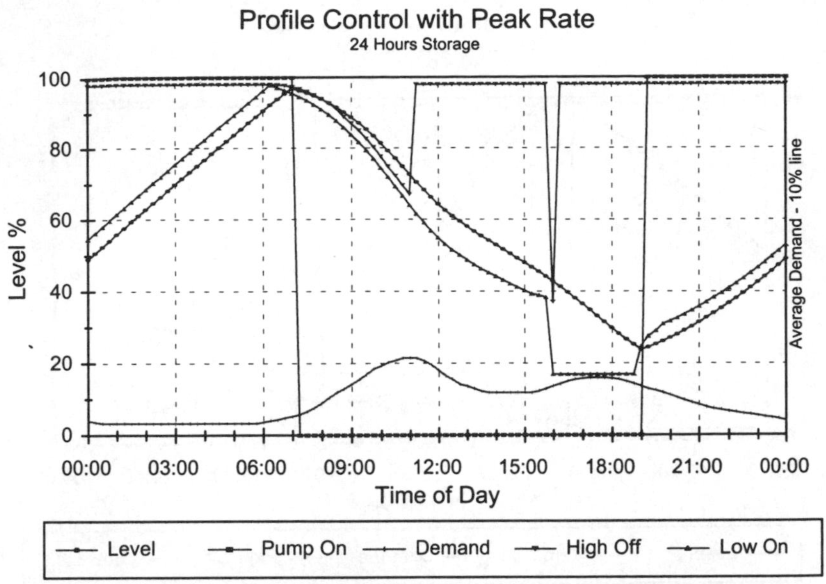

Figure 3

Figure 3, Profile Control with Peak Rate (24 Hours Storage), shows control profiles for a system with the same tariff parameters as figure 1 and a peak rate tariff between 16:00 and 19:00. This profile is probably operationally unacceptable, as the reservoir level will fall below 10% when only the peak rate slot is left available for further pumping, though this would be an exceptional situation. The profiles in figure 4 have the same parameters as figure 3 and provide the modifications to raise the minimum level to around 21%. This has been achieved by not using all evening units before starting day units and using some peak rate units before all the day rate units have been fully used.

The natural curves tend to be more resilient to large changes in demand and the period required for the control process to compensate is less. The optimisation can be adversely affected during the period of compensation. Incremental daily demand variation is unlikely to effect the optimisation as the period of compensation will be correspondingly small. System events which cause large changes in demand will affect the optimisation more, depending on how involved and evolved the curves are.

The natural curves tend to be more resilient to large changes in demand and the

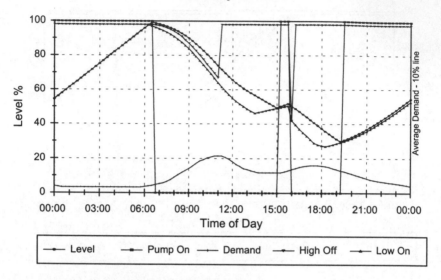

Figure 4

period required for the control process to compensate is less. The optimisation can be adversely affected during the period of compensation. Incremental daily demand variation is unlikely to affect the optimisation as the period of compensation will be correspondingly small. System events which cause large changes in demand will affect the optimisation more depending on how involved and evolved the curves are.

5   EXTENSION

The effect of this type of control will not be limited to just the system to which it is applied. A booster fed service reservoir may be supported from boreholes through a contact tank. If the tank is relatively small or has tight level band the borehole pumps will follow the operation of the booster, also being optimised. This type of knock-on effect can be extensive and should be extended where possible.

This technique can be adapted the other way round. A supply booster could feed a header tank too small to be optimised itself, which then feeds further supply service reservoirs, which may be controlled with float valves. These reservoirs, if fitted with actuated valves in-line with the float valves, can be controlled on their own individual profiles giving the equivalent of pump controls, but opening and closing values instead of starting and stopping pumps.. This approach allows the storage downstream to be used for optimising the pump, which in reality is controlled at the head of the system by level.

Other system constraints can be modified to increase the benefit, such as changing booster control from Duty-Standby to Duty-Assist or Duty-Duty to increase throughput, hydraulics allowing. To do this on systems where the hydraulic losses are small (10% say) compared to the lift, will produce savings even though hydraulic losses increase. Low head systems have to be examined and determined on merit.

6   PRACTICAL IMPLEMENTATION

There are many forms that this control system can take in its implementation. The reservoir's level needs to be monitored against a real time clock, so that it can be compared with the profiles which also relate to real time. The result of the comparison sets and resets a control to start and stop the pump.

Telemetry and SCADA systems can be used, but they need the capability of control and the facility for evaluating the reservoir level against the control profiles. Standalone PLCs, programmable pump controllers and programmable outstations are also devices which can be used to achieve this control.

The real time clock needs to be available continuously or uninterrupted (independent of mains power). A pump will not operate if the mains power is off, but the clock must still increment and show the correct time when the power is restored. Ordinary mains powered timers are not sufficient.

7   COST SAVING

The cost savings that can be achieved depend on many factors. The tariff structure that is available, the configuration of the system and the operational flexibility are all factors that apply.

However, users in the UK who have reached a significant level of application negotiate their own tariffs. The cost saving that has been achieved is in the order of 25% of the clean water budget for electricity.

Criteria for Applying Optimisation in Water Distribution Networks

J. Quevedo, G.Cembrano, G.Wells, R. Pérez *and* R.Argelaguet

ABSTRACT
In this paper, the applicability of optimisation methods to general water distribution systems has been analysed. When, where and why an optimisation tool can be interesting for the planning of water networks has been discussed here. This paper describes the features of a new optimisation tool, designed and developed in the frame of an European ESPRIT IV project named WATERNET "Knowledge Capture for Advanced Supervision of Water Distribution Networks". The paper concludes with the necessary steps to implement the proposed optimisation tool in an arbitrary water network.

1   OPTIMIZATION OF WATER DISTRIBUTION NETWORKS

Optimisation in water networks refers to the problem of improving the management of water resources in a supervisory control system, to achieve specific operation goals. Water networks are generally composed of a large number of interconnected pipes, reservoirs, pumps, valves and other hydraulic elements which carry water to demand nodes from the supply areas, with specific pressure levels to provide a good service to consumers.

The hydraulic elements in a network may be classified into two categories: active and passive. The active elements are those which can be operated to control the flow and the pressure of water in specific parts of the network, such as pumps, valves and turbines. The pipes and reservoirs are passive elements, insofar as they receive the effects of the operation of the active elements, in terms of pressure and flow, but they cannot be directly acted upon.

A supervisory control system in a water network generally includes a telemetry system which periodically updates some information from a selected set of passive elements and from most of (ideally all) the active elements. This information is generally composed of pressure and flow readings, as well as status of the active elements, and it reflects the instantaneous operation condition of the water network. It also includes the mechanisms to actuate the active elements in the network and to control their performance. However, it is not always obvious how to derive appropriate control strategies for the active elements, in order to achieve an efficient use of the resources and to meet the specific pressure and flow needs in the whole network, at all times.

Optimisation in water networks deals with the problem of generating control strategies ahead in time, to guarantee a good service in the network, while achieving certain management goals. According to the needs of a specific utility, these goals may include one or more of the following: minimisation of cost, including supply and power for pumping operations; maximisation of water quality, in terms of dissolved materials and residence time; minimising deviations of pressure and/or flow values from specific desired levels for leak prevention, etc. This work attempts to show where, when and why optimisation methods can be successfully applied to water distribution operation. The benefits of the optimisation are introduced in section 2 and the interaction of the optimisation module with the general supervisory control system is outlined. Section 3 deals with the posing of the optimisation problem in water utilities and section 4 is concerned with the factors influencing the computational complexity of the optimisation problems arising in water management.

2 WHERE AND WHEN IS OPTIMIZATION APPLICABLE?

The benefits of using this subsystem depend to a great extent on a number of characteristics of the system in question, such as: the dimension and the structure of the network, its geographic distribution, the number of pump stations and automatic valves, the number and the capacity of the reservoirs, the power consumption at the pump stations, the electrical tariff(s) of the facility and finally, the degree of automation of the water distribution system: sensors, actuators, telemetry, monitoring, supervision, availability of models, such as GIS, hydraulic models, operational models, expert systems, etc.

2.1 Dimension and Structure

The dimension and the structure of the network are important aspects to take into account in order to decide whether or not to implement an optimal daily operational scheduling. A small water distribution network (for instance, the networks of communities of less than 5,000 inhabitants) or larger ones with very simple structures, such as networks with just one source of supply and a tree-like structure, are quite easily amenable to a manual or heuristic definition of the appropriate control strategies and do not require an optimisation procedure. Conversely, networks with several sources of supply and a more redundant or mesh-like structure (Figure 1) are very hard to optimise manually and generally benefit from the use of an automatic optimisation procedure.

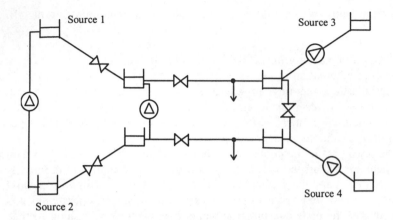

Fig. 1 An example of mesh–like structure of the network

2.2 Active Components and Infrastructure

An optimisation module is typically concerned with optimising the power consumption caused by pumping operations of water acquisition and elevation. In this context, it is clear that a flat area with practically the same elevation in all the nodes of the network does not require much effort from the pump components to maintain the pressures at the consumer nodes. Consequently the electrical consumption probably does not justify the use of an optimisation subsystem.

It is also useful to analyse the capacity of the reservoirs relative to the consumption volume of water. For example, if the capacity of the reservoirs is twice or more times larger than the global daily consumption of water in the network, the optimal strategy is trivial: store water in the reservoirs during the nights, when the electrical tariff is lower, and stop during the day, when the electrical tariff is higher.

However, in many European cities, the terrain is hilly, the building heights differ from one area to another and so do the pressure requirements to give a good service to the consumers, so that extensive use is made of pumping and booster stations. Furthermore, in many cases, the water systems were designed a long time ago and with the increase in population and the extension of the cities towards other (possibly not flat either) areas, the storage capacity has remained largely insufficient. Water networks in these conditions are very good potential candidates for the implementation of optimisation of the pumping operations.

2.3 Variable electricity Tariffs

During the last few years in European countries, the power supply market is free for large consumers, so that from one day to the next, a company can change its electrical supplier. So far, suppliers have been offering a variable tariff with 3 or 4 time zones; now, they may offer a more complex tariff structure. In this case, the optimisation subsystem would be all the more useful in minimising the electrical costs.

2.4 Level of Automation: Hardware and Software

The level of automation in the water distribution system (Figure 2) is a key factor. The optimisation requires an advanced telemetry system to read the node consumptions, the reservoir levels and the pressures at some nodes at least hourly. Once the optimal strategy has been computed and validated at the supervision system level, it should be automatically sent to the remote units.

The two main functions considered in the WATERNET optimisation subsystem are a short-term (1 day) forecasting of the demands in the nodes of the operative model and the computation of optimal strategies, 24 hours ahead, for the pumping stations and servo-valves so as to minimise the operational cost. The forecasting of the node consumption works in a stand-alone fashion so that it can also be used as a tool for other applications, such as: fault detection (in particular, faults in the field-sensors), generation of a manual strategy on special days, infrastructure design, etc. It requires the storage of historic consumption data.

Additionally, for its second important function, the optimisation requires information about the interactions of the different active and passive components in the network. This means that a model must be supplied to the optimisation module. This is the so-called operative model.

A reasonable computer configuration for the implementation of the optimisation module for a medium-size water network is a PC pentium with 16Mb RAM memory and Windows 95. Table 1 summarises the conditions of applicability of an optimisation module with potential success.

Table 1. Conditions of applicability of an optimisation module

condition	status
dimension	>5000 habitants
structure of the network	redundant
geografic distribution	variable piezometric height
pump stations	>3 pumps of 100Kw or more
electrical tariff	variable
electrical consumption	>1.000.000 Kw-H per year
nodes consum sensors	yes
reservoirs level sensors	yes
telemetry system	yes
remote control of pumps	yes
remote control of valves	yes
hydraulic model	yes

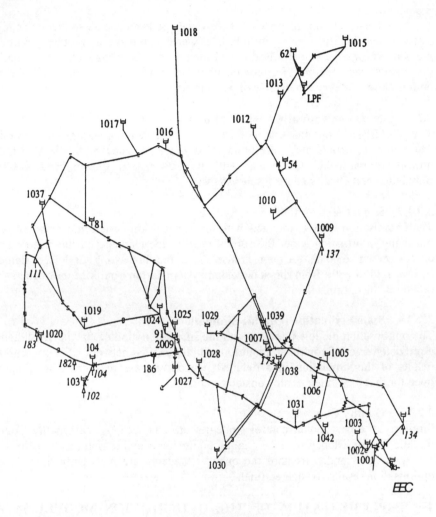

Fig. 2 A representative hydraulic network of SINTRA water distribution system

3 WHY USE THE WATERNET OPTIMISATION SOLVER?

3.1 A Robust General-Purpose Tool

The WATERNET optimisation solver relies on powerful commercial optimisation tools (CONOPT), which can efficiently deal with linear or non-linear cost functions and constraints. It incorporates several optimisation procedures and it can provide information on the optimisation process, as required by the user, with different degrees of detail. Additionally, the WATERNET module incorporates a commercial interface to the CONOPT program (GAMS) which makes it easier for a user who is knowledgeable in optimisation to introduce the definition of the problem to the CONOPT program.

A higher-level user interface has been developed and added, so that it is simple for a water facility user to introduce a water optimisation problem, with no previous knowledge in optimisation and then to run the optimisation and analyse the results very easily. This combination constitutes a powerful tool for the optimisation of a very broad class of water systems.

3.2   Integration with other Water Management Tools
The WATERNET optimisation module can be executed in co-operation with other water management tools, corresponding to other modules in the WATERNET project. In particular it interacts with the Supervision system, and with the Simulation and Quality Control subsystems.

3.2.1   Supervision system
The optimisation module interacts with the Supervision system most obviously in retrieving information about the current state of the elements in the remote units and providing setpoints for them. It also shares the forecasting module, which can be accessed directly from the supervision system for any purpose, not necessarily related to the optimisation.

3.2.2   State Estimation (On line Simulation)
The optimisation module interacts with the simulation module for the validation of a provisional set of control strategies in a more detailed network. It is not until the results of the simulation in a detailed network appear reasonable to a human operator that the strategies are considered correct.

3.2.3   Quality Control
Similarly, the provisional control strategies are analysed by the Quality Control module for consistency with the quality requirements and re-optimisations must be run with new conditions until the quality standards are met. Only then are the optimisation results considered final.

4   IMPLEMENTATION OF THE OPTIMIZATION MODULE IN AN ARBITRARY WATER NETWORK

4.1   Modelling the Network
The first step is to model the network. Assuming the water facility has already developed a detailed model for simulation with any custom-made or commercial simulator, the task consists of generating the reduced operative model of the network dynamics. This is the model of the effect produced by the control actions used in the network over time and in general it must express the evolution of the water volumes stored in the reservoir caused by the water transfers produced by the valves and the pumping stations.

In practice, this model must contain only those reservoirs, valves and pumping stations that can be automatically operated, i.e., those which correspond to remote units. The elements connecting the automatically controlled reservoirs, pumping stations and valves must be designed with extensive use of a simulator, such as the WATERNET simulation tool, and they may correspond to real pipes of the infrastructure or equivalent pipes obtained by model reduction. A method for model reduction has been proposed and successfully used for the Sintra network in this project.

4.2 Posing the Optimisation Problem

The statement of the optimisation problem in a water network is a complex task that should be carried out with the person(s) in charge of the exploitation of the facility in question, so that the appropriate optimisation objectives are detected and the required constraints are defined. The mathematical characteristics of the problem statement determine the complexity of the task to be carried out by the optimisation solver and therefore the speed (and the feasibility) of the computation of the control strategies. In this context, it is very important to attempt to produce the simplest possible statement that is meaningful for expressing the objectives to be optimised and the physical and operational limitations of the water system.

4.2.1 The Cost Function

The cost function expresses the desired exploitation objectives. In many cases, the optimisation will deal with the pumping and acquisition costs, to be minimised by the optimiser. In this case, a sum of linear functions of the flows is usually tested initially. Then, more complex functions may be used to take into account the pressure-flow relationship and the characteristic equations of pumps and other hydraulic elements.

Additionally, it may be useful at some stage of the optimisation study to incorporate some of the constraints of the problem in the cost function, in the form of penalty terms. These are usually designed as quadratic functions that penalise the violation of constraints.

4.2.2 Constraints

The operational limits of the water network impose a certain number of constraints on the flows, the reservoir volumes and (in the general case) the pressures. Firstly, the dynamic mass balance in the network node produces a set of constraints which is usually a set of linear functions of the flows and the reservoir volumes. Furthermore, the physical and operational limits of pumps and valves may be expressed as a set of bounds on the control variables (usually flows). Additionally, the physical and operational limits of the reservoir volumes impose bounds on these variables and finally, operational requirements may impose some bounds on pressures at specific consumer nodes and/or flows in specific pipes of the network (at sites which are not remote units of the telecontrol system). All these constraints imply different degrees of complexity in the optimisation module.

4.2.3   Mass balance constraints

Mass balance constraints are usually a set of linear combinations of flows and reservoir levels and as such, they do not impose a special difficulty for the optimisation.

4.2.4   Bounds on the control variables

The upper and lower bounds on the control variables, subject to which the optimisation is carried out, are easily expressed as a set of inequalities and they do not impose difficulties for the optimisation.

4.2.5   Bounds on the reservoir volumes

The upper and lower bounds on the reservoir volumes, which are present in all cases, do impose some added complexity on the optimisation problem, since they are constraints on variables which cannot be directly acted upon; they are obtained indirectly as a result of a control action. Again, they can be expressed as a set of inequalities; in this case, it is usually helpful to allow for some tolerance in the fulfilment of the constraints.

4.2.6   Bounds at sites other than remote units

In some cases, other bounds may be required at sites which do not correspond to remote units, i.e., they do not correspond to either reservoirs or control elements, but the pressure or flow must be limited nonetheless. In these situations, non-linear equations appear as a result of using the pressure-flow relationships in the network to express the values of the pressures or flows at the desired sites in terms of the variables accessible at the remote units. This type of constraint imposes a serious complication of the optimisation problem, but they can be efficiently dealt with by optimisers such as the WATERNET module.

4.3   Treatment of the discrete-valued variables

In pumping stations where the individual pumps are of fixed speed, the flow produced by the station can only take a few discrete values, corresponding to the different combinations of pumps on and off. This is a common situation in a large class of water systems and it has been dealt with in literature using two main approaches.

The straightforward approach is to use discrete non-linear mathematical programming (e.g. modified scheduling problem (Brdys, 1993), branch and bound method (Minoud, 1993...)), but the computational complexity of discrete problems is exponential and is therefore much higher than continuous ones which are polynomial (Ulanicki, 1997). The second approach is based on the relaxation of the optimal scheduling problem in a continuous problem assuming that all the pump stations are continuous variables. This assumption means that the number of pumps switched on can vary continuously. If a given operating point is not reachable instantaneously (e.g. npumps=4.55) it can be achieved as an average over a time period (Pulse Width Modulation). This common approach (Ulanicki, 1997, Andersen, 1997) is useful if the pump station has more than 4 or 5 pumps working in parallel scheme and this situation is not always true. Another problem of the

PWM approach in pump scheduling is the real maintenance problems when the switching time for pump is short (e.g. less than 1 hour).

In the WATERNET optimisation module a third new possibility has been incorporated. The idea is to use penalty terms in the cost function to be minimised in the optimisation execution. These penalty terms are sinusoidal functions of the pump variables with a period of 1cycle/pump and a big value of amplitude (e.g. 10^6). The cost of these terms contributes significantly to producing a large value of the cost function when the pump variables are not close to integer values $(0,1,2,...)$. This new approach has given good results (Quevedo, 1998) for the SINTRA network, where the number of pumps is very limited in each station. In any event, it is possible to use any of the three possibilities using the WATERNET optimisation software, due to the general-purpose nature of the CONOPT solver.

5 CONCLUSIONS

The WATERNET optimisation module provides a very useful tool for the supervisory control of a broad class of water systems. The main advantages of this software are as follows.

Generality: because it is based on a general-purpose optimisation solver, a broad class of optimisation problems may be dealt with including linear and non-linear solution methods.

Specific User Interface: the specific user interface developed in the project guides the user for a clear description and easy input of the problem equations and data; for the editing and correction of data for different optimisation trials; for the execution and for the analysis of the results.

A very important issue in the optimal control of a water network is to what extent the strategies and set points computed by a program should be automatically sent from the optimisation procedures to the real system actuators. This decision should be taken carefully by the water facility managers and some key factors to be taken into account therein are the structure of the network (tree-like or mesh) and previous experience in controlling the network.

It is common in large facilities to require that an experienced operator validate the optimised strategy before sending it to the remote units. This is the approach assumed in the WATERNET project.

In addition to the current features of the optimisation module, a number of improvements are envisaged as on-going research. The first one is to allow for an asynchronous execution of the optimisation. Alongside an optimisation procedure run periodically every 24 hours to generate a one-day ahead optimised strategy, most water control systems will benefit from running other partial optimisations at different times during the day, in order to cater for deviations in the demands or other elements with respect to the models used by the 24-hour optimisation.

Furthermore, research on the subject of generating the reduced operative model from the detailed network data must be continued, based on that proposed in the WATERNET project, to produce a more general method. Similarly, improvements for the translation of the data provided by the optimisation for the set points of control elements in the reduced operative model into more detailed data to be used in each remote unit must be studied more deeply to provide more general solutions.

ACKNOWLEDGEMENTS

This work has been partly supported by the Commission of European Union ESPRIT project n° 22.186 project WATER-NET and the University and Research Commission of the Generalitat de Catalunya (group SAC ref 1997SGR00098). Authors are members of CERCA (Collective for the Study and Research of Automatic Control) and LEA-SICA (Associated European Laboratory on Intelligent Systems and Advanced Control).

REFERENCES

A. Brooke, D. Kendrick, A. Meeraus, R. Raman, "GAMS Language Guide". GAMS Developement Corporation, 1997(USA).

M.A.Brdys, "Water systems. Structure, algorithms and applications", Prentice Hall 1994.

B. Ulanicki, B. Coulbeck and K. Ulanicka, "Generalised techniques for optimisation of water networks", Computing and Control for the Water Industry, Research Studies Press Ltd, UK, 1997.

J.H. Andersen and R.S. Powell, "The use of continuous decision variables in an optimising fixed speed pump scheduling algorithm", Computing and Control for the Water Industry, Research Studies Press Ltd, UK, 1997.

M.A. Brdys, "A method for optimal scheduling of general class of water supply network with strong non linear couplings", Proc. Of the IFAC Congress, Sydney, 1993.

H. Minoux, "Mathematical programming", John Wiley, 1986.

J. Quevedo, R. Pérez, G. Wells, G. Cembrano and R. Argelaguet, "New design of an optimisation tool for water distribution systems", draft paper presented to 14th World Congress of IFAC, Beijing, 1998.

Models for the Optimal Scheduling of Pumps to Meet Water Quality

A.B. Sakarya, F. E. Goldman *and* L. W. Mays

ABSTRACT
Two new models have been developed for determining the optimal operation of water distribution systems for water quality purposes. These models are based upon describing the operation as a discrete time optimal control problem that can be used to determine the optimal schedules of the pumps in the distribution systems. One solution methodology is based upon a mathematical programming approach and the second methodology is based upon a simulated annealing approach. Each model is interfaced with the US EPA simulation model EPANET. The two models have been applied to sample water distribution systems for the sake of performing comparisons. The results are compared along with their respective advantages and disadvantages.

1 FORMULATIONS TO SATISFY WATER QUALITY

Computer models that simulate the hydraulic behavior of water distribution systems have been available for many years. More recently these models have been extended to analyze the quality of water as well as the hydraulic behavior. These models are capable of simulating the transport and fate of dissolved substances in water distribution systems and can be used to predict water quality at the point of delivery to the customer. The Safe Drinking Water Act in 1974 and its Amendments in 1986 (SDWAA) requires the measurement of residual chlorine concentration, lead and copper at the point of delivery. Previously regulatory concerns have been focused on water as it leaves the treatment plant before entering the distribution system (Clark, 1994), disregarding the variations in water quality which occur in the water distribution systems.

Several optimization models have been developed: (1) to determine the optimal operation schedule of the pumps in a water distribution system for a predefined time horizon for water quality purposes and (2) to minimize pumping costs while satisfying the hydraulic constraints, the water quality constraints, and the bound constraints on pump operation times, pressures, and the tank water storage heights.

Several objective functions can be formulated that address the optimal operation of water distribution systems for water quality purposes. Three such objective functions are: (1) the minimization of deviations of the actual concentrations of a constituent from the desired concentration values; (2) the minimization of the total pump operation times; and (3) the minimization of the total energy cost. The constraints that have to be considered for all time periods for all cases are basically the hydraulic constraints, the water quality constraints, and the bound constraints.

Consider a water distribution system with M pipes, K junction nodes, S storage nodes (tank or reservoir), and P pumps, which are operated for T time periods.

The objective function that minimizes the deviations of concentrations from the desired range of concentrations is

$$Min\,Z_I = Minimize \sum_{t=1}^{T} \sum_{n=1}^{N} \; min\left(0, min\left(C_{nt} - \underline{C}_{nt}, \overline{C}_{nt} - C_{nt}\right)\right)^2 \tag{1}$$

where N is the total number of nodes (junction and storage); C_{nt} is the substance concentration at node n at time t (mass/ft^3); $\underline{C}_{nt}$ and $\overline{C}_{nt}$ are the lower and upper bounds, respectively on substance concentration at node n at time t (mass/ft^3).

The objective function that considers the minimization of the total pump durations is

$$Min\,Z_{II} = Minimize \sum_{p=1}^{P} \sum_{t=1}^{T} D_{pt} \tag{2}$$

where D_{pt} is the length of time pump p operates during time period t (hr).

The objective function for the minimization of the total energy cost is

$$Min\,Z_{III} = Minimize \sum_{p=1}^{P} \sum_{t=1}^{T} \frac{UC_t \, 0.746 \, PP_{pt}}{EFF_{pt}} D_{pt} \tag{3}$$

where UC_t is the unit energy or pumping cost during time period t ($/Kwhr), PP_{pt} is the power of pump p during time period t (hp), D_{pt} is the length of time pump p operates during time period t (hr), EFF_{pt} is the efficiency of pump p in time period t. The efficiency of the pump and the unit energy cost are considered to be constant for all time periods.

The distribution of flow throughout the network must satisfy the conservation of mass and the conservation of energy, which are defined as the hydraulic constraints. The conservation of mass at each junction node, assuming water is an incompressible fluid, is

$$\sum_{i} (q_{ik})_t - \sum_{j} (q_{kj})_t - Q_{kt} = 0 \quad k = 1, \ldots, K \text{ and } t = 1, \ldots, T \tag{4}$$

where $(q_{ij})_t$ is the flow in the pipe m connecting nodes i and j at time t (cfs) and Q_{kt} is the flow consumed (+) or supplied (-) at node k at time t(cfs);

The conservation of energy for each pipe m connecting nodes i and j, in the set of all pipes, M is,

$$h_{it} - h_{jt} = f\left(q_{ij}\right)_t \ \forall \ i, j \in M \text{ and } t = 1, \dots, T \tag{5}$$

where h_{it} is the hydraulic grade line elevation at node i (equal to elevation head, E_i plus pressure head, H_{it}) at time t (ft) and $f(q_{ij})_t$ is the functional relation between headloss and flow in a pipe connecting nodes i and j at time t (ft).

The total number of hydraulic constraints is (K+M)T, and the total number of unknowns is also (K+M)T, which are the discharges in M pipes and the hydraulic grade line elevations at K nodes. The pump operation problem is an extended period simulation problem. The height of water stored at a storage node for the current time period, y_{st}, is a function of the height of water stored from the previous time period which can be expressed as,

$$y_{st} = y_{s(t-1)} + \frac{q_{s(t-1)}}{A_s} \Delta t \ s = 1, \dots, S \text{ and } t = 1, \dots, T \tag{6}$$

The water quality constraint which is the conservation of mass of the substance within each pipe m, connecting nodes i and j in the set of all pipes, M is,

$$\frac{\partial\left(C_{ij}\right)_t}{\partial t} = -\frac{\left(q_{ij}\right)_t}{A_{ij}} \frac{\left(\partial C_{ij}\right)_t}{\partial x_{ij}} + \theta\left(C_{ij}\right)_t \ \forall \ i, j \in M \text{ and } t = 1, \dots, T \tag{7}$$

where $(C_{ij})_t$ is the concentration of substance in pipe m connecting nodes i and j as a function of distance and time (mass/ft^3); x_{ij} is the distance along pipe (ft); A_{ij} is the cross-sectional area of pipe connecting nodes i and j (ft^2); and $\theta(C_{ij})_t$ is the rate of reaction of constituent within the pipe connecting nodes i and j at time t (mass/ft^3/day).

All the constraints considered up to now are equality constraints. These constraints are handled by the network simulator. In addition, there also exist inequality constraints which are the bound constraints. The lower and upper bounds on pump operation time is given as,

$$\Delta t_{min} \leq D_{pt} \leq \Delta t_{max} p = 1, \dots, P \text{ and } t = 1, \dots, T \tag{8}$$

where D_{pt} is the length of the operation time of pump p at time t, and Δt_{min} and Δt_{max} are the lower and upper bounds on D_{pt}, respectively. Δt_{min} can be zero in order to simulate an idle pump and Δt_{max} is equal to the length of one time period.

The nodal pressure head bounds are

$$\underline{H}_{kt} \leq H_{kt} \leq \overline{H}_{kt} \ k = 1, \dots, K \text{ and } t = 1, \dots, T \tag{9}$$

where $\underline{H}_{kt}$ and $\overline{H}_{kt}$ are the lower and upper bounds, respectively on pressure head at node k at time t, H_{kt}. There are no universally accepted values for the lower and upper bound values. The range of 20 to 40 psi is acceptable for minimum pressure for average loading conditions, but may be lowered during emergency situations such as a fire. High pressures are usually 80 to 100 psi depending on the distribution system circumstances.

The bounds on the height of water storage are

$$\underline{y}_{st} \leq y_{st} \leq \overline{y}_{st} \quad s = 1, \ldots, S \text{ and } t = 1, \ldots, T \tag{10}$$

where $\underline{y}_{st}$ and $\overline{y}_{st}$ are the lower and upper bounds, respectively on the height of water stored at node s at time t, y_{st}. These limits are due to physical limitations of the storage tank or the reserve required for fire protection.

Cohen (1982) stated that, if there is not a requirement for the periodicity in operation of a network then the optimization of the operation has no meaning. Hence, to achieve this, traditionally, final storage bounds are tightened so that the storage in the tanks will be more or less the same as the initial states. However, it has been found that the system reaches steady state if the daily pump operation schedules repeat themselves for a certain period of time. The tank water levels at the beginning and at the end of the day are equal to each other, when the steady state conditions are reached and have adjusted to the time that the pumps are operating during the simulation. Hence, the constraint that forces the tank water levels at the end of the day to be more or less equal to the initial state does not need to be met and should not be used in simulations that include water quality since it takes many days for the system to reach steady state.

The minimization of pump operation times and pump operation cost consider the same constraint set defined by Equations (4)-(10) as the minimization of the deviations of concentrations from the desired values, with an additional constraint for the substance concentration values to be within their desired limits.

The bounds on substance concentrations are

$$\underline{C}_{nt} \leq C_{nt} \leq \overline{C}_{nt} \quad n = 1, \ldots, N \text{ and } t = 1, \ldots, T \tag{11}$$

where $\underline{C}_{nt}$ and $\overline{C}_{nt}$ are the lower and upper bounds, respectively on substance concentration at node n at time t, C_{nt}.

The above formulation, results in a large scale nonlinear programming problem with decision variables, $(q_{ij})_t$, H_{kt}, y_{st}, D_{pt}, and $(C_{ij})_t$. One method of developing an optimization algorithm is to partition the decision variables into two sets; control (independent) and state (dependent) variables. The pump operation times are the control variables. The problem is formulated above as a discrete time optimal control problem. Equations 1 - 3 are examples of objective functions for the minimization of the deviations of concentrations from the desired range of values, the minimization of the pump operation times and the minimization of the energy cost, respectively. Equations 9 – 11 define typical hydraulic equality constraints and water quality inequality constraints. The inequality constraints are bounds in the state variables: pressure, tank elevation and nodal contaminant concentrations. Equation 8 is the bound constraint for the control variable, the duration of pumping during a time period.

2 SOLUTION METHODOLOGIES AND APPLICATIONS FOR WATER QUALITY PURPOSES

Two methodologies will be described for determining the optimal operation of water distribution systems for water quality purposes. These methodologies are based upon describing the operation as a discrete time optimal control problem that can be used to determine the optimal operation schedules of the pumps in distribution systems. One solution methodology is based upon a mathematical programming approach and the second methodology is based upon a simulated annealing approach. The following sections describe these two methodologies and present example applications with comparisons.

2.1 Mathematical Programming Approach

The mathematical approach reformulates the problem using an optimal control framework resulting in linking a simulation code, EPANET (Rossman, 1994) with an optimization code, GRG2 (Lasdon and Waren, 1986) to find the optimal solution. The decision variables are partitioned into control variables and state variables in the formulation of the reduced problem. The control variables are the amount of time that the pump operates during each time period. The series of period operating times results in a pump operation schedule. The control variable values (pump operation schedule) are determined by the optimizer and they are given as input to the simulator which solves for the state variables (pressure, water quality, and tank levels). Hence, the state variables are obtained as implicit functions of the control variables. This results in a large reduction in the number of constraints as the hydraulic and water quality equality constraints are solved by the simulator and only the inequality bound constraints are left to be solved by the optimizer.

Improvements in the objective function of the nonlinear programming problems are obtained by changing the control variables of the reduced problem. NLP codes restrict the step size by which the control variables change so that the control variable bounds are not violated. The state variables, which are implicit functions of control variables, are not taken into consideration in the determination of step size. If the bounds of the state variables are violated, more iterations would be needed to obtain a feasible solution. The penalty function method is used to overcome this problem. The state variable bound constraints are included in the objective function as penalty terms. The application of this technique is also beneficial because it reduces the size of the problem and the number of the constraints.

There are many kinds of penalty functions that can be used to incorporate the bound constraints into the objective function. In this application, two different penalty functions, the bracket and the augmented Lagrangian are used.

The bracket penalty function method (Mays, 1997) uses a very simple penalty function which has the following form

$$PB_j(V_{j,i}, R_j) = R_j \sum_i \left[\min(0, V_{j,i}) \right]^2 \tag{12}$$

and the augmented Lagrangian method (Fletcher, 1975) uses the following penalty function

$$PA_j(V_{j,i}, \mu_{j,i}, \sigma_{j,i}) = \frac{1}{2} \sum_i \sigma_{j,i} \min\left[0, \left(V_{j,i} - \frac{\mu_{j,i}}{\sigma_{j,i}}\right)\right]^2 - \frac{1}{2} \sum_i \frac{\mu_{j,i}^2}{\sigma_{j,i}} \tag{13}$$

where the index j is the representation of H for pressure head, C for concentration, and y for water storage height bound constraints. The index i is a one dimensional representation of the double index (k, t) for the pressure head penalty term, (n, t) for the concentration penalty term, and (s, t) for the storage bound penalty term. PB_j and PA_j define the bracket and the augmented Lagrangian penalty functions for bound constraint j, respectively. $V_{j,i}$ is the violation of the bound constraint j. R_j is a penalty parameter used in the bracket penalty method and $\sigma_{j,i}$ and $\mu_{j,i}$ are the penalty weights and Lagrangian multipliers used in the augmented Lagrangian method, respectively. A detailed description of the bracket and the augmented Lagrangian penalty function methods and the methods to update the penalty function parameters can be found in Sakarya (1998).

The violation of the pressure head constraint is defined as

$$V_{H,kt} = \min[(H_{kt} - \underline{H}_{kt}), (\overline{H}_{kt} - H_{kt})] \tag{14}$$

Similarly the violations of the substance concentration and the water storage height bound constraints can be defined.

The reduced problem for minimizing the deviations of the concentrations from the upper and lower bounds is

$$Min L_I = P_C(V_{C,nt}, F_C) + P_H(V_{H,kt}, F_H) + P_y(V_{y,st}, F_y) \tag{15}$$

subject to

$$0 \le D_{pt} \le \Delta tp = 1, \ldots, P \text{ and } t = 1, \ldots, T \tag{16}$$

where P_C, P_H, and P_y define the bracket or the augmented penalty terms associated with the concentration, pressure head, and storage bound constraints, respectively, depending on the penalty function method used. Similarly, F_C, F_H, and F_y define the penalty function parameters which are the penalty parameters for the bracket or the penalty weights and the Lagrangian multipliers for the augmented Lagrangian penalty method, associated with the concentration, pressure head, and storage bound constraints, respectively.

For minimizing pump operation time, the reduced objective function is subjected to the same constraint defined by Equation (16), and has the following form

$$Min L_{II} = \sum_{p=1}^{P} \sum_{t=1}^{T} D_{pt} + P_C(V_{C,nt}, F_C) + P_H(V_{H,kt}, F_H) + P_y(V_{y,st}, F_y) \qquad (17)$$

The reduced objective function for the minimization of pump energy is defined as

$$Min L_{III} = \sum_{p=1}^{P} \sum_{t=1}^{T} \frac{UC_t \, 0.746 PP_{pt}}{EFF_{pt}} D_{pt} + P_C(V_{C,nt}, F_C) + P_H(V_{H,kt}, F_H) + P_y(V_{y,st}, F_y) \qquad (18)$$

subject to the constraint defined by Equation (16).

The solution of the final form of the problem is obtained by the two step optimization procedure described in Lansey and Mays (1989), Brion and Mays (1991), and Mays (1997). Finite difference approximations were used to calculate the derivatives of the objective function with respect to the control variables, the pump operation times. These derivatives are the reduced gradients which the optimization code needs to find the optimal solution. Figure 1 shows the flow chart of the optimization model.

2.2   Simulated Annealing Approach

Simulated annealing is a combinatorial optimization method that uses the Metropolis algorithm to evaluate the acceptability of alternate arrangements and slowly converge to an optimum solution. The method does not require derivatives and has the flexibility to consider many different objective functions and constraints. Simulated annealing uses concepts from statistical thermodynamics and applies them to combinatorial optimization problems. Kirkpatrick, Gelatt, and Vecchi (1983) explained the simulated annealing methodology and applied the method to the "Traveling Salesman" and computer design problems. Kirkpatrick (1984) provided additional insights and applications including graphical partitioning, useful in the electronics industry for the design of circuits.

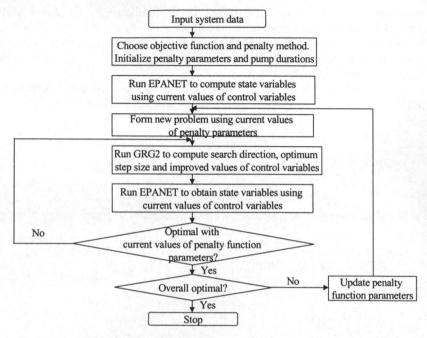

Figure 1. Flow Chart of the Optimization Model

Combinatorial optimization requires that the decision variables be restricted to a set of discrete values. The set of all possible combinations is called the configuration space. For example, consider a pump that operates for 6 hours divided into 1 hour periods, where the pump can either be on or off for any period. The number of pump operation combinations is $2^6 = 64$. A pump which operates for 24 hours divided into one hour periods has $2^{24} = 16,777,216$ combinations. The six hour example can be solved by trial and error but the 24 hour example is prohibitively large to solve by trial and error. For large combinatorial optimization problems, simulated annealing provides a manageable solution strategy. Refer to Goldman (1998) for details of the application of the simulated annealing approach.

3   EXAMPLE APPLICATION

To illustrate the two solution approaches the primary zone of the North Marin Water Distribution (MNWD) system shown in Figure 2 was used (Rossman, 1994 and Vasconcelos, et al, 1996). This system contains 115 pipes, 91 junction nodes, 2 pumps, 3 storage tanks, and 2 reservoirs. The minimum and maximum pressures at demand nodes were set at 20 psi and 100 psi, respectively. The desired minimum water storage height at all tanks was 5 ft. The minimum and maximum allowable concentration limits at all demand nodes were set at 50 µg/L and 500 µg/L, respectively. The bulk and the wall rate coefficients used in the simulation were -0.1 day⁻¹ and -1 ft/day, respectively. The simulation was conducted for a total of 12 days where the values at the last day were used to evaluate the objective function and constraints. The unit cost of energy was assumed to be constant at 0.07 \$/Kwhr for all time periods, and the efficiency of both pumps was a constant 0.75.

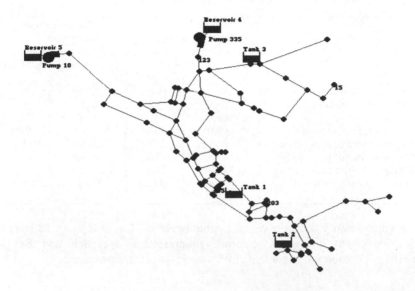

Figure 2 Water Distribution System of North Marin Water District Zone I (Source: Rossman, 1994)

Table 1 lists the optimized solutions obtained by using the mathematical programming and the simulated annealing approaches with a 500 µg/L concentration at both reservoirs, constant throughout the simulation for minimizing pumping time and minimizing pumping cost applications. The final results obtained by both approaches had no violations of any penalty term for both cases. Pump 335 is larger than pump 10, hence closing pump 335 for a certain time period has more effect than closing pump 10. Since the strongest gradient resulted from closing Pump 335, it was always closed first for the mathematical programming approach which solves for the optimum solution by using reduced gradients. The simulated annealing approach had no bias since it randomly chose a pump and period to change to obtain a trial pump operation schedule. The total pump operation times obtained from the mathematical programming approach were greater than the ones obtained from the simulated annealing approach. However, the total pump operation times of pump 335 and the total 24 hr energy cost obtained from the mathematical programming approach were less than the ones obtained from the simulated annealing approach since the mathematical programming approach preferred to close the larger pump 335 first. The total number of EPANET calls were 425 and 356 for minimizing pump operation time and pump operation cost, respectively for the mathematical programming approach, whereas 1500 calls were made for both cases when the simulated annealing approach was used.

Table 1 Optimized solutions obtained by using different solution approaches for 500 µg/L of concentration at both reservoirs

	Minimize Pump Operation Time		Minimize Pump Operation Cost	
	Mathematical Programming	Simulated Annealing	Mathematical Programming	Simulated Annealing
Original objective function	34.47	24.00	399.95	429.53
Concentration violation	0.00	0.00	0.00	0.00
Pressure violation	0.00	0.00	0.00	0.00
Tank water level violation	0.00	0.00	0.00	0.00
Operation time of pump 10 (hr)	19.97	5.00	24.00	14.00
Operation time of pump 335(hr)	14.50	19.00	13.62	17.00
Total pump operation time (hr)	34.47	24.00	37.62	31.00
Total 24 hr energy cost ($)	401.06	433.63	399.95	429.53

Figures 3 and 4 show the optimal pump operation times of pump 10 and pump 335 obtained by using mathematical programming approach and simulated annealing approach for minimizing pump operation cost, respectively

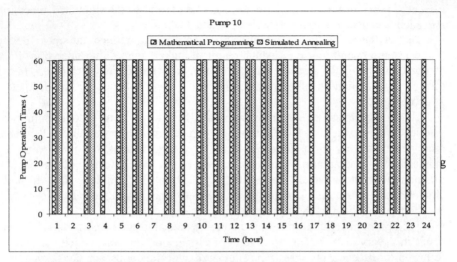

Figure 3. Optimal Pump Operation Times of Pump 10 Used in NMWD for Minimizing Pump Operation Cost

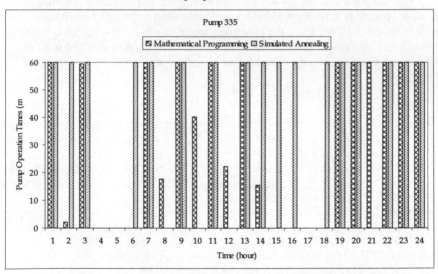

Figure 4. Optimal Pump Operation Times of Pump 335 Used in NMWD for Minimizing Pump Operation Cost

4 ADVANTAGES AND DISADVANTAGES

Both solution approaches are capable of finding the optimum solution for the pump operation problem for water quality purposes. The mathematical programming approach requires the calculation of the derivatives of the objective function with respect to pump operation times (reduced gradients) and so the simulated annealing approach is considered to be more flexible and adaptable.

The mathematical programming approach tries to reduce the most costly (larger) pump operation times, which result in higher total pump operation times compared to simulated annealing approach. There were distinct differences in the total pump operation times whereas similar results were obtained for the total 24-hr energy costs.

The mathematical programming approach tries to find one "global optimum" solution, whereas the simulated annealing approach finds many solutions that have total penalties very close to each another. The solutions obtained from the mathematical programming approach depend on the initial values of the penalty function parameters which makes finding the global optimum more difficult. Global optimum solution can not be guaranteed, since convexity of the objective function can not be proven.

The optimum solution found by the mathematical programming approach may result in a very short operating time during one time interval, which can not be followed for practical purposes. The simulated annealing approach considered one-hour periods so this problem did not occur.

The mathematical programming approach finds the optimum solution in much shorter operating times than the simulated annealing approach. Each optimization trial took about 10 seconds; the simulated annealing approach required three times the number of trials as the mathematical-programming approach and took over six hours computer time versus about two hours for the mathematical programming approach.

REFERENCES

Brion, L. M., and L. W. Mays, "Methodology for Optimal Operation of Pumping Stations in Water Distribution Systems," Journal of Hydraulic Engineering, 117(11): 1551-1569, 1991.

Clark, R. M., "Applying Water Quality Models," in M. H. Chaudry and L. W. Mays, eds., Computer Modeling of Free-Surface and Pressurized Flows, Chap. 19, Kluwer Academic Publishers, Netherlands, pp. 581 – 612, 1994.

Cohen, G., "Optimal Control of Water Supply Networks," in S. G. Tzafestas ed., Optimization and Control of Dynamic Operational Research Models, Chap. 8 in Vol. 4, North-Holland Publishing Company, Amsterdam, 1982.

Goldman, F. E., "The Application of Simulated Annealing for Optimal Operation of Water Distribution Systems," Ph.D. dissertation, Department of Civil and Environmental Engineering, Arizona State University, Tempe, Arizona, 1998.

Kirkpatrick, S., "Optimization by Simulated Annealing: Quantitative Studies," Journal of Statistical Physics, 34(5/6), 975-986, 1990.

Kirkpatrick, S., Gelatt, C.D., and Vecchi, M.P., "Optimization by Simulated Annealing," Science, 220(4598), 671-680, 1983.

Lansey, K. E., and L. W. Mays, "Optimization Model for Water Distribution System Design," Journal of Hydraulic Engineering, 115(10): 1401-1418, 1989.

Lasdon, L. S., and A. D. Waren, "GRG2 User's Guide," Department of General Business, The University of Texas at Austin, Austin, Texas, 1986.

Mays, L. W., Optimal Control of Hydrosystems, Marcel Dekker Inc., New York, 1997.

Rossman, L. A., "EPANET Users Guide," Drinking Water Research Division, Risk Reduction Engineering Laboratory, Office of Research and Development, U.S. Environmental Protection Agency, Cincinnati, Ohio, 1994.

Sakarya, A. B., "Optimal Operation of Water Distribution Systems for Water Quality Purposes," Ph.D. dissertation, Department of Civil and Environmental Engineering, Arizona State University, Tempe, Arizona, 1998.

Vasconcelos, J.J., Boulos, P.F., Grayman, W.M., Kiene, L., Wable, O., Biswas, P., Bhari, A., Rossman, L.A., Clark, R.M., and Goodrich, J.A. (1996), Characterization and modeling of chlorine decay in distribution systems, AWWA Research Foundation and American Water Works Association, Denver, Colorado pp.258-271.

Optimal Regeneration Scheduling for Reverse Osmosis Membranes in Water and Wastewater Treatment

H. J. See, V. S. Vassiliadis *and* D. I. Wilson

ABSTRACT

A systematic approach is developed to determine optimal cleaning and replacement schedules for reverse osmosis (RO) network configurations subject to fouling. The approach is based on a discrete time interval analysis and is solved as a mixed-integer non-linear programming (MINLP) problem assuming an exponential decay in membrane permeability. This approach is capable of identifying the optimal membrane regeneration and replacement schedules over the entire membrane lifetime under both perfect and imperfect cleaning scenarios. The methodology is demonstrated by solving the scheduling problem for various network configurations of two previously water treatment case studies. The results generated show that maintenance considerations can have significant effects on the design of optimal RO network superstructures.

1 INTRODUCTION

Reverse osmosis (RO) membranes are now widely used in treating water and wastewater to meet the increasing demand for inexpensive clean water for potable and industrial purposes. The relatively low energy requirement involved, and the ability to achieve high solute separations over wide pH and temperature ranges, have promoted their use over other conventional treatment methods [1]. Furthermore, RO membranes are less complicated to use owing to the compactness of the modules' design, and the absence of phase changes [2].

A typical RO plant comprises RO membranes, which perform the separation task, booster pumps to overcome the osmotic pressure of the solutes in solution, and turbines to recover excess energy from the high pressure brine (reject) streams. The success of the network not only depends on the configuration of these units, but also on its ability to maintain the product requirements continuously [3]. As RO membrane performance is bound to deteriorate over time due to fouling [4, 5], regular cleaning is required to recover performance [1]. Since the additional operating costs due to cleaning in these systems are significant, there is therefore a need for robust methods for generating optimal cleaning schedules.

Various methods have been used to design RO plants, such as the short-cut method using graphical-analytical technique [6, 7], the mathematical simulation approach [8, 9], the superstructure approach [10, 11, 12, 13 and 14] and the object-orientated approach [14]. Although these methods were capable of identifying the optimum RO network structure and number of RO units required, little consideration was given to cleaning considerations, except in [11], where a pre-determined scheduling approach was used.

In this study, a systematic analysis based on a multiple discrete time interval approach is used to identify not only the optimal membrane cleaning schedule, but also the unit replacement schedule for several seawater desalination and wastewater (pulp and paper dephenolisation) networks. It is assumed that membrane performance decays exponentially. The effect of non-perfect regeneration is considered. The effects of RO network configuration on cleaning and membrane replacement actions are also discussed.

2   SCHEDULING PROBLEM FORMULATION

The membrane cleaning and replacement schedule problem is approached by discretising the total time horizon into P intervals (or periods) of equal duration for each membrane module unit u ($u \in U$, where U is the total number of RO units), as shown in Figure 1. The 'operating day' parameter ($T_{u,p}$) at each period p for RO unit u denotes the length of time since the last cleaning action was performed, while $t_{u,p}$ represents the length of time elapsed since the operation was started.

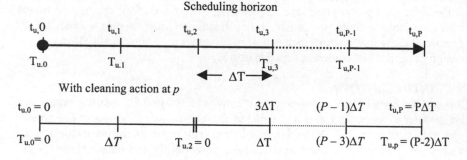

Figure 1   Scheduling time horizon representation illustrating $t_{u,p}$ and $T_{u,p}$ relationships over the horizon with a single cleaning action at $p = 2$.

Since membrane regeneration and replacement actions incur only a few hours of downtime, the amount of downtime is taken to be insignificant compared to the duration (in term of months) of each period. Without downtime, the length of an interval (ΔT) represents the non-cleaning duration between two subsequent periods. A membrane cleaning decision, denoted by $Z_{u,p}^{clean}$ is therefore only considered at the boundaries of the chosen time interval, $T_{u,p}$,

$$T_{u,p} = \left(1 - Z_{u,p}^{clean}\right)\left(T_{u,p-1} + \Delta T\right) \qquad \forall u \text{ and } p, Z_{u,p}^{clean} = \{0,1\} \qquad (1)$$

where $Z_{u,p}^{clean}$ is a binary integer that takes the value of 0 when no cleaning is required and takes the value of 1 when cleaning is deemed necessary for module unit u at the end of period p. $Z_{u,p}^{clean}$ also takes the value of 1 whenever the corresponding RO unit is replaced with a new set of RO modules ($Z_{u,p}^{new} = 1$, where $Z_{u,p}^{new} = \{0,1\}$).

An exponential decay in membrane permeability over time is assumed [11] and the permeability of the membranes ($K_{u,p}$) in unit u at period p is given by

$$K_{u,p} = K_{u,p}^{o} \cdot \exp\left(-\frac{T_{u,p}}{\Gamma}\right) \qquad \forall \ u \text{ and } p \qquad (2)$$

where $K_{u,p}^{o}$ is the permeability of the clean membrane and Γ is a decay constant. For the perfect cleaning scenario, membrane permeability is assumed to be restored fully to its initial condition (K_{u}^{new}) after each cleaning action ($K_{u,p}^{o} = K_{u}^{new}$).

In order to account for imperfect cleaning, a constant loss in membrane permeability (ΔK_{loss}^{o}) is assumed for any membrane used after a cleaning action. Under these circumstances, cleaning of the membranes will eventually become unfavourable and a new unit will have to be installed to meet the production requirements. When membrane replacement occurs, the permeability is restored to K_{u}^{new}. This case is implemented via,

$$K_{u,p}^{o} = \left(1 - Z_{u,p}^{new}\right)\left(K_{u,p-1}^{o} - Z_{u,p}^{clean} \Delta K_{loss}^{o}\right) + Z_{u,p}^{new} K_{u}^{new}$$
$$\forall \ u \text{ and } p, Z_{u,p}^{clean} = \{0,1\}, Z_{u,p}^{new} = \{0,1\} \qquad (3)$$

In order to ensure that membrane replacement can only take place when membrane regeneration is considered, the following constraint is imposed on both integer variables,

$$Z_{u,p}^{clean} \leq 1 + Z_{u,p}^{replace} \qquad \forall \ u \text{ and } p, Z_{u,p}^{clean} = \{0,1\}, Z_{u,p}^{new} = \{0,1\} \qquad (4)$$

3   REVERSE OSMOSIS PERMEATOR

The performance of the RO membrane units is modelled using the approximated model based on membrane permeability and solute transport parameters proposed by [6] for hollow fiber modules. The model predicts the permeate flowrate ($F_{u,p}^{P}$) and the concentration ($C_{u,p}^{x,P}$) of each solute species x ($x \in X$, where X is the total species of solute in the solution) in the permeate of unit u at period p,

$$F_{u,p}^{P} = N_u K_{u,p} S\left(\Delta P_{u,p} - \Delta \pi_{u,p}\right)\gamma_{u,p} \qquad \forall \ u \text{ and } p \qquad (5)$$

$$C_{u,p}^{x,P} = \frac{D_x C_{u,p}^{x,m}}{K_{u,p}\left(\Delta P_{u,p} - \Delta \pi_{u,p}\right)\gamma_{u,p}} \qquad \forall \ u, p \text{ and } x \qquad (6)$$

where N_u is the number of membrane modules in unit u, $K_{u,p}$ is the permeability of the membrane, S is the membrane surface area, $\Delta P_{u,p}$ is the effective trans-membrane operating pressure, $\Delta \pi_{u,p}$ is the trans-membrane osmotic pressure, $\gamma_{u,p}$ is the membrane geometry correction factor, D_x is the solute transport parameter of species x, and $C_{u,p}^{x,m}$ is the concentration of species x at the high pressure side of the membrane wall.

The following constraint is imposed to ensure that the average flow per module ($F_{u,p}^{ave}$) is within the manufacturer's recommended range of the lower (F_L^{ave}) and the upper (F_U^{ave}) bounds for the modules involved,

$$F_L^{ave} \le F_{u,p}^{ave} = \frac{F_{u,p}^f}{N_u} \le F_U^{ave} \qquad \forall \ u \text{ and } p \qquad (7)$$

where $F_{u,p}^f$ is the flowrate of the feed stream that enters unit u at period p.

The pressure drop (ΔP) for flows across the membrane units is assumed to be constant depending on the membrane types,

$$P_{u,p}^f = P_{u,p}^r - \Delta P \qquad \forall \ u \text{ and } p \qquad (8)$$

where $P_{u,p}^f$ and $P_{u,p}^r$ are the pressures of the feed and reject streams respectively.

4 SOLUTION STRATEGY

4.1 The Objective Function
The objective function employed is the total cost, TC, based on that used by [11]. It is taken as the sum of the initial and replacement membrane module costs (*MFC*), the capital cost of the pumps (*PFC*) and turbines (*TFC*), the membrane regeneration cost (*MOC*), and the pumping energy cost of all the pumps (*POC*) including energy cost savings generated by recovery turbines (*TOC*), viz.

$$TC = C_M \sum_{u=A}^{U} \sum_{p=1}^{P} \left(N_u + Z_{u,p}^{new} N_u \right) + \sum_{b=1}^{B} 2590 \, PWR_b^{0.79} + \sum_{t=1}^{T} 830 \, PWR_t^{0.47}$$

$$+ \sum_{p=1}^{P} \sum_{u=A}^{U} \left[10^3 \, Z_{u,p}^{clean} + 450 \, N_u \left(Z_{u,p}^{clean} - Z_{u,p}^{new} \right) \right]$$

$$+ \frac{C_E}{\eta} \sum_{p=1}^{P} \sum_{b=1}^{B} KWH_{b,p} + C_E \, \eta \sum_{p=1}^{P} \sum_{t=1}^{T} KWH_{t,p}$$

$$(9)$$

The sum of *MFC*, *PFC* and *TFC* represents the total fixed cost (*FC*) and the sum of *MOC*, *POC* and *TOC* represents the total operating cost (*OC*) of the network. It is assumed that the membrane modules cost, C_M, \$2700/module, is the same for both initial and replacement modules. The membrane regeneration cost involves a fixed cost of \$10k for each period in which cleaning occurs regardless of the number of units to be cleaned in any period, and a variable cost of $450/module. Both *PFC* and *TFC* are calculated from the maximum power requirements in kW as indicated by PWR_b ($b \in B$, where B is the total number of booster pumps used) and PWR_t ($t \in T$, where T is the total number of turbines used) respectively. $KWH_{b,p}$ and $KWH_{t,p}$ are the electrical duties of the pumps and turbines respectively; the efficiency, η, is 65% and the cost of electricity, C_E, is $0.07/kWh.

4.2 Solution Technique

Solutions are obtained by solving the scheduling problem as a mixed-integer non-linear programming (MINLP) problem using DICOPT++ under the General Algebraic Modelling System (GAMS) environment [16] on a SunSparc 10 workstation. The outer approximation/extended relaxation (OA/ER) method is used to solve the problem by decomposing it into a series of non-linear (NLP) sub-problem and mixed integer (MIP) master problem [17]. The NLP sub-problem is solved using MINOS and the MIP master problem is solved using CPLEX by employing several starting points to obtain the best possible solution. This is because there is no guarantee of obtaining a global optimum solution due to the non-convexity of the MINLP problem. The applicability of the approach in identifying the effect of network configuration on operational cost is demonstrated using the following two case studies. A three year operating horizon is used here as this period represents the average lifetime of a membrane module.

5   CASE STUDY 1: SEAWATER DESALINATION

Seawater desalination is one of the most established applications of RO membranes in water treatment technology for potable and industrial uses [13]. Here, we consider scheduling of the optimal network configurations obtained by Zhu *et al.* [11] (DSL-1) and four other network configurations (DSL-2 to -5) shown in Figure 2. The aim is to determine the effects of different staging arrangements on the optimal cleaning schedule. DSL-1, -2 and -5 are two-stage systems with 2:2, 2:1 and 3:1 configurations respectively, while DSL-3 and -4 are single stage systems with 2 and 3 parallel units respectively.

All of the configurations are optimised using the same feed flowrate (45-51 m³/hr), raw seawater concentration (34 800 ppm) and booster pump operating pressure (62-70 atm) to generate permeate with flowrate ≥ 21.6 m³/hr and salt content ≤ 570 ppm. Table 1 summarises the operating parameters used. The geometrical properties of the Du Pont B-10 hollow fibre membranes used are described by [11].

398

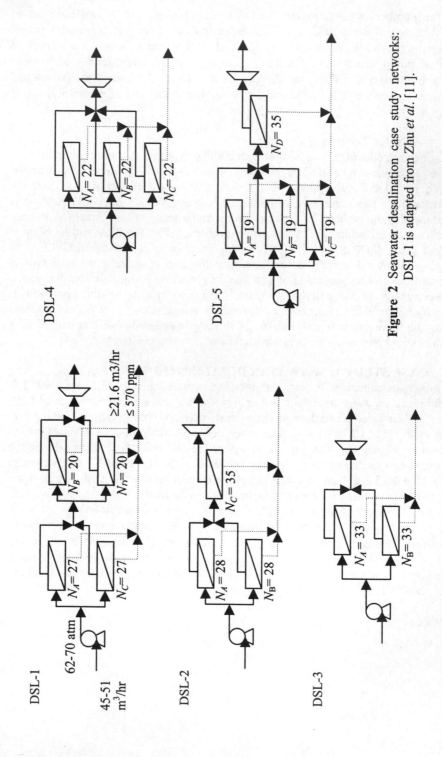

Figure 2 Seawater desalination case study networks: DSL-1 is adapted from Zhu *et al.* [11].

Table 1 Seawater desalination case study operating parameters (after [11]).

F_L^{ave}	$= 0.756 \ \mathrm{m^3/hr}$	F_U^{ave}	$= 0.972 \ \mathrm{m^3/hr}$
ΔK_{loss}^o	$= 0.03 \times 10^{-10} \ \mathrm{kg/s \ N}$	K_u^{new}	$= 3.0 \times 10^{-10} \ \mathrm{kg/s \ N}$
ΔP	$= 0.22 \ \mathrm{atm}$	ΔT	$= 60 \ \mathrm{days}$
Γ	$= 328 \ \mathrm{days}$	D	$= 4.0 \times 10^{-6} \ \mathrm{kg/s \ m^2}$

Table 2 Cleaning schedules for seawater desalination case study: perfect cleaning scenario. Dark areas denote periods ending in cleaning. Costs are in $.

Day ($t_{u,p}$)	DSL-1 A B C D	DSL-2 A B C	DSL-3 A B	DSL-4 A B C	DSL-5 A B C D
0					
60					
120					
180					
240					
300					
360					
420					
480					
540					
600					
660					
720					
780					
840					
900					
960					
1020					
1080					
TCL	14	13	18	34	17
FC	**358 700**	**341 600**	**280 500**	**279 000**	**348 700**
MFC	253 800	245 700	178 200	178 200	248 400
PFC	99 300	90 800	96 800	95 400	94 900
TFC	5 600	5 100	5 500	5 400	5 400
OC	**390 500**	**426 500**	**637 000**	**704 100**	**433 800**
MOC	208 100	246 400	437 300	506 600	254 200
POC	235 200	231 600	262 600	259 500	231 200
TOC	52 800	51 500	62 900	62 000	51 600
TC	**749 200**	**768 100**	**917 500**	**983 100**	**782 500**

Table 2 shows the membrane regeneration and replacement schedules, presented as Gantt charts, for the perfect cleaning scenario. The shaded periods indicate periods where the membrane unit is cleaned at the end of that period. The fixed cost (*FC*)

values for the single stage networks (DSL-3 and -4) are significantly smaller than those of the two-stage networks. The proportion of this cost due to membrane units (*MOC*) is smaller for the single stage systems
(64% vs. 71%). The operating costs show the reverse pattern, as may be expected, where the single stage networks prove to be unable to cope with the effects of membrane fouling. The operating costs are significantly larger than the capital costs in this case study, rendering the (cheaper) single stage networks economically unattractive. This result is not unexpected, as the feed pressure limit is active more often in these less flexible systems (as indicated by the power costs). Analogous situations in heat exchanger networks suggest that oversized or distributed systems are more robust if fouling is likely to occur. The inflexibility of the single stage networks is demonstrated by the number of cleaning actions performed; this number would be reduced by adding more modules to each unit, thereby increasing the *FC*. There is obvious scope for further iteration of the DSL-3 and -4 designs to try and optimise the total operating cost of such configurations. The most effective design is that reported by Zhu *et al.*, although the advantage over DSL-2 is not large. Methods for combining optimal scheduling within a superstructure design algorithm are the subject of ongoing work.

The optimised schedules in Table 2 show several notable features. The inflexibility of the single stage networks is very evident in the repetitive cycles obtained for DSL-3 and -4. The two-stage systems feature regular patterns of cleaning more than one unit; this feature is driven by the significant fixed cost associated with any cleaning action. Each schedule shows 'end effects' where no cleaning is carried out near the start (modules are clean) or end (insufficient payback time) of the horizon. Inspection of the DSL-1 schedule shows that the pattern is quasi-regular, with a cleaning interval varying between 120 and 180 days. This variation suggests that the time discretisation used here may have been too coarse. Nevertheless, the non-regular pattern obtained is significantly different from that employed by Zhu *et al.* The schedule used by these workers featured cleaning of unit pairs A-C and B-D alternatively every three months. This regular schedule would give a TC of $882k, compared with the $749k obtained here. Quasi-regular patterns are evident in the schedules obtained for DSL-2 and -5.

The results for the imperfect cleaning scenario given in Table 3 show similar rankings in *OC* as in Table 2, except that DSL-5 is as favourable as DSL-2; DSL-1 remains marginally the best. More cleaning actions are required due to the degradation model, and the cost model again encourages several units to be cleaned simultaneously. The pumping costs (*POC*) do not change appreciably between the two scenarios; the increase in cleaning costs is caused by membrane degradation, which cannot be countered by the allowable range in pump pressures. Only the single stage systems require unit replacement, which causes their capital costs (*FC*) to exceed those of the two-stage networks. This result again suggests that the configuration is sub-optimal and could be improved upon.

The quasi-regular cleaning patterns for the two-stage schedules in Table 2 are largely replaced by complete plant cleaning in Table 3. This change is promoted by the absence of downtime or limits on the amount of cleaning which can be performed in any period, the form of the cost function used and the size of the discretisation intervals. More intervals could be used, at the cost of computing time: the DSL-1 case, for example, features 72 binary variables, 2701 continuous variables, 4464 constraints and 191 CPU seconds.

Table 3 Cleaning schedules for seawater desalination case study: imperfect cleaning scenario. Dark and grey areas denote periods with cleaning and replacement actions respectively. All costs are in $.

Day $(t_{u,p})$	DSL-1 A	B	C	D	DSL-2 A	B	C	DSL-3 A	B	DSL-4 A	B	C	DSL-5 A	B	C	D
0																
60																
120																
180																
240																
300																
360																
420																
480																
540																
600																
660																
720																
780																
840																
900																
960																
1020																
1080																

	DSL-1	DSL-2	DSL-3	DSL-4	DSL-5
TCL	17	17	19	39	20
FC	**356 900**	**341 000**	**462 000**	**343 600**	**349 600**
MFC	253 800	245 700	356 400	237 600	248 400
PFC	97 600	90 200	99 900	100 300	95 800
TFC	5 500	5 100	5 700	5 700	5 400
OC	**428 900**	**468 900**	**633 400**	**976 800**	**459 400**
MOC	248 200	290 000	432 200	776 100	279 800
POC	232 500	229 900	265 100	264 300	231 100
TOC	51 800	51 000	63 900	63 600	51 500
TC	**785 800**	**809 900**	**1 095 400**	**1 320 400**	**809 000**

6   CASE STUDY 2: WASTEWATER DEPHENOLISATION

The pulp and paper industry consumes large volumes of water and produces a number of complex waste streams from the different pulping and bleaching processes. The task of treating these effluents to meet environmental regulations is becoming more challenging. The capability of RO membranes to concentrate such wastes without generating chemical impacts has prompted the application of RO technology in treating these wastewaters.

This case study compares three optimised configurations reported by El-Halwagi [10] for the treatment of Kraft mill wastewaters contaminated with mono-chlorophenol (*MCP*) and tri-chlorophenol (*TCP*). The networks are based on the Du Pont B-10 RO membrane and are summarised in Table 4 and Figure 3. The geometrical properties of the membranes can be found in [10]. The characteristics of the feed and product streams are given in Table 5. Table 6 shows the optimal cleaning schedule results obtained for the perfect cleaning scenario.

Table 4 Wastewater dephenolisation case study operating parameters (after [10]).

F_L^{ave}	= 0.21 kg/s	F_U^{ave}	= 0.46 kg/s
D_{MCP}	= $2.43{\times}10^{-4}$ kg/s m^2	D_{TCP}	= $2.78{\times}10^{-4}$ kg/s m^2
ΔP	= 0.40 atm	K_u^{new}	= $5.90{\times}10^{-10}$ kg/s N
Γ	= 2190 days	ΔT	= 60 days

Table 5 Characteristics of the feed and product streams (after [10]).

Stream	A	B	C	D
Flow rate (kg/s)	25	6	≥ 9	≥ 4.5
MCP concentration (ppm)	12	26	≤ 8.8	≤ 8.8
TCP concentration (ppm)	4	3	≤ 1.4	≤ 1.4

El-Halwagi identified DPH-1 as the least expensive option. The FC values in Table 6 confirm this result, where the FC of DPH-1 is 6% and 50% less than that of DPH-2 and -3 respectively. The total cost (TC) results, however, show that DPH-3 was the most favourable network. This network has the lowest operating cost and requires very few cleaning actions due to its use of permeate staging. The primary unit (A) has to be cleaned regularly in order to maintain the total permeate flow; there is sufficient capacity in units B and C to counter the effect of the relatively slow decay in permeability in this case. This result again demonstrates that OC considerations can reverse the trends shown in capital costs alone.

7   CONCLUSIONS

A systematic approach for determining optimal regeneration and replacement schedules for RO membranes in water and wastewater treatment has been implemented successfully. The results obtained from the two case studies presented here have shown that the approach has a significant contribution to the design of the optimal RO networks. The approach provides more accurate and realistic fixed and operating costs of these networks as it simulates the cleaning process over the entire membrane lifetime. The different perfect and imperfect cleaning scenarios demonstrate the flexibility of the approach. As the approach uses discrete time intervals, the approach can also be used for long term planning encompassing the plant lifetime by extending the time horizon of interest.

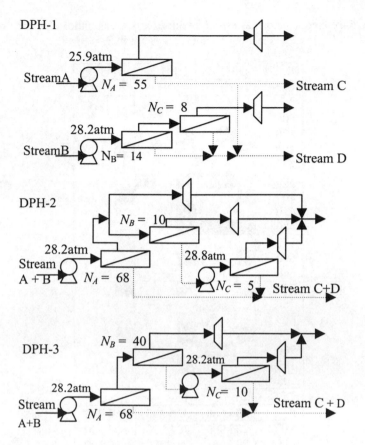

Figure 3   Dephenolisation case study networks (after [10]).

Table 6 Dephenolisation case study results. Dark areas indicate periods with cleaning actions. Costs are in $.

Day ($t_{u,p}$)	DPH-1			DPH-2			DPH-3		
	A	B	C	A	B	C	A	B	C
0									
60									
120		■		■					
180	■			■				■	
240									
300		■	■	■		■			
360	■							■	
420				■					
480	■	■	■	■					
540								■	
600	■			■					
660									
720	■			■				■	
780		■	■						
840			■	■					
900	■			■				■	
960		■	■						
1020				■					
1080									
TCL	16			11			5		
FC	**306 600**			**326 200**			**460 700**		
MFC	207 900			224 100			337 500		
PFC	92 200			94 800			116 100		
TFC	6 500			7 300			7 100		
OC	**479 800**			**602 400**			**261 100**		
MOC	308 000			408 300			203 000		
POC	222 200			248 200			288 800		
TOC	50 400			54 100			47 900		
TC	**786 400**			**928 600**			**721 800**		

REFERENCES

[1] Buckley, C.A., C.J. Brouckaert, and C.A. Kerr, 1993. RO Application in Brackish Water Desalination and in the Treatment of Industrial Effluents. In *Reverse Osmosis: Membrane Technology, Water Chemistry and Industrial Applications*, (Z. Amjad, ed.), Van Nostrand Reinhold, New York, pp:275-299.

[2] Rautenbach, R., and R. Albrecht, 1989. *Membrane Processes*. John Wiley and Sons, Chichester, UK.

[3] Qi, R., and M.A. Henson, 1998. Optimal Design of Spiral-Wound Membrane Networks for Gas Separations. *J. Membrane Science*, 148:71-89.

[4] Ohya, H., 1978. An Expression Method of Compaction Effects on Reverse Osmosis Membranes at High Pressure Operation. *Desalination*, 26:163-174.

[5] Al-Qahtany, H.I., and N.M.S. Al-Bastaki, 1995. Effect of Aging on the Performance of RO Hollow Fiber Membranes in a Section of an RO Plant. *Desalination*, 101:177-183.

[6] Evangelista, F., 1985. A Short Cut Method for the Design of Reverse Osmosis Desalination Plants. *Ind. Eng. Chem. Process Des. Dev.*, 24(1):211-223.

[7] Evangelista, F., 1986. Improved Graphical-Analytical Method for the Design of Reverse Osmosis Plants. *Ind. Eng. Chem. Process Des. Dev.*, 25(2):366-375.

[8] Rautenbach, R.,R. Knauf, A. Struck, and J. Vier, 1996. Simulation and Design of Membrane Plants with AspenPlus. *Chem. Eng. Techol.*, 19:391-397.

[9] Bansal, I.K., and A.J. Wiley, 1973. A Mathematical Model for Optimizing the Design of Reverse Osmosis Systems. *Tappi J.,* 56(10):112-115.

[10]. El-Halwagi, M.M., 1992. Synthesis of Reverse Osmosis Networks for Waste Reduction. *AIChE J.*, 38(8):1185-1198.

[11] Zhu, M., M.M. El-Halwagi and M. Al-Ahmad., 1997. Optimal Design and Scheduling of Flexible Reverse Osmosis Networks. *J. Membrane Science*, 129:161-174.

[12] Voros, N., Z.B. Maroulis, and D. Marinos-Kouris, 1996. Optimization of Reverse Osmosis Networks for Seawater Desalination. *Computers Chem. Engng.*, 20(Suppl):S345-S350.

[13] Voros, N.G., Z.B. Maroulis, and D. Marinos-Kouris, 1997. Short-cut Structural Design of Reverse Osmosis Desalination Plants. *Journal of Membrane Science*, 127:47-68.

[14] Galan B., and I.E. Grossmann, 1998. Optimal Design of Distributed Wastewater Treatment Networks. *Ind. Eng. Chem. Res.*, 37:4036-4048.

[15] Voivontas, D., K. Papafotiou, D. Assimacopoulos, and E. Mitsoulis, 1998. A Hybrid Expert-Procedural System for the Design of Reverse-Osmosis Desalination Plants. *Can. J. Chem. Eng.*, 76:1102-1109.

[16] Brooke A., D. Kendrick, and A. Meeraus, 1992. *GAMS: A User's Guide*. Scientific Press, Palo Alto, CA.

[17] Quesada, I., and I.E. Grossmann, 1992. An LP/NLP Based Branch and Bound Algorithm for Convex MINLP Optimisation Problems. *Computers Chem. Engng.*, 16:937-947.

Optimal Control of Pumping at a Water Filtration Plant Using Genetic Algorithms

A.R. Simpson, D.C. Sutton, D.S. Keane *and* S.J. Sheriff

ABSTRACT

At many water filtration plants water is delivered by pumps from a clear water storage to nearby tanks in other parts of the water distribution system. The control of the pumps in terms of startup and shut-down is usually achieved by trigger levels based on low and high tank water levels. This paper investigates the use of genetic algorithm optimisation to optimise the operation of pumps with the objective of minimising the pumping cost whilst staying within the operating constraints of minimum and maximum water levels for the tanks. Results for a case study are presented. Two different formulations of the genetic algorithm are considered. In both cases, it is assumed that the demand for the day has been forecast and that the water treatment output has been set to this value. Penalty costs are applied to degrade the fitness of solutions in a genetic algorithm population when the tank level drops below or rises above contracted minimum and maximum levels. In addition, a penalty cost is applied when a pump starting limitation is violated. In the first formulation, the trigger levels for the lower tank are optimised. The optimised trigger levels were found to be dependent on the starting level in the upper tank. In the second formulation, the trigger levels in the upper tank are optimised. The outcome of the genetic algorithm optimisation for the second case shows that the lower tank is too small to enable any feasible selection of trigger levels for the upper tank. The results show that there is potential for savings in pumping costs by using genetic algorithm optimisation for real-time operation of water filtration plant pump operations.

1   INTRODUCTION

This paper explores the application of genetic algorithms to the optimisation of pump operation at a water filtration plant. At many water filtration plants, water is delivered by pumps from a clear water storage adjacent to the plant to other parts of the water distribution system. Often water is taken from the clear water storage to another tank at a higher elevation. The control of the pumps in terms of start-up and shut-down is usually achieved by trigger levels based on low and high water levels in either the upstream (upper) or downstream (lower) tank. This paper investigates the use of genetic algorithm optimisation to optimise the operation of pumps with the objective of minimising the pumping cost whilst staying within the

407

operating constraints of minimum and maximum water levels for the tanks. Results of a case study for a water filtration plant located in the Adelaide hills in South Australia are presented. Water is assumed to be supplied by gravity feed from the water filtration plant at a constant rate to the clear water storage. A demand loading condition is considered with two lateral offtakes of demand from the pipeline leading from the lower clear water storage tank to the upper tank.

Two different formulations of the genetic algorithm are considered. In both cases, it is assumed that the demand for the day has been forecast and that the water treatment output has been set to this value. Penalty costs are applied to degrade the fitness of solutions in a genetic algorithm population when the water level in the upper Balhannah tank drops below or rises above contracted minimum and maximum levels. In addition, a penalty cost is applied to ensure the number of pump starts does not exceed 6 per hour over the 24-hour period.

2   LITERATURE REVIEW

The focus of this paper is the use of genetic algorithms for optimisation. Previous papers have considered the optimisation of the sizing of elements in water distribution systems (Simpson et. al 1994; Dandy et al. 1996). The cost of electricity accounts for a large part of the total operating cost for water-supply networks (Jowitt and Germanopoulos 1992). Over the last 25 years a number of authors have considered optimisation of pumping. Tarquin and Dowdy (1989) presented a heuristic method where the best pump or pump combination was selected. Previous attempts at optimisation of pumping using the more traditional optimisation approaches include: linear programming (Jowitt and Germanopoulos 1992); dynamic programming (Sterling and Coulbeck 1975a, Sabet and Helweg 1985, Ormsbee et al. 1989; Zessler and Shamir 1989; Lansey and Awumah 1994; Nitivattananon et. al 1996); hierarchical decomposition methods (Sterling and Coulbeck 1975b); and adaptive search (Pezashk and Helweg 1996).

3   GENETIC ALGORITHM FORMULATION APPLIED TO PUMPING OPTIMISATION

Genetic Algorithms (GAs) are search based optimisation algorithms driven by operators that attempt to replicate the processes of natural selection and population genetics. A string of numbers is used to represent the decision variables and their possible choices. In the genetic algorithm application developed here for pump optimisation, the fitness of each string is taken as proportional to the inverse of the total cost of pumping per kL of water pumped. The total cost is determined by the total electricity cost of pumping during the 24-hour period plus penalty costs associated with violating any of the constraints defined (e.g violation of tank levels, exceeding number of pump starts).

In the GA process, survival of the fittest and mixing by crossover drives the genetic algorithm procedure to obtain fitter and fitter solutions. GAs represent an efficient search method for non-linear optimisation problems. The GA search starts off with the random seeding of a population of strings, each of which represents a possible solution to selection of factors that determine the pumping cost. A simple genetic algorithm formulation may comprise: (i) tournament selection that selects a mating pool for the next generation giving preference to strings that are fitter (or of lower cost) (ii) crossover to permit mixing together of the strings from the mating pool and (iii) mutation to prevent the loss of genetic diversity. Over the last 15 years genetic algorithms have been applied to many engineering problems. Genetic algorithm or evolutionary strategy methods have been applied to pumping optimisation problems during the past 5 years (Murphy et al. 1994; Mackle et al. 1995; Savic et al. 1997; Illich and Simonovic, 1998).

4   CASE STUDY AND RESULTS

The case study to which the genetic algorithm (GA) has been applied is a pump station at the Summit Storage Water Filtration Plant (WFP) located approximately 4 km north of Littlehampton in the Adelaide hills, South Australia. A schematic is shown in Fig. 1. Trigger levels in the Clear Water Storage (CWS) tank are currently used to control when the pumps are turned on and off as shown in Fig. 2. The focus of the investigation is the optimisation of the operation of three fixed speed Clear Water Storage pumps (two duty and one stand-by) to minimise pumping costs. Two formulations have been investigated. A detailed analysis of the first formulation is given.

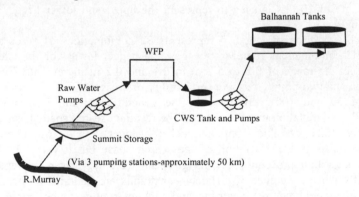

Figure 1. Water supply from the River Murray to the Balhannah tanks

4.1   Formulation 1

In the first formulation, the trigger levels for the operation of the pumps (based on levels in the Clear Water Storage or lower tank) are optimised. In this case, the upper trigger level signals when the pumps should turn on while the lower level signals when the pump should shut down. The diameter of the tank is 16.745 m.

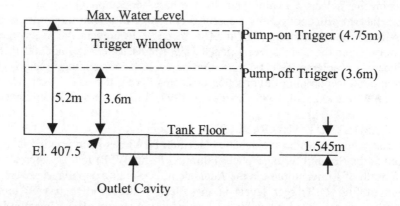

Figure 2. Clear Water Storage (CWS) tank existing pump trigger levels

The genetic algorithm formulation for the optimisation of the trigger levels uses a binary coded string to represent the two decision variables. The string of total length of 10 bits is used to represent the upper and lower trigger levels as shown in Fig. 3.

Each sub-string of length 5 represents 32 combinations of zeros and ones. These 32 combinations represent the possible trigger levels for the Clear Water Storage in the range of 0.5 to 5.2 m in 0.1 and 0.2 m increments. Tournament selection and two-point crossover (probability of crossover = 1.0) with population sizes of 50 and 100 were used in the genetic algorithm optimisation. The probability of mutation was set to zero. The resultant search space is small – there are only 1024 possible combinations of trigger levels. An alternative to optimisation is full enumeration of the search space but it was found that the genetic algorithm procedure has a significant advantage over full enumeration especially when a number of different constraints are included as penalty costs. Full enumeration requires a sorting and ranking routine to be implemented to determine the lowest cost solution that satisfies the constraints.

The genetic algorithm routine calls EPANET (a hydraulic simulation model) to simulate the variation of demands and tank levels at a one-minute time intervals over the 24-hour period. In each EPANET model, the tanks are provided with a 5-metre depth buffer above and below the true physical maximum and minimum levels for each tank. This enables satisfactory hydraulic evaluation of solutions that may cause the tank to drain empty or overflow. The costs of pumping were determined by computing the power costs as shown in Eq. 1.

$$P = \frac{\gamma Q H_P}{e_p e_m} \qquad (1)$$

where power (P) used by the pump is a function of the specific weight of fluid being pumped (γ), flow (Q), pump head (H_p), pump efficiency (e_p) and its motor efficiency (e_m). The motor efficiency for the pumps was set at 97%. An equation was fitted to the efficiency versus discharge curve for the pumps. Two power tariffs were available: (i) 14.5c/kWh for peak (7am-11pm) and (ii) 6.1 c/kWh (11pm-7am). The range of pumping heads is between 28 to 34 m while the range of discharges (each pump) is 475 to 490 L/s. The rising main between the two tanks is around 1000 m long with an internal diameter of 826 mm.

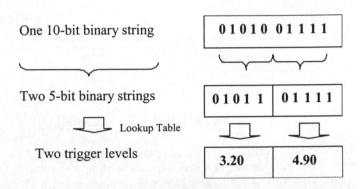

Figure 3. The conversion of the 10 bit string to two trigger levels

The overall pumping cost is around $400 per day for a constant water filtration plant output of 25.7 ML/day. To avoid a distortion in the results because of varying finishing levels in both tanks the objective function that is optimised is the unit cost per kL of water pumped. This in turn controls the fitness of the strings in the genetic algorithm optimisation process.

The constraints in this problem that have been included in the genetic algorithm procedure as penalty costs include:

1.   The upper trigger level (for turning pumps on) should not be lower than the lower trigger level (for turning pumps off). If this is detected in the genetic algorithm computer run, a large penalty is added to the power cost.

2.   The Balhannah water surface level in the upper tank (as shown in Fig. 4) is not permitted to go below the minimum contract level (49%) or above the maximum contract level (96%). A penalty cost is applied whenever this is detected.

3.   There must not be more than an average of 6 pump starts per hour (less than 144 for the day). If this is detected a large penalty cost is applied.

412

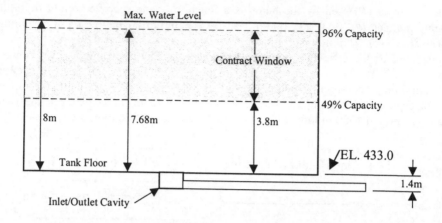

Figure 4. Balhannah tank levels

The correct operation of the genetic algorithm optimisation for selecting optimum trigger levels was confirmed by completing a run where the Balhannah tank penalty factor was set to zero (cases no. 1 to 4 in Table 1). For this penalty factor, one would expect the CWS to operate as close as possible to the top of the possible range of allowable trigger levels. In this way the power cost would be minimised. The results of the GA optimisation runs are shown in Table 1. Case no. 1 has a trigger level range of 0.24 m and is near the top of the CWS, thus the costs of operating the pumps will be minimised. If the range is increased to 0.74 m as for case no. 2 the cost increases by about $11 for the day while the number of pump starts drops from 118 to 39. Cases 3 and 4 show the results for a higher starting level in the Balhannah tanks of 6.9 m versus 6.4 m. Again the trigger levels are near the top of the CWS as expected.

Cases 5 to 8 consider pump operation with a Balhannah tank level penalty cost of $50 per incursion. A starting level for the Balhannah tank of 6.4 m was used for cases 5 and 6. The plot of the variation in water level in both tanks is shown in Fig. 5. Clearly, the Balhannah tank starting level is too low. The CWS is drawn down initially to increase the water level of the Balhannah tank and the trigger levels are set about 1.3 m below the top of the CWS tank. As a result the pumping costs are not as low as they could be. Cases 7 and 8 remedy this situation by raising the starting level in the Balhannah tank to 6.9 m. For case no. 7 the Balhannah tank level drops slightly at the beginning and the trigger levels for the CWS are very close to the top of the tank as shown in Fig. 6. A saving of 4.8% in pumping per kL of water occurs when moving from case no. 5 to case no. 7. The GA in all cases limited the number of pump starts to less than 144.

Thus the optimised trigger levels were found to be dependent on the starting level in the upper tank.

Table 1. Results for GA optimisation of formulation no. 1

Lower Trigger Level (m)	Upper Trigger Level (m)	Balhannah Tank Penalty ($/m)/ Balhannah starting level (m)	Cost (cents per kL)	Cost for 24-hours ($)	No. of pump starts
4.955	5.195	0 / 6.4	1.4035*	351.66	119
4.155	5.195	0 / 6.4	1.4302	360.24	28
4.955	5.195	0 / 6.9	1.4283	355.74	118
4.455	5.195	0 / 6.9	1.4382	362.67	39
3.255	3.855	50 / 6.4	1.5085	387.62	46
2.455	3.855	50 / 6.4	1.5228	392.68	14
4.755	5.195	50 / 6.9	1.4355	359.50	65
4.055	5.0555	50 / 6.9	1.4630	371.58	29

*=The tank level goes well below the Balhannah minimum

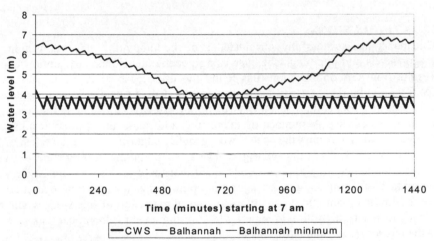

Balhannah starting water level 6.4 m, Penalty $50

Figure 5. Tank level variation for GA optimised trigger levels (case 5)

4.2   Formulation 2

In the second formulation, the trigger levels in the upper tank (Balhannah) are optimised. The opposite now occurs with the trigger levels compared to the first formulation. In this case, the lower trigger level signals when the pumps should turn on while the upper level signals when the pump should shut down. The outcome of the genetic algorithm optimisation for the second case shows that the lower tank is too small to enable any feasible selection of trigger levels for the upper tank. Thus it would not be possible for the system to be operated based on trigger levels in the upper tank.

414

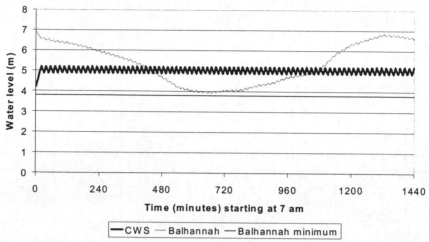

Figure 6. Tank level variation for GA optimised trigger levels (case 7)

5 CONCLUSIONS

This paper has successfully formulated a methodology for the optimisation of trigger levels and pumping schedules using genetic algorithm (GA) optimisation. Results from GA optimisation runs found the use of trigger levels in the lower CWS tank to be the least cost alternative for the demand loading case considered. This research found the use of GAs to be an effective method optimisation of pumping schedules. A number of interesting and some unexpected results are produced when comparing the two genetic algorithm formulations. For formulation no. 1, the optimised trigger levels depend on the starting water surface level in the upper storage. The results show that there is potential for savings in pumping costs by using genetic algorithm optimisation for real-time operation of water filtration plant pump operations. A logical extension of this work would be to use the genetic algorithm to optimise the starting level for the upper tank as well as the trigger levels in the lower tanks.

REFERENCES

Dandy, G.C., Simpson, A.R. and Murphy, L.J. (1996). "An improved genetic algorithm for pipe network optimisation." *Water Resources Research*, Vol. 32, No. 2, Feb., 449-458.

Illich, N. and Simonovich, S.P. (1998). "Evolutionary algorithm for minimization of pumping cost." *Journal of Computing in Civil Engineering*, ASCE, Vol. 12, No. 4, Oct., 232-240.

Jowitt, P.W. and Germanopoulos, G. (1992). "Optimal pump scheduling in water-supply networks." *Journal of Water Resources Planning and Management*, ASCE, Vol. 118, No. 4, 406-422.

Lansey, K.E. and Awumah, K. (1994). "Optimal pump operations considering pump switches." *Journal of Water Resources Planning and Management*, ASCE, Vol. 120,

No. 1, Jan./Feb., 17-35.

Mäckle, G., Savic, D.A. and Walters, G.A. (1995). "Application of Genetic Algorithms to Pump Scheduling for Water Supply," *Conference on Genetic Algorithms in Engineering Systems: Innovations and Applications*, GALESIA '95, IEE Conference Publication No. 414, Sheffield, UK, pp. 400-405.

Murphy, L.J., Dandy, G.C. and Simpson, A.R. (1994). "Optimum Design and Operation of Pumped Water Distribution Systems." *1994 International Conference on Hydraulics and Civil Engineering, Hydraulic working with the Environment*, The Institution of Engineers, Australia, Brisbane,15-17 Feb., 149-155.

Nitivattananon, V., Sadowski, E.C. and Quimpo, R.G. (1996). "Optimization of water supply system operation." *Journal of Water Resources Planning and Management*, ASCE, 122 (4), Sept./Oct., 374-384.

Ormsbee, L.E., Walski, T.M., Chase, D.V., and Sharp, W.W. (1989). "Methodology for improving pump operation efficiency." *Journal of Water Resources Planning and Management*, ASCE, Vol. 115, No. 2, 148-164.

Pezeshk, S. and Helweg, O.J. (1996). "Adaptive search optimisation in reducing pump operating costs." *Journal of Water Resources Planning and Management*, ASCE, Vol. 122, No. 1, Jan/Feb., 57-63.

Sabet, M.H. and Helweg, O.J. (1985). "Cost effective operation of urban water supply system using dynamic programming." *Water Resources Bulletin*, Vol. 21, No. 1, 75-81.

Savic, D.A., Walters, G.A. and Schwab, M. (1997). ""Multiobjective genetic algorithms for pump scheduling in water supply, *Evolutionary Computing workshop*, AISB '97, Manchester, 7-8 April.

Simpson, A.R., Dandy, G.C. and Murphy, L.J. (1994). "Genetic algorithms compared to other techniques for pipe optimisation," *Journal of Water Resources Planning and Management*, ASCE, 120 (4), July/August, 423-443.

Sterling, M.J. and Coulbeck, B. (1975a). "Dynamic programming solution to optimization of pumping costs." *Proceedings of the Institution of Civil Engineers*, Part 2, Dec., 789-797.

Sterling, M.J. and Coulbeck, B. (1975b). "Optimization of water pumping costs by hierarchical methods." *Proceedings of the Institution of Civil Engineers*, Part 2, Dec., 813-818.

Tarquin, A.J. and Dowdy, J. (1989). "Optimal pump operation in water distribution." *Journal of Hydraulic Engineering*, ASCE, Vol. 115, No. 2, February, 158-169.

Zessler, U. and Shamir, U. (1989). "Optimal operation of water distribution systems." *Journal of Water Resources Planning and Management*, ASCE, Vol. 115, No. 6, 735-752.

PART VIII

GENETIC ALGORITHMS

The Development of an Optimal Strategy to Schedule Main Replacements

M.O. Engelhardt *and* G.C. Dandy

ABSTRACT

Water distribution systems, like all other physical systems, deteriorate with time. This deterioration increases the frequency of burst events, increasing the cost required to maintain the system and the number of interruptions to service experienced by the consumer. The best way to minimise the occurrences of these events is to use a decision model that will enable the optimal schedule for the replacement of these mains to be found, subject to budget constraints. A decision model based on genetic algorithm (GA) optimisation is developed in this paper in two stages. Firstly, a purely economic decision model is developed. This model allows for the mains to be scheduled for replacement over a twenty-year period. Not only is the GA able to identify and schedule those mains that should be replaced due to their physical degradation, but it can also schedule mains due to pressure deficiencies caused by increases in demands.

The second stage of the decision model is the inclusion of a reliability measure into a multi-objective framework. The use of GAs enables a large set of non-dominated solutions to be generated. This facilitates the identification of a trade-off between reliability and the economic criteria. The final decision model enables the optimal replacement schedule to be identified for a given level of service. The ability of the GA to solve for both single and multiple objective problems means that the optimal schedule on any point along this curve can be identified.

1 INTRODUCTION

Rehabilitation strategies play an important role in the operation of a water distribution system. These rehabilitation strategies provide the framework through which the water companies can define and identify the trade-offs that must be made. The trade-off of most interest is that of performance and the cost required to meet these service levels. Of equal importance to a water company is the incorporation of the operating and maintenance costs to allow the network to be run at its most efficient. The optimisation of the operation and rehabilitation of reasonable sized networks may require large computation times, especially if extended period simulations are deemed necessary. To overcome this, a number of simplifications, and hence trade-offs may be required. Some of the trade-offs

419

required are demonstrated through the setting up of a rehabilitation strategy for a typical network in Adelaide, Australia.

This case study system will be introduced in Section 2. Section 3 provides a discussion on genetic algorithms and the unique alternatives that they offer as a decision support system for the rehabilitation problem. This leads into the two principal formulations that have been developed to demonstrate this. The first concentrates on providing the least cost design over an extended period given the structural deterioration that will occur and expected increases in demands. The second of the formulations is a multiple objective approach. A level of service performance criteria based on a reliability measure is introduced into the problem formulation. This enables the trade-off between the cost of rehabilitation and performance levels to be identified.

2   CASE STUDY

The first step in the formulation process was the decision of the case study system. The chosen system was an average sized subsystem in the city of Adelaide. The model used in the rehabilitation strategy consisted of over 200 km of mains over 150 mm in diameter. The supply is gravity fed through 5 storage reservoirs. In the 1960s the majority of the cast iron mains were lined in-situ. Therefore, the only rehabilitation option that was included was replacement. The break history of the network was obtained, with break prediction curves developed through multiple regression. Neither leakage nor water quality were considered in the decision to replace a main in the region.

3   GENETIC ALGORITHMS

GAs are one of the new evolutionary computation techniques. They utilise many of the concepts of nature, simulating a "survival of the fittest" regime. Each solution is represented by a chromosome, which contains the decision variables for the problem. Each individual, based on the values contained in the chromosome is given an objective value, in this case the cost of the solution. Based on its objective value each individual is assigned a fitness value. This is a measure of the worth of an individual to its surrounding environment. A population of these individuals is used to search the solution space. Each competes to mate and form a new population (based on its fitness). This process uses a stochastic process to share and exchange the chromosome values from the parent individuals in the hope of forming fitter offspring.

In recent years GAs have been shown to be successful in finding near optimal solutions in a variety of water network design problems (Dandy et al. 1996). Their use throughout the design and operation of water distribution systems has increased, including pump scheduling (Mackle, et al. 1995), calibration (Vitkovsky and Simpson 1997), leakage reduction (Wood and Reddy 1996) and previously for rehabilitation based problems (Halhal, et al. 1997). The ability of the GA to handle these problems makes them ideal to use for a rehabilitation problem.

It should be noted that the use of GAs does not guarantee the global optimal solution. As indicated by the examples presented above, they can handle highly non-linear problems better than other techniques, however, like many optimisation techniques, this comes at a high computational cost through using a high number of trial evaluations. As a rule of thumb, the more trials, generally the higher the likelihood that the optimal solution is found. Naturally, the more complicated the problem, the larger the search space and the more trials one would expect to obtain near optimal solutions. Equally, the more complicated the problem the higher simulation requirements, further increasing the computational requirements of the GA. To reduce this computer time, simplifications to the simulation are generally adopted.

4   SINGLE OBJECTIVE PROBLEM FORMULATION

The single objective framework was formulated so that the optimal rehabilitation schedule over a twenty-year time period is obtained. The GA modelled the planning horizon by splitting it into five, five-year time periods. A budget constraint is applied to model the available funds for rehabilitation works. The size of this budget was assigned through consultation with the supervising water authority (South Australian Water Corporation). Demand increases were assumed to occur over the planning horizon. The single objective formulation can be given mathematically as,

$$\min(SystemCost) = \sum_{i=0}^{NL} Cost_i$$

(1)

where,

$$Cost_i = PVOB_i + PVNB_i + NPR_i$$

(2)

with

$$NPR_i=0 \text{ and } PVNB_i=0 \text{ if } T_{ri} \geq 5,$$

(3)

subject to,

$$\sum_{i=1}^{N_r} Rep(d_{ni}) * \delta(t) * l_i \leq k * Fund, \quad \forall t \in T_{ri}, N_r \in (1,...N)$$

(4)

$$H_j = f(Q,d) \geq H_{min}$$

(5)

$$v_i = f(Q,d) \leq V_{max}$$

(6)

where $SystemCost$ is the total cost of the rehabilitated system, $Cost_i$ is the cost of replacing and/or maintaining link i, NL is the number of links in the system, $PVOB_i$ is the net present value of the burst costs in the existing main, $PVNB_i$ is the net present value of the expected burst costs in the new main, NPR_i is the present value of the costs of replacing main i, T_{ri} is the time step for replacement for main $i \in (0,1,..,4)$ with the actual time of replacement being $5*T_{ri}$, N_r is the number of pipes that are to be replaced, $Rep(d_{ni})$ is the replacement cost of main i with diameter d_{ni}, l_i is the length of main i, $\delta(t)$ is 1 if $t=T_{ri}$, or 0 otherwise, $Fund$ is the allowed budget for each year, $k=2.5$ for $t=0$ or 5 otherwise, H_j is the pressure head at node j, H_{min} is the minimum allowable pressure head, Q is the system demand, d is the diameters of the mains in the system, v_i is the velocity in link i and V_{max} is the maximum allowable velocity.

The constraint given by Equation (4) was included to limit the funds that can be spent in any one time step. Equations (5) and (6) are the hydraulic performance constraints. It was shown in Engelhardt (1999) that the GA was able to obtain near optimal schedules for a variety of different formulations of varying complexities. A typical solution provided by the GA using the representation chosen provides an output as given in Tables 1 and 2. Table 1 identifies the mains that require replacement and the funds required to implement these works over a 20 year planning horizon. Table 2 provides the diameters of the new (duplicate) and replaced mains.

Table1: The Mains Identified for Replacement.

Time (Years)	Replacement Costs ($1000)	Mains	
		Replaced	Duplicated
0	123.72	152, 153, 162, 164, 166	
5	249.955	155, 165, 167, 249, 251, 252, 469	
10	225.8	14, 154, 168, 179, 181, 250, 357, 470	
15	0		
20	205.05	24, 183, 186, 356	174, 175, 248

Table 2: New Mains Required to Overcome Pressure Deficiencies.

Main No.	Time Step	Orig. Diam.	Rep Diam.	Dupl. Diam.	Length (m)	Main No.	Time Step	Orig. Diam.	Rep. Diam.	Dupl. Diam.	Length (m)
14	10	150	250	-	356	179	10	200	450	-	46
24	20	150	150	-	154	181	10	200	375	-	467
152	0	150	150	-	206	183	20	200	100	-	159
153	0	150	150	-	213	186	20	200	100	-	25
154	10	150	150	-	71	248	20	200	-	200	782
155	5	150	150	-	101	249	5	200	100	-	589
162	0	150	150	-	185	250	10	200	100	-	238
164	0	150	150	-	139	251	5	200	100	-	1156
165	5	150	150	-	104	252	5	200	100	-	324
166	0	150	150	-	288	356	20	300	450	-	54
167	5	150	150	-	124	357	10	300	100	-	44
168	10	150	150	-	41	469	5	150	150	-	116
174	20	150	-	150	119	470	10	150	150	-	103
175	20	150	-	150	263						

The above formulation requires amendments to ensure its practicality and acceptability. Firstly the costs are considered. This formulation is driven purely by the burst costs, hence the mains identified for replacement (mains 152-155, 162, 164-168, 249-252, 469, 470) are the sections of mains that had the highest probability of failure. Other costs were not included as part of this research, although any cost can be included that can be assigned to individual main lengths. This limitation is created primarily due to the representation used in the GA chromosome that is based on discrete sections of mains.

Care must also be taken on when the hydraulic model is run as part of the formulation. It can be seen that all replacements to overcome the pressure deficiencies created by the increased demands are either in the 10 or 20 year time periods. This is due to the fact that the demand increase was modelled in two stages. The demand was assumed to increase by half the expected amount in the 10 year time step with the total expected increase modelled at the 20 year time step. As was shown in Engelhardt (1999), a single hydraulic analysis results in all main replacements required to overcome the pressure deficiencies being scheduled in the last time step. Scheduling replacements in this time step has the least effect on the net present value of the system cost. Ideally, the hydraulic model should be run at each time step, but this comes at a computational cost.

One other problem that needs to be addressed is that of the practicality of the resulting schedule. Some solutions are actually cheaper than others due to the spatial connectivity of the identified mains. A good example of this is the scheduling for mains 152-155, 162, 164-168, 469 and 470. All of these mains are connected along a road, but are scheduled for replacement over a ten year period, with just 71 metres of main replaced in the 10 year time step. Clearly, it would be cheaper to replace all of the mains at the same time.

4.1   Multiple Objective Economic-Reliability Formulation

The single objective model given above does not take into account multiple service performance criteria. These are now being increasingly applied to water distribution systems. The one level of service criteria that was addressed above is the hydraulic one. Any other performance indicators can easily be applied as other constraints, similar to these hydraulic constraints. This, however, does not answer the question of the financial requirements to meet levels of service indicators. This requires a multi-objective framework.

The level of service measure that is investigated here is reliability. Numerous measures of reliability have been proposed in the literature, including some borrowed from traditional probability theory. These include the probability that all nodes are connected to a source (Goulter and Coals 1986), and the probability that the demand is met (Hobbs and Beim 1988). Surrogate measures have also been utilised, including the number of days' outage (Walski 1987) and the volume of unserved demand (Wagner, et al. 1988). Index based measures have also been utilised, such as the ratio of the served demand to the total demand (Fujiwara and Tung 1992), and the complement of the ratio of the unserved demand over the total demand (Fujiwara and De Silva 1990).

None of these reliability measures was considered as the ideal measure for a rehabilitation strategy. The measure that is proposed is the number of customer interruptions that can be expected in a year. This can be given mathematically as:

$$\sum_i TENCI(i,t) = \sum_i (\text{Probability of link } i \text{ failing at time } t) *$$

(number of customer interruptions associated with the failure of link i) (7)

where $TENCI(i,t)$ total expected number of customer interruptions for main i at time t. The probability of link i will burst is obtained using the Poisson process, given that the break prediction equations developed for the case study system provide the burst rate/km of length. This burst rate is assumed to be constant over the five-year time intervals. The connectivity of the system and location of valves determine the number of customers interrupted. When a main bursts, valves either side of it are closed to allow repair. This has a localised effect, and depending on its location within the network, may have network wide effects. In the former case, knowledge of the location of valves and service connections is paramount. The latter interruptions are obtained by taking each main out of operation successively and observing the pressures throughout the system. If the pressure falls below a set level at a node, then an interruption to this node's supply was assumed.

Therefore the rehabilitation formulation becomes,

$$\min(SystemCost) = \sum_{i=0}^{NL} Cost_i \tag{8}$$

and

$$\min \sum_i TENCI(i,t) \tag{9}$$

The question though is at what time step should the reliability level be used as the objective in Equation (9). This depends on the planning horizon of the economic criteria. The simplest of the economic models is to only allow the replacement of the mains to occur in the first time step. Therefore, the most appropriate reliability measure is that of the current time step. Also to simplify the computational requirements it was assumed that the main was replaced by one of an equivalent size, thereby not affecting the hydraulic properties, or more importantly the number of customers effected by a burst event.

Figure 1 provides the trade-off curve generated by converting the single objective GA into a multi-objective GA using a method proposed by Fonseca and Fleming (1993). Five curves are provided here. They enable determination of the effect that spending money now has on the future expected number of customer interruptions at various times in the future. These curves clearly provide a means by which the supervising water authority can quantify the required expenditure to meet a level of service that is required in the system's operation into the near future.

5 CONCLUSION

The research presented here provides the various alternatives available to the water design engineer in the application of GAs to the rehabilitation problem. The GA can be used in a variety of steps throughout the decision making process. This can occur in two stages. The GA can obtain the optimal solution for a complex, long term planning strategy. Traditionally the major concern is to ensure the correct hydraulic delivery of the supply. However, increasingly other levels of service performance criteria are included. The GA can be used to identify the trade-off between the introduced performance criteria and the required funds to be investigated. More importantly, once the decision is made with regard to the required level of service, the GA can be used to identify the required works to meet this level of service.

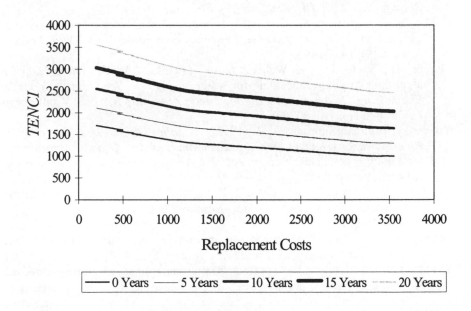

Fig. 1 The Reliability-Economic Trade-Off for a Water Distribution System

REFERENCES

Dandy, G. C., Simpson, A. R. and Murphy, L. J. (1996). "An improved genetic algorithm for pipe network optimisation." *Wat. Resour. Res.*,32 (2), 449-458.

Engelhardt, M. O. (1999). *"Development of a Strategy for The Optimum Replacement of Water Mains."* PhD, Department of Civil and Environmental Engineering,University of Adelaide, 514.

Fonseca, C. M. and Fleming, P. J. (1993). "Genetic algorithms for multi-objective optimisation: formulation, discussion and generalisation." *Fifth International Conference on Genetic Algorithms*, Morgan Kaufman, 416-423.

Fujiwara, O. and De Silva, A. U. (1990). "Algorithm for reliability based optimal design of water networks." *Jour, of Env. Eng. Div. Am. Soc. Civ. Eng.*, 116 (3), 575-587.

Fujiwara, O. and Tung, D. H. (1992). "Reliability improvement for water distribution through increasing the pipe size." *Water Resources Journal*, 28 (3), 29-36.

Goulter, I. C. and Coals, V. (1986). "Quantitative approaches to reliability assessment in pipe networks." *Jour. of Trans. Div. Am. Soc. Civ. Eng.*, 112 (3), 287-301.

Halhal, D., Walters, G. A., Ouzar, D. and Savic, D. A. (1997). "Water network rehabilitation with a structured messy genetic algorithm." *Jour. of Wat. Resour. Plan. Man. Div. Am. Soc. Civ. Eng.*, 123 (3), 137-146.

Hobbs, B. F. and Beim, G. K. (1988). "Analytical simulation of water system capacity reliability - 1. modified frequency duration analysis." *Wat. Resour. Res.*, 24 (9), 1431-1444.

Mackle, G., Savic, D. and Walters, G. A. (1995). "Application of genetic algorithms to pump scheduling for water supply." *Genetic Algorithms in Engineering Systems: Innovations and Applications*, IEE Conference Publication No. 414, 400-405.

Vitkovsky, J. P. and Simpson, A. R. (1997). "*Calibration and Leak Detection in Pipe Networks Using Inverse Transient Analysis and Genetic Algorithms.*" University of Adelaide, Department of Civil & Environmental Engineering, Research Report R157,

Wagner, J. M., Shamir, U. and Marks, D. H. (1988). "Water distribution reliability: simulation methods." *Jour. of Wat. Resour. Plan. Man. Div. Am. Soc. Civ. Eng.*, 114 (3), 276-293.

Walski, T. M. (1987). "Discussion on: multi-objective optimisation of water distribution networks by I.C. Goulter." *Civ. Eng. Sys.*, 4 (4), 215- 217.

Wood, D. J. and Reddy, L. S. (1996). "Using variable speed pumps to reduce leakage and improve performance." *Improving Efficiency and Reliability in Water Distribution Systems*. E. Crabrera and A. F. Vela. Kluwer Academic Publisher. 135-164.

Discrete Optimisation of Water Distribution Networks Using Genetic Algorithms

R. Farmani, R.G.S. Matthew *and* A. A. Javadi

ABSTRACT

A computer programme using genetic algorithms has been developed and used for optimum design of water distribution networks. The programme includes both simulation and optimisation in the context of a single programme. The programme can consider both discrete and continuous design variables or a combination of the two.

A sine function has been introduced for discretisation of the variables which are not taken as the unknown design variables. The programme has been applied to test case studies and the obtained improvements have been discussed. It has been shown that the method provides an efficient tool in analysis, design and optimisation of water distribution networks.

1   INTRODUCTION

Water distribution networks are one of the most expensive civil engineering structures and so searching for the optimum network can lead to a considerable saving in the construction cost.

Many researchers have studied the optimisation of water distribution networks. Many different formulation and solution methods have been proposed. For example Alperovits and Shamir (1977) presented a linear programming gradient method. Eiger et al. (1994) used a two stage decomposition model. Simpson et al. (1994) and Savic and Walters (1997) used Genetic Algorithms (GAs) for optimal design of water networks. This paper looks at the application of different methods for discrete optimisation of water distribution networks including GAs.

2   HYDRAULIC FORMULATION

Hydraulic formulation of the problem is based on the continuity equation and conservation of energy.

Continuity of the flow dictates that the inflows must be equal to the outflows at every node of the network. The continuity equation can be written as:

$$\sum Q_{in} - \sum Q_{out} = Q_e \tag{1}$$

where Q_e represents the external inflow or demand at the junction node.

Conservation of energy dictates that the algebraic sum of all increments of head around a loop must be equal to zero. The equation can be written as:

$$\sum h_{ij} = \sum E_p$$

(2)

where h_{ij} is the head loss in pipe ij and E_p is the energy head put into the liquid by pumping. For head loss calculations the Hazen-Williams formula is used, which can be expressed as:

$$h_{ij} = \alpha L_{ij} \left(\frac{q_{ij}}{c_{ij}} \right)^{1.852} \frac{1}{D_{ij}^{4.87}}$$

(3)

where L_{ij}, q_{ij}, c_{ij} and D_{ij} are length, flow rate, Hazen-Williams coefficient and diameter of the pipe between nodes i and j respectively and α is a conversion factor to account for units. In this paper $\alpha = 10.5088$ was used.

3   OPTIMISATION

The main objective of optimum design is the selection of project alternatives, which are compared to find the most economical alternative, according to the present value of cost for the design model. The total cost of a project involves a number of different components such as equipment costs (pipes, fittings, pumps, reservoirs etc), construction cost, operation cost and energy cost.

The cost of installing pipes in the network is usually considered as the construction cost. This cost is a function of the pipe diameter and length and can be written as:

$$c_{ij} = k\, D_{ij}^n\, L$$

(4)

where c_{ij} is cost of installing the pipe between nodes i and j, k is the cost per unit length of the pipe between i and j with a diameter D_{ij}, and n usually varies between 1 and 2.

In a water distribution network with a reservoir, the additional storage elevation cost can be considered as follows:

$$c_{res} = \beta\, x_{res}$$

(5)

where c_{res} is the cost of raising the location of the reservoir, β is the unit cost for raising the reservoir by one metre and x_{res} is the additional elevation to be selected by optimising the objective function.

In a water distribution network with a pump, the operational cost of the pump is considered as a function of its capacity and its rated horsepower. The power is given by:

$$h_p = \gamma.Q.X_p/\eta \qquad (6)$$

where X_p is the head added by the pump (m), Q is flow rate (m^3/s), $\gamma = \rho g/1000$, ρ is density of water (kg/m^3), g is gravity acceleration $9.81 (m/s^2)$, η is the efficiency of pump.

4   GENETIC ALGORITHMS

Genetic algorithms are methods used in function optimisation. They offer a much higher probability of locating the global optimum in the design space for complex optimisation problems. The main steps of the GAs method can be summarised as follows (Goldberg, 1989):

a)   The initial population (*lpop*) is created as binary random numbers of 0s and 1s with a fixed length (*lcrom*). In this study *lpop* = 200 and *lcrom* = 10 were used.

b)   The binary strings are decoded into parameter values. In this work, the design variables are pipe diameters (which are discrete) and pressure heads (which are continuous). For continuous variables, having the upper and lower levels of variables (which in this work are the maximum and minimum pressure levels in pipes) and by using linear scaling, the values of decoded binary strings are transferred to the domain of design variable. For discrete variables each decoded binary string value represents a discrete pipe size.

c)   The objective function, which is cost of the network, is evaluated from the established design variable values. The feasibility of the solution can be verified by a simulation program which is linked to the optimisation program. The simulation model is based on the sets of non-linear equations (1) and (2). For a constrained minimisation problem in GAs, firstly constraints should be converted into unconstrained form by using a penalty function.

d)   Selection: the fitter members in the population are selected to produce new members for the next generation.

e)   Crossover: the chosen individuals of the members of the population are exchanged to improve the fitness of the next generation. The probability of crossover (P_c) shows whether selected chromosomes are used in reproduction or not. The most suitable value for P_c from the literature is in the range of 0.5-1 (Goldberg et al. (1987)).

f) Mutation: this operator works based on probability of mutation (p_m) which varies from 0.001-0.025.

g) Convergence: at the end of each generation, convergence is checked and the procedure is repeated until no further improvements can be obtained.

4.1 Sine Penalty function

In water distribution networks the discrete design variables are pipe diameters, ($D_i, i = 1,2,...n$), that have to be chosen from a list of available commercial pipe sizes. There are different methods for finding discrete design variables in search space, such as rounded continuous optimum, segmental optimisation, dynamic rounding, discrete optimisation using penalty function. In this work, a sine penalty function has been used in discrete optimisation of water distribution networks. The application and efficiency of the method have been illustrated by examples. The sine penalty function was used to discretise those pipe diameters, which were not taken as the design variables. The penalty term is defined to take on a zero value at the discrete points. At non-discrete points the penalty term assumes a non-zero value. This forces the optimal solution to be at the selected discrete points. The method can be explained in detail as follows:

Consider the following optimisation problem:

$$\text{minimise} \quad f(x) \tag{7}$$
$$\text{subject to} \quad g_j (x) \geq 0 \qquad j=1,2,..., n_g$$

$$\text{where} \quad x=(x_1, x_2,..x_n)^t,$$
$$x_i =(d_{i1}\ d_{i2}...d_{iq})^t \qquad i=1,2,..., n_d,$$

n_d is the number of discrete design variables, n_g is the number of constraints, d_{ik} the kth are discrete value of the ith design variable and q is the number of discrete values for each variable.

To account for the discrete design variables, the objective function is modified to a pseudo-objective function φ, which includes the penalty terms due to constraints, $p(g_j)$, and the non-discrete values of the design variables as:

$$\varphi(x,r,s) = f(x) + r\sum_{j=1}^{n_g} p(g_j) + s\sum_{i=1}^{n_d} \phi_d^i(x) \tag{8}$$

where $\phi_d^i(x)$ denotes the penalty term for non discrete values of the ith design variable. Different forms for ϕ are possible, but Dong, et al. (1990) recommended the following function:

$$\varphi_d^i(x) = \frac{1}{2}\left(\sin\frac{2\pi\left[x_i - \left(d_{i(j+1)} + 3d_{ij}\right)/4\right]}{d_{i(j+1)} - d_{ij}} + 1\right)$$

(9)

where $\qquad d_{ij} \prec x_i \prec d_{i(j+1)}$

The multiplier, r, in equation 8 controls the contribution of the constraint penalty terms. It is clear that r must be chosen so as to increase the weight of constraint violations. After several iterations the constraint penalty (φ -f) is within the desired range of the objective function to activate the penalty terms for non-discrete values of the design variables.

A criterion for the activation of the non-discrete penalty multiplier, s, is the same as the convergence criterion, that is:

$$\left|\frac{\varphi - f}{f}\right| \le \varepsilon_c$$

A typical value for ε_c is 0.01. The magnitude of the non discrete penalty multiplier, s, at the first discrete iteration is calculated such that the penalty associated with the discrete valued design variables - that are not at their allowed values - is of the order of 10 percent of the constraint penalty, i.e., $s \approx 0.1\ r\ p(g)$ (Haftka, and Gurdal, 1992). As the iteration for discrete optimisation proceeds, the non-discrete penalty multiplier for the new iteration is increased by a factor of the order of 10. It is also important to decide how to control the penalty multiplier for the constraints, r, during the discrete optimisation process. If r is decreased for each discrete optimisation iteration, as in the continuous optimisation process, the design can be stalled due to high penalties for constraint violation. Thus, it is suggested that the penalty multiplier r be frozen at the end of the continuous optimisation process. However, the nearest discrete solution at this response surface may not be a feasible design, in which case the design must move away from the continuous optimum by moving back to the previous response surface. This can be achieved by increasing the penalty multiplier, r, by a factor of 10.

5   EXAMPLE

5.1   Example 1
In this example, a two loop network with a pump added at the source and a reservoir linked to node 7 by an additional pipe, studied by Alperovits and Shamir (1977), is considered. The operation of the network is studied under two loading conditions (see Figure 1). The cost for the pump as a function of its horsepower is given in Table 1. Table 2 shows the cost data for pipes.

Table 1 Cost function of pumps, (Alperovits and Shamir (1977))

h_n	1	11	21	31	41	51	100
Cost	3000	3000	1800	1300	1000	850	500

Table 2 Cost data for pipes

Diameter (in)	1	2	3	4	6	8	10	12	14	16	18	20	22	24
Cost	2	5	8	11	16	23	32	50	60	90	130	170	300	550

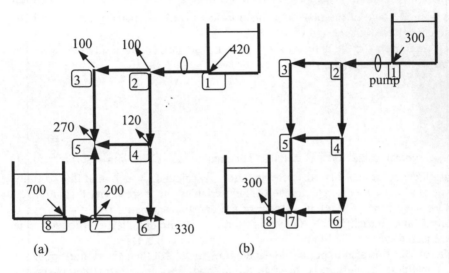

Fig. 1 A two loop network with a pump and a reservoir.
(a) Peak demand   (b) Low demand

The additional storage elevation cost (per unit of elevation) was 2000 (*$/m*). Table 3 shows the solutions for the network of Figure 1 presented by Alperovits and Shamir (1977) and the results of GA analysis. It is seen that the GA method for discrete optimisation gives a slightly higher value in the total cost for the optimum network compared to the result by Alperovits and Shamir (1977) which was obtained by segmental optimisation. By considering segmental optimisation in the GA procedure, the total cost of the network is reduced by (4.7%) (GA segmental (I)). Also by accepting a slight infeasibility (0.17%) at two nodes, 3 and 6, further reduction (6.4%) in the total cost can be obtained (GA segmental (II)).

Table 3 Comparison of the minimum cost solution for the network with pump and reservoir

Pipe	Alperovits et al. (1977) L(m)	D(in)	GA (discrete) D(in)	GA (segmental I) L(m)	D(in)	GA (segmental II) L(m)	D(in)
1	33.89	10	12	914.04	12	44.02	10
	966.08	12		85.96	14	955.98	12
2	374.46	5	8	231.81	6	260.9	6
	625.51	8					
3	1000.0	10	10	768.19	8	739.1	8
				331.91	8	174.69	8
4	816.14	6	1	668.09	10	825.31	10
	183.85	8		1000.0	1	1000	1
5	999.97	10	8	1000	10	1000	10
6	999.99	14	12	376.8	12	461.02	12
				623.2	14	538.98	14
7	929.62	6	10	469.11	8	884.43	8
	70.36	4				115.57	10
8	825.67	10	12	530.89	10	67.43	8
	174.28	8		72.81	8	932.57	10
				927.19	10		
9	100.0	16	14	2.27	14	1.01	14
				97.73	16	98.99	16
Additional elevation of the reservoir(m)	2.2		3.95	2.32		2.29	
head added by pump above head at reservoir (m)	6.3		4.58	3.545		3.62	
Cost of raising the reservoir(unit)	4330		7891	4645		4575	
pumping cost(unit)	28408		28759	22252		22730	
pipeline cost(unit)	267113		268000	259534		254600	
total cost(unit)	299851		304650	286432		281905	

434

5.2   Example 2

The network shown in Figure 2 has been used in previous studies by Awumah and Goulter (1992). The network consists of 12 links, 8 demand nodes, one source node and four loops. The demand and minimum pressure data for the nodes are given in Table 4. The height of all nodes above datum was taken as zero. The results of discrete optimisation using the sine function method are presented in Table 5.

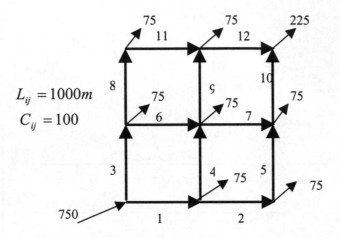

Fig. 2 A four-loop network for model application

Table 4 Demand and minimum head at nodes in the network

Node	Demand (m^3/h)	Minimum pressure head (m)
1	-750(*source*)	65(*maximum*)
2	75	25
3	75	25
4	75	25
5	75	25
6	75	25
7	75	25
8	75	25
9	225	25

In this example, a four-loop water distribution network has been analysed using different discrete and continuous optimisation methods. The design variables were considered as the pipe diameters and head losses in the pipes. The results of the analyses using different methods have been summarised in Table 6. Comparison of the results obtained from different optimisation methods shows that discrete optimisation using the proposed sine penalty function provides the least cost while maintaining a high value of entropy.

Table 5 Results of the network layout (discrete diameter)

Pipe	Flow rate (m^3/h)	Diameter (*mm*)	Head loss (*m*)
1	374.59	300	11.29
2	157.70	200	15.44
3	375.41	300	11.23
4	141.89	200	13.2
5	82.70	200	5.01
6	141.65	200	13.25
7	102.21	200	7.25
8	158.76	200	15.99
9	106.33	200	7.82
10	109.91	200	8.27
11	83.76	200	5.09
12	115.09	200	7.7

Table 6 Comparison of the different methods in terms of cost

Method	Entropy	Cost
Continuous	Infeasible	Infeasible
Rounded	Infeasible	Infeasible
Segmental	2.7993	486.87
Dynamic rounding	2.7776	473.208
Sine function	2.7863	469.89

6   CONCLUSION

A Genetic Algorithm based program has been developed for particular application to the optimisation of water distribution networks. The GA program has been applied to a number of case studies and the results have been compared to those previously obtained by other authors using traditional optimisation techniques as well as GAs.

A sine penalty function was introduced for discretisation of the variables which are discrete in nature, but are not taken as unknown design variables. Multiple loading case has also been solved for the network and the results have been compared to those obtained using traditional optimisation techniques. Comparison shows that the GA provides better (lower cost) and more practical (feasible) results. Beside giving the lowest cost network compared to other techniques, one of the other important advantages of the GA technique is that in the search space, it gives more than one solution. This gives the designer more than one choice to select the best optimum.

436

REFERENCES

1   Alperovits, A., Shamir, U. (1977) Design of optimal water distribution systems. Water Resources Research, 885-900.

2   Awumah, K., Goulter, I. C. (1992) Maximising entropy defined reliability of water distribution networks. Engineering Optimisation, Vol. 20, pp. 57-80.

3   Dong, K. S., Gurdal, Z. and Griffin, O. H., Jr (1990) A penalty approach for nonlinear optimisation with discrete design variables. *Engineering optimisation,* vol. 16, pp. 29-32.

4   Eiger, G., Shamir, U., Ben-Tal, A. (1994) Optimal design of water distribution networks. Water Resources Research, 2637-2646.

5   Goldberg, D., Kuo, C. H. (1987) Genetic Algorithms in pipeline optimisation. Journal of Computing in Civil Engineering, 128-141.

6   Goldberg, D. E. (1989) Genetic algorithms in search, optimisation and machine learning. Addison-Wesley Publishing Co., Reading, Mass.

7   Savic, D. A., Walters, G. A. (1997) Genetic Algorithm for least-cost design of water distribution networks. Journal of Water Resources Planning and Management, 67-77.

8   Simpson, A., Dandy, G. C., Murphy, L. J. (1994) Genetic algorithms compared to other techniques for pipe optimisation. Journal of Water Resources Planning and Management, 423-443.

Optimal Phasing of Water Distribution Systems Rehabilitation

D. Halhal, G.A. Walters, D.A. Savic *and* D. Ouazar

ABSTRACT

A Multi Objective Structured Messy Genetic Algorithm (SMGA) based method dealing with the optimal staging of water distribution systems rehabilitation is discussed. Its principle consists of developing an optimal and progressive improvement of a water distribution network according to the different horizon data, by finding then decomposing the final optimal or near optimal solution (which has to be completely achieved at the end of the planning period) into a sequence of improvement events, of relatively small costs, forming intermediary and complementary solutions which are in themselves optimal or near optimal solutions in terms of benefit yielded to the network with regard to their cost. The method permits the network to take full advantage of the different improvement events as their implementation proceeds. The proposed method has been implemented and tested on examples, yielding very encouraging results.

1   INTRODUCTION

Ageing water distribution systems experience, with time, many problems such as structural failures, drop in carrying capacity, poor water quality, and disruption in the service provided, which, in the absence of any adequate rehabilitation programme, are aggravated over the years. These systems are often large sized networks for which an immediate and complete rehabilitation involves prohibitively high financial outlays, which makes rehabilitation phasing the only practical approach in most cases.

Often, utilities allocate a fixed amount of money for the rehabilitation and improvement of their water distribution systems over a period of time. Improvement in the system's performance is achieved through replacing, rehabilitating, duplicating and/or repairing some of the pipes in the network, and also by adding completely new pipes. As the planning period is usually long, it is important that the network takes full advantage of the different improvement events as implementation proceeds, not only after achievement of the totality of the work i.e.at the end of the planning period. There is an imperative need for the development of a methodology for an optimal and economically feasible

437

438

rehabilitation phasing which defines the different solution events to be prioritised and implemented in steps over a planning horizon according to the availability of the required money.

2   THE APPROACH

The purpose of the present approach is to invest a limited amount of money gradually over a period of time such that after each investment, the benefit yielded is maximised. The approach aims to find a set of improvement sequences forming complementary solutions to be implemented hierarchically along a planning horizon, such that the execution of each one constitutes a step towards achievement of the final solution, and produces the maximum improvement to the system. The curve D1 in Fig.1 shows the evolution of the network improvements achieved along the planning period expressed in terms of the benefits yielded with respect to the money invested.

After each step in the improvement of the network, the solution achieved constitutes an optimal or near optimal solution, called here the partial cumulative solution. The purpose is to find the optimal set of partial cumulative solutions, the implementation of which leads to the final optimal or near optimal solution and produces in each step the maximum benefit to the system.

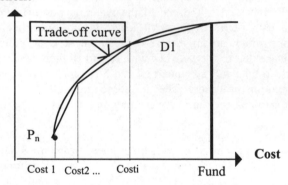

Fig. 1 : Optimal Planning curve D1

The optimal set of partial cumulative solutions forming the final solution consists of a series of non dominated solutions situated on the trade-off curve between the benefit and the cost as represented by the curve D1 in Fig. 1. Unfortunately, solutions along this trade-off curve are often not cumulative, and partial cumulative solutions belonging to the best planning set may not all be non-dominated solutions. Therefore, it is sensible to find a compromise such that the resulting planning set is an optimal or near optimal set.

Curve D2 in Fig. 2 shows the real planing curve with Δi as the benefit difference between partial cumulative solution i in the planning set and the optimal solution having the same cost.

3   PROBLEM FORMULATION

The problem is defined here as: find a set of optimal or near optimal partial cumulative solutions forming a planning set for a progressive rehabilitation, replacement and/or expansion of an existing water distribution system according to a fixed amount of money.

Benefit

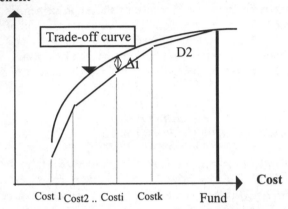

Fig. 2 : Real planning curve D2

The problem can be formulated analytically as:

$$\text{Minimise}\quad f(i) = \frac{1}{N}\sum_{i=1}^{N}\alpha_i\Delta_i \tag{1}$$

$$\text{Subject to}\quad Cost(i) \le Fund$$

Where N is the number of solutions in the planning set;
Δi is the difference in the benefit between solution i and the optimal solution having the same cost.
$Cost(i)$ is the cost of the solution i;
$Fund$ is the fixed available amount of money
α_i is a weighting coefficient for solution i, which is set equal to:

$$\alpha_i = \frac{Cost(i)}{Fund} \tag{2}$$

The value Δi measures how far the chosen solution is from the optimal solution of the same cost. Δi has different impacts on the network improvement depending on the order of execution of the corresponding solution in the planning period. The network improvements produced during the first steps of the rehabilitation programme are minor as the costs involved are relatively small, but they become more and more important during the process, as the increase in benefit from the improvements becomes increasingly costly. So, the influence of Δi on the « quality » of the solution becomes more and more significant with the order of execution of the solution in the planning period. The introduction of the weighting coefficient α_i in the formula regulates the value of Δi with respect to the solution cost, and favours the development of planning sets containing high order (to be executed late in the programme), « good » solutions.

The benefit attributed to a solution is a measure of the improvement in the level of service experienced by consumers combined with a measure of the decrease in costs associated with system operation and maintenance. For simplicity, the method combines four factors, with weights chosen by the user. The total benefit of a solution i is therefore expressed as the weighted sum of four benefits :

$$Benefit(i) = w_h Benefit\_hyd(i) + w_p Benefit\_phy(i) + \qquad (3)$$
$$w_f Benefit\_flx(i) + w_q Benefit\_qual(i)$$

where $Benefit\_hyd(i)$ is the pressure benefit derived from better hydraulics
$Benefit\_phy(i)$ is the maintenance benefit derived from better physical condition of the pipes.
$Benefit\_flx(i)$ is the operational benefit derived from greater network flexibility
$Benefit\_qual(i)$ is the quality benefit derived from pipe replacement and/or cleaning and lining,
wh, wp, wf and wq are the respective weightings.

4   STRUCTURED MESSY GENETIC ALGORITHM (SMGA)

The SMGA is a type of GA with flexible coding. Each gene carries both its name and value. Extendible strings begin with short lengths and grow longer in a structured manner throughout the process until they reach a maximum length or until the process meets a termination criterion. A full description of SMGA can be found elsewhere [1,2]. Fig 3 shows a flowchart representation of SMGA.

The use of SMGA permits the easy handling of large and complex systems with less computing time and memory space than conventional GA. In fact, SMGA encodes only active decision variables. So, in the present case, only the relevant arc numbers and the rehabilitation decision for each of those arcs are stored in the string, which increases in length through the evolution process until reaching its maximum length, which is much smaller than when all the arcs are considered. The total search space is thereby explored with strings of small maximum length.

5   MULTI-OBJECTIVE OPTIMISATION (MOO)

The optimisation of existing water distribution systems consists of choosing the best combination of the different alternatives which maximises the benefit resulting from the improvement operation on the one hand and minimises the cost, which must be less than or equal to the available fund. However, the maximisation of the benefit is contradictory to the minimisation of the cost, the two criteria being in competition with one another. The problem is better tackled by a multi-objective optimisation procedure, where the fitness function is split into two separate terms: the benefit and the cost, and the solutions are ranked according to the two criteria. In addition to that, the solution yielded by the process is not a single one but a set of non-dominated solutions, which are neither superior nor inferior to each other.

Actually, GAs are very well suited to multi-objective optimisation problems, as they search for many non-dominated solutions at the same time. The procedure adopted uses the concept of Pareto optimality ranking and fitness sharing. Pareto optimality ranking [3] consists of attributing the same rank to all non-dominated individuals in the population and removing them from contention. The next rank is attributed to the next set of non-dominated individuals, and so on until the entire population is ranked. A technique of niche formation and speciation is implemented through a sharing scheme. The total fund is divided into several intervals. Each individual is assigned to the interval (class) corresponding to its cost, and its shared fitness is calculated by dividing its fitness by the total number of individuals in its class. The result obtained is a set of non-dominated solutions which when plotted according to benefit and cost on a Cartesian graph, yield a trade-off curve of the form shown in Fig 2.

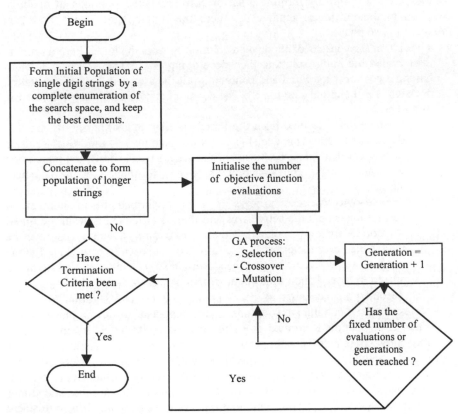

Fig. 3: Flowchart representation of SMGA

6   SMGA AND MOO

SMGA builds up the complexity of solutions in a progressive way, starting with short strings of one digit length and increasing them during the process until reaching a maximum length. It needs a large range of "building blocks" of different costs and benefits to generate useful solutions with increasingly longer strings throughout the process. The multi-objective approach provides these "building blocks" with a range of costs and benefits, ideal for subsequent generation of more complex solutions, while SMGA provides an effective technique for generating and developing a wide range of good solutions for the MOO. Hence the link between SMGA and MOO is mutually beneficial, and forms an efficient model capable of handling complex water distribution systems. In addition, the model provides solutions of different ranges of costs and benefits from which are chosen the partial cumulative solutions, which suits very well the type of problem being considered.

7   THE METHOD

The method is based on the idea of developing planning sets containing at least one non-dominated solution, around which are developed all the partial cumulative solutions. It starts first by forming a set of non-dominated solutions of different costs up to the maximum fund, called here the initial Pareto front, and then proceeds in two phases.

- The first phase consists of choosing a solution S_0 from the initial Pareto optimal front called the initial solution, then developing it to obtain the best solution having a cost near the Fund and containing all the arcs and their rehabilitation decisions for the initial solution S_0. The development operation is performed in two steps :

  - a population is formed from the initial solution S_0 by altering its inactive decision variables (the decision of 'doing nothing'). If the initial solution S_0 has less than two inactive arcs (its string is quite short), the population is created by the concatenation process to form individuals of longer strings.

  - a multi-objective GA process is then performed for a number of generations where the active arcs and their decisions present in the initial solution S_0 are kept unaltered. The best solution S_1 having a cost near the fund, called here the original solution, is then chosen to be included in the planning set, and to constitute the starting point of the second phase of the procedure. Fig. 4 shows the development process of the method.

- The second phase consists of decomposing the original solution S_1 into a series of partial cumulative solutions of different costs to form a planning set. First, a population is formed from the original solution S_1 by inactivating randomly chosen decision digits of its string, i.e. setting them equal to 1 ('doing nothing').   Then a multi-objective GA process is performed for a number of generations, during which different classes of solutions are created. The best solution S_2 of the nearest lower cost class is chosen to be added into the planning set. The process continues by forming a population from this new solution S_2 by inactivating some randomly chosen digits. The multi-objective

GA process succeeds for a number of generations in yielding different classes of solutions from which the nearest lower cost class best solution S_3 is chosen to increase the number of solutions in the planning set, and so on, until reaching the least cost best solution. Each partial cumulative solution thus formed is added to the planning set (represented by D3 in Fig. 5), and its distance Δ from the trade-off curve is calculated, weighted and accumulated with the values of the previous solutions of the same planning set.

When the planning set has been formed and its objective function has been calculated, another set is developed from another solution in the initial Pareto optimal front in the same way, and the process continues until covering all the non-dominated solutions of the initial Pareto front.

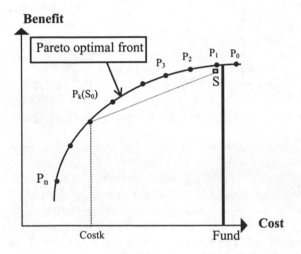

Fig. 4: Development of Solution P_k chosen from the initial Pareto front to form the best solution S_1 having a cost near the fund

444

The objective function (OF) of each planning set, thus formed, is evaluated, and the set having the least OF value is chosen as the best planning set containing partial cumulative solutions from which exists at least one solution belonging to the trade-off curve.

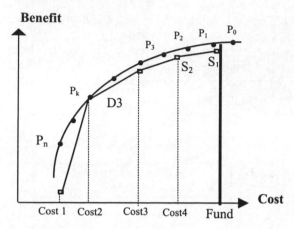

Fig. 5: Planning curve D3 containing solution P_k, and formed by decomposing solution S_1

8   EXAMPLE

The example is a looped network containing 92 nodes and 155 pipes, fed from 2 reservoirs, as shown in Figure 6. Two time horizons are considered. In the short-term horizon, an expansion of the system is planned to supply areas located in the North West. The minimum desired pressure is fixed at 20 m above the ground. In the long-term horizon, the nodal demand is changed and the minimum required pressure is equal to 25 m above the ground, while new areas are to be created in the South East and supplied with potable water. The total fund of the whole operation is fixed at 2,500,000 from which 1,000,000 is taken as the budget allocated to the short-term improvement planning programme.

The number of decisions for each pipe and the available pipe sizes and costs are given in Tables 1 and 2.

Table 1: Option for each decision variable

Decision code	Option for an existing pipe	Option for an new pipe
1	Leave as it is	Diameter 80 mm
2	Clean and line	Diameter 100 mm
3	Duplicate with Diameter 150 mm	Diameter 150 mm
4	Duplicate with Diameter 200 mm	Diameter 200 mm
5	Duplicate with Diameter 300 mm	Diameter 300 mm
6	Duplicate with Diameter 400 mm	Diameter 400 mm
7	Renew with the same Diameter	Diameter 500 mm
8	Renew with the next largest Diameter	Diameter 600 mm

Table 2 : Available pipe sizes and alternative solution costs

Pipes Diameter (mm)	Cost of		
	New pipe	Clean & Line	Replace existing pipe
80	100	85	110
100	175	100	190
150	220	150	240
200	320	200	350
300	550	300	600
400	780	410	850
500	980	500	1050
600	1350	630	1500

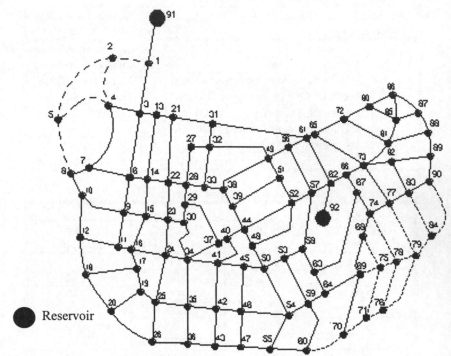

Fig. 6: Example - Layout of the network

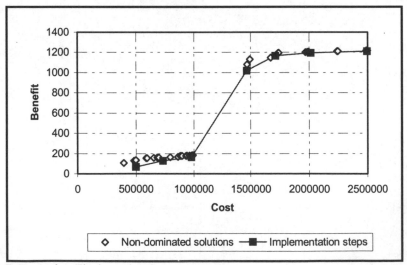

Fig. 7: Example - The best planning programme

Table 3: Example : the best planning set details

Arcs	Solution of Step 1	Solution of Step 2	Solution of Step 3	Solution of Step 4	Solution of Step 5	Solution of Step 6	Solution of Step 7
1	-	-	-	-	-	-	300**
2	200*	&	&	&	&	&	&
4	80	&	&	&	&	&	&
5	80	&	&	&	&	&	&
9	80	&	&	&	&	&	&
11	80	&	&	&	&	&	&
12	-	-	-	-	150*	&	&
29	-	-	-	Clean	&	&	&
30	-	-	-	100**	&	&	&
41	-	-	-	-	150*	&	&
52	-	150*	&	&	&	&	&
61	-	-	-	-	-	200*	&
62	-	-	-	-	300*	&	&
67	-	200*	&	&	&	&	&
68	-	-	-	-	150*	&	&
70	-	-	-	-	200*	&	&
73	200*	&	&	&	&	&	&
79	-	-	300*	&	&	&	&
93	-	300*	&	&	&	&	&
94	-	-	-	-	200*	&	&
101	-	-	-	-	-	200*	&
102	-	400*	&	&	&	&	&
109	-	-	-	-	200*	&	&
110	300*	&	&	&	&	&	&
117	150*	&	&	&	&	&	&
120	-	-	-	-	-	150*	&
121	x	x	x	80	&	&	&
123	x	x	x	80	&	&	&
124	x	x	x	80	&	&	&
125	x	x	x	80	&	&	&
126	x	x	x	80	&	&	&
129	-	-	-	150*	&	&	&
132	-	300*	&	&	&	&	&
133	x	x	x	80	&	&	&
134	x	x	x	80	&	&	&
135	x	x	x	80	&	&	&
136	x	x	x	80	&	&	&
137	X	x	x	80	&	&	&
139	X	x	x	80	&	&	&
140	X	x	x	80	&	&	&
147	X	x	x	80	&	&	&
149	X	x	x	80	&	&	&
Cost	497500	233400	247500	486700	247800	301800	480000
Accumulated cost	497500	730900	978400	1712900	1712900	2014700	2494700
Benefit	68	128	163	1018	1167	1195	1211
Min pressure (m)	-0.15	7.55	18.36	-0.10	13.22	16.23	19.06
node	6	6	6	71	71	5	6

448

* : Add parallel pipe of diameter shown
** : Replace existing pipe by diameter shown
& : Implemented in the previous steps
x : Not existing during the period

The detail of the different solutions for the best planning set is given in Table 3, and plotted together with the Pareto optimal front in the Cartesian graph in Fig. 7. It can be seen that the implementation steps closely follow the Pareto optimal front.

9   CONCLUSION

A multi-horizon optimisation procedure has been developed to find the best way to invest optimally and progressively a limited amount of money in improving the performance of a water distribution system, according to different horizon data and funds. The method is based on a multi-objective optimisation procedure using SMGA. It involves finding a set of optimal or near optimal complementary solutions, where each solution is a part of the final solution. The method permits a progressive and optimal improvement of the network according to the availability of money and time. It implements optimally and progressively along the planning period an optimal or near optimal solution that has to be completely achieved by the end of the time horizon.

REFERENCES

Halhal, D., Walters, G.A & Ouazar, D.(1995): " Structured Messy Genetic Algorithm Approach for the Optimal Improvement of Water Distribution Systems". GALESIA 95, University of Sheffield,   12-14 September 1995.

Halhal, D., Walters, G.A, Ouazar, D and Savic D.A.(1997): "Water Network Rehabilitation with a Structured Messy Genetic Algorithm". J. Water Resource Planning and Management, ASCE, 123/3, pp 137-146.

Goldberg, D.E.(1989): « Genetic Algorithms in Search, Optimisation, and Machine Learning>> Addison-Wesley, 1989.

A Genetic Algorithm for Wastewater Treatment Planning

C. G. Wang *and* D. G. Jamieson

ABSTRACT

This paper outlines the possible use of a genetic algorithm in conjunction with a process-based model and an artificial neural network for solving the optimisation problem associated with regional wastewater-treatment planning. To that end, a model has been developed to determine the optimal configuration of individual wastewater-treatment plants and associated waste-load allocations within a regional wastewater disposal system. The objective is to minimise the overall cost of such a system, subject to ensuring that the required river-quality standard is achieved. These procedures have been applied to part of the upper Thames basin in southern England, based on seven possible wastewater-treatment sites. Whilst an optimal solution cannot be guaranteed, the results show that at least for fixed-emission standards, it can be approximated to a degree that makes no practical difference. The approach adopted also indicates that the combination of GA optimisation and ANN simulation provides a powerful tool to assist engineers in determining not only the number of sites and optimal locations of regional facilities, but also the waste-load allocation and effluent-discharge standards.

1   INTRODUCTION

Regional wastewater-treatment planning aims to find the optimal location of wastewater-treatment facilities and possibly effluent-discharge standards, so that potential cost-savings might be achieved through economies of scale. However, regional wastewater-treatment planning is significantly more difficult than separate, case-by-case studies since it is a highly-combinatorial problem which requires the analysis of a large number of alternative configurations. This has led to a considerable amount of research effort being spent on applying different mathematical-programming techniques to identify the optimal solution on a regional base. These techniques include mixed-integer programming (Graves, 1972; Wanielista and Bauer, 1972; Joeres et al. 1974), extreme-point ranking (Deininger and Su, 1973), geometric programming (McNamara, 1976), combinatorial programming (Brill and Nakamura, 1978; Nakamura et al. 1981) and heuristic programming (Nikolay and Boulos, 1993). However, none of these techniques has gained widespread acceptance owing to their complexity and the amount of user effort required.

The purpose of this paper is to present an approach which based on the combination of a genetic algorithm (GA) and an artificial neural network (ANN), the latter having been used to replicate a complex, process-based simulation model. The objective is to minimise the total cost (capital and operating) of wastewater treatment subject to either a fixed-emission standard or satisfying in-stream water-quality requirements. The difference between the two is that when a fixed-emission standard is assumed, the solution is restricted to selecting the optimal sites, whereas if the sites selected are allowed to have individual effluent-discharge standards, the problem also involves waste-load allocation. Either way, the proposed approach provides a powerful search procedure based on the mechanics of natural selection which has been shown to perform well in recent optimisation applications related to water-resources engineering (Wang, 1991; Ritzel and Eheart, 1994). The case study presented here is based on the upper Thames basin in southern England, where seven possible treatment sites have previously been identified.

2 REGIONAL WASTEWATER TREATMENT PLANNING

2.1 Evaluation model

In this section, a general optimisation model for a regional wastewater-treatment planning is outlined based on a fixed-emission standard. In other words, all treatment sites selected are assumed to have the same predetermined effluent discharge standard. The system is envisaged to have N wastewater sources with N possible treatment plants. If each source can be connected by a trunk sewer to any other source, then each source has N different ways of disposing of its waste (pipe to N-1 other sources or built a treatment plant). Therefore, the number of possible combinations, where one of N options is chosen at each of the N sources, is N^N, without even considering different levels of partial treatment at each site.

The evaluation model comprises:

Objective function:

$$\text{Min} \sum_{i=1}^{N} f_i(Q_i, x_i, y_i) \tag{1}$$

$$f_i(Q_i, x_i, y_i) = \begin{cases} C_i(Q_i, x_i) & y_i = 0, i \\ T_{ij}(Q_i, L_{ij}, E_i, E_j) & y_i = j, \ j \neq i \end{cases} \tag{2}$$

Constraints:

$$WQ_k(\mathbf{Q}, \mathbf{X}, \mathbf{Y}) \leq WO_k \tag{3}$$

$$Q_i = QE_i + \sum_{j \in J_i} Q_j \quad \text{for } i=1,2,...,N \tag{4}$$

$$\sum_{j \in J_0} Q_j = \sum_{i=1}^{N} QE_i \tag{5}$$

$$QE_i \geq 0 \tag{6}$$

$$0 \leq y_i \leq N \tag{7}$$

$$Q_{\min} \leq Q \leq Q_{\max} \tag{8}$$

$$x_{\min} \leq x_i \leq x_{\max} \tag{9}$$

$$\text{if } y_i = j \text{ then } a_{ij} = 1 \quad a_{ij} * a_{ji} = 0 \text{ for } i = 1,2,...,n; j = 1,2,...,n; \quad i \neq j \tag{10}$$

where:

y_i = disposal choice at source i; $y_i = 0$, i indicates that disposal choice is to treat; $y_i = j$ indicates that disposal choice is to pipe to source j;

a_{ij} = parameter to indicate selection of piping option: if $y_i = j$, $a_{ij} =1$, otherwise $a_{ij} =0$;

$f_i(Q_i, x_i, y_i)$ = cost to dispose of wastewater at source i as a function of the volume Q_i, removal rate x_i and disposal choice y_i;

$C_i(Q_i, x_i)$ = total cost of treatment option which is the function of capacity Q_i and removal rate x_i including capital, operation and maintenance costs at site i;

$T_{ij}(Q_i, L_{ij}, E_i, E_j)$ = total cost of piping wastewater Q_i from source i to source j including capital, operation and maintenance;

$WQ_k(\mathbf{Q}, \mathbf{X}, \mathbf{Y}) \leq WO_k$ requires that the water quality WQ_K at any check points must meet the specified water-quality objective WO_k. $\mathbf{Q}, \mathbf{X}, \mathbf{Y}$ are a set of discharge volumes, removal rates and disposal choices, respectively;

QE_i = amount of wastewater generated at source i;

E_i = elevation at source I, E_j = elevation at source j;

J_0 = the set of all sources j that are treating;

J_i = the set of all sources j that are piping to source i;

L_{ij} = distance between source i and source j (length of the trunk sewer);

i = index of sources, j = alternative index of sources;

N =number of sources;

Q_i = amount of wastewater accumulated at source i.

The objective is to minimise the sum of the disposal costs at each source by selecting either treatment or piping to another source as the means of disposal. The constraint (3) ensures the specified water-quality objective to be met. Equation (4) ensures that the wastewater disposed at source i is equal to the quantity generated there, plus the sum of any wastewater piped there from other sources. Equation (5) requires that all wastewater generated in the system is treated. Equation (10) prohibits source j from piping wastewater to source i while source i is piping to source j, a condition referred to as "opposing flows".

It is clear that the complexity of this problem is due to its highly-combinatorial features, the number of possible system configurations increasing exponentially with the number of alternatives considered. From the above model structure, it can be seen that:

a) piping costs $T_{ij}(Q_i,L_{ij},E_i,E_j)$ as the function of the design flow Q_i, are concave;

b) the same applies for the treatment costs in the sense that the functions $C_i(Q_i,x_i)$ are concave in Q_i for given x_i;

c) if all x_i were fixed, the objective function would be concave and the problem would then consist of minimising a concave function subject to linear constraints: this problem equates to the same level of complexity as a general integer-programming problem and hence, would not normally be expected to be solvable in a reasonable time.

2.2 Optimisation Process

Recently, a relatively-new search procedure, namely genetic algorithms (GA) has received much attention in various research fields as a result of their ability to deal efficiently with optimisation problems. Although GAs can be classified as a branch of heuristic programming, they are very different in their philosophy and the way they work. Details of their attributes and the way they differ from other forms of optimisation can be found in the publications by Holland (1975) and Goldberg (1989). For the purpose of this paper, it suffices to say that GAs have been successfully applied to a diverse range of scientific, engineering and economic problems including: structural optimisation (Goldberg and Samtani, 1986); pipeline optimisation (Goldberg and Kuo, 1987; Simpson et al., 1994; Savic and Walters, 1997); calibrating conceptual rainfall-runoff model (Wang, 1991); optimising groundwater-management models (Mckinney and Lin, 1994); and groundwater pollution control (Ritzel and Eheart, 1994; Rao and Jamieson, 1997).

In GAs, the set of decision variables appears as a coded string of fixed length (Goldberg, 1989) which in this particular case, is assumed to be binary. The decision variable for each possible treatment site y_i is represented as an integer value $(0,1,...,N)$, with a $l-$bits binary variable being used to represent each parameter y_i. The integer of the decoded binary variable ranges from 0 to $2^l - 1$ and can be mapped linearly onto the parameter range $[0, N]$. The parameter range is discretized into 2^l points with a discretization interval of:

$$\Delta x_i = \frac{N}{2^l - 1} \tag{11}$$

3   APPLICATIONS

3.1   Upper Thames Basin

To demonstrate the capability and efficiency of the proposed GA-based search procedure for optimising the wastewater-treatment-planning model, it has been applied to part of the upper Thames basin near Oxford in southern England. Each of the seven sites where wastewater is generated can be either a potential location for wastewater treatment or has an existing plant. These can also be linked by pipeline in the manner shown in Figure 1, so that the wastewater could be conveyed to any other site. A pumping station needs to be built at site i if its elevation is lower than that to which its wastewater is to be conveyed. It is assumed that all selected wastewater treatment plants have to achieve the prescribed effluent-quality standard before being allowed to discharge to the river and that the maximum capacity for any one treatment site is 100000 m^3/d. The amounts of wastewater generated at each site, together with their elevations are summarised in Table 1.

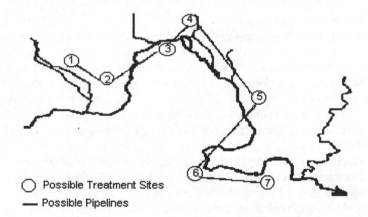

○ Possible Treatment Sites
— Possible Pipelines

Figure 1 Locations of possible treatment sites

3.2 Cost Functions

The cost functions used in this research have previously been developed by the Water Research Centre (WRC, 1985). They are taken from UK tender prices for different elements of water and wastewater treatment and related to the physical characteristics, such as dry weather flow, maximum design flow, area of screen, volume of storm tank, biological loading, retention time etc., by means of regression analysis. In essence, these relationships cover civil and mechanical engineering construction costs, together with operating costs, which can be combined to give an average annual cost for each process. Similarly, the conveyance-cost equation is also an aggregate function in terms of pipe length, diameter and depth underground, with the diameter being determined by the maximum flow rate. For pumping costs, again the costs are related to the flow rate and normal operating head. All costs are calculated in terms of 1979 £ sterling which, of course, can be converted to present-day costs by using the appropriate indices.

Table .1. Characteristics of Each Possible Wastewater Treatment Site

Sewage Treatment site	#1	#2	#3	4#	5#	6#	7#
Distance from the upstream site (km)	0.00	3.36	6.09	3.40	11.40	8.59	6.03
Dry Weather Flow (m^3/d)	82	436	12000	12600	90000	30000	136
Elevation (m)	80	70	90	55	60	50	55

3.3 Process Options

For each site, there was a choice of four levels of wastewater treatment, these being:

i. Pre-treatment (raked screens, comminutors, detritus basins and storm tanks);
ii. Pre-treatment plus Primary Sedimentation (sedimentation tanks and sludge-dewatering);
iii. Pre-treatment, Primary Sedimentation plus Secondary Treatment (percolating filters or activated sludge, final separating tanks and sludge-dewatering);
iv. Pre-treatment, Primary Sedimentation, Secondary Treatment plus Tertiary Treatment (mechanical filtration or slow sand filters or land treatment or lagooning).

The selection of different combinations of processes was largely determined by the scale of operation but no site was allowed to discharge to the river unless its effluent achieved a standard of 10 mg/l Biochemical Oxygen Demand and 10 mg/l Suspended Solids. Obviously, partly-treated effluent could be conveyed to other sites provided the cost of the pipeline was included.

3.4 Optimal Configuration Assuming a Fixed Emission Standard

By initially giving a random set of numbers to the decision variables and evaluating the objective function after each iteration, the GA was used to generate successive populations by means of a weighted random selection process. The search procedure quickly converged to provide the optimal configuration of treatment processes at each site, pipeline linkage and discharge points to the river. In order to examine the rate of progress towards identifying the optimal configuration, some four possible options have been selected, based on different maximum number of generations. For the purpose of this exercise, all costs have been converted to annual costs (using a discount rate of 8 percent), over the economic life of the wastewater-treatment facilities and pipelines which were assumed to be 20 and 30 years, respectively.

It can been seen from Table 2 that the optimal configuration is Option 1 which is depicted in Figure 2. With this arrangement, sites #4, #5 and #6 have been selected as discharge points to the river where the effluent has to achieve the required the water-quality standards. As a consequence, the wastewater would need to be conveyed from sites #1, #2 and #3 to site #4 where it would be given final treatment. Similarly, with site #7, where the wastewater would be piped to site #6 for final treatment. The total annual cost for this preferred option would be £1,795,497. Comparing this with Option 2, which has total annual cost of £1,799,808, the only significant different is that #7 site has been selected for full treatment. Even this small difference in cost demonstrates the effectiveness and for that matter, the robustness of the GA search procedure.

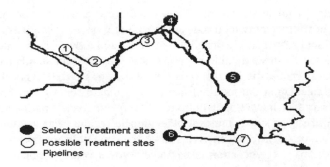

Figure 2 Optimal configuration

Table 2 Summary of Feasible Solutions

Site	# 1	# 2	# 3	# 4	# 5	# 6	# 7	Total Annual Cost (£)
Option 0: Generation=300								
Decision Variable	2	3	4	0	0	0	6	
DWF (m³/d)	82	518	12,518	25,118	90,000	30,136	136	
Annual Cost (£)	3,675	13,182	7,970	381,680	943,652	439,037	6,301	1,795,497
Option 1: Generation=30								
Decision Variable	2	3	4	0	0	0	6	
DWF (m³/d)	82	518	12,518	25,118	90,000	30,136	136	
Annual Cost (£)	3,675	13,182	7,970	381,680	943,652	439,037	6,301	1,795,497
Option 2: Generation=25								
Decision Variable	2	3	4	0	0	0	0	
DWF (m³/d)	82	518	12,518	25,118	90,000	30,136	136	
Annual Cost (£)	3,675	13,182	7,970	381,680	943,652	437,508	12,141	1,799,808
Option 3: Generation=20								
Decision Variable	2	0	4	0	0	0	0	
DWF (m³/d)	82	518	12,518	25,118	90,000	30,136	136	
Annual Cost (£)	3,675	27,939	7,964	375,632	943,652	437,508	12,141	1,808,511
Option 4: Generation=19								
Decision Variable	0	1	2	0	0	0	6	
DWF (m³/d)	82	518	12,518	25,118	90,000	30,136	136	
Annual Cost (£)	224,983	7,893	13,419	226,090	943,652	439,037	6,301	1,861,375

It is also apparent from Table 2 that improved cost-effective solutions were being found as the maximum number of generations was increased, until the global optimal had been identified. In this particular case, the global optimal was achieved after 30 generations which took just 80 seconds to compute on a modern PC. This confirmed the rapid convergence towards the optimal solution, a characteristic of the GA approach which in part, is due to the avoidance of becoming entrapped by a local optimum. Global optimality was confirmed by completing some 300 generations which resulted in no further improvement.

3.5 Regional Wastewater Treatment Planning Assuming Individual Discharge Standards

Obviously, there is no guarantee that prescribed fixed-emission standards will achieve the river water quality required for the different downstream users such as public water supply, irrigation, fisheries etc. Therefore, there is an increasing tendency towards to a more flexible approach in which consent standards are set individually for various discharges rather than being all the same. In these circumstances, calculating the required treatment standard for a given discharge, which is necessary to achieve a specific river water quality, is not particularly easy for a number of reasons since, amongst other things, the size and location of the discharge are important. Moreover, rivers have the capability to absorb some of the pollution load as re-oxygenation takes place. Further complications arise with the "carry-over" effect from other upstream reaches which may necessitate higher treatment standards than those required for the immediate reaches to meet more stringent river water-quality requirements downstream. The combination of these effects, together with the attenuation which takes place and the differing degrees of in-river dilution caused by tributaries, require discharge consent standards to be set in the context of an integrated river system rather than in a piecemeal fashion.

However, estimating the effects of different combinations of effluent discharge standards on river water quality implies the use of a process-based model which is capable of simulating the cause-and-effect relationship. The problem with this approach is that it is impractical to include some form of optimisation since the computational burden of running a simulation model for each parameter set is excessive even for modern computers. Therefore, if optimisation is to be included, a more computationally-efficient form of simulation is required. Nevertheless, the knowledge contained in process-based modelling needs to be retained since there is no alternative for estimating the consequences of modifying decision variables. To that end, the approach adopted in this study to improve the computational efficiency of a process-based model was to replicate its predictive capability using an artificial neural network which has been detailed by Wang and Jamieson (1999).

Integrating the ANN water-quality simulator into optimisation model and using the above GA search procedure with a maximum generation of 100, the optimal configuration which considers the waste-load allocation is shown Figure 3. In this particular case, the river water quality should meet class 1B, which requires that BOD concentration should not be greater than 5 mg/l and DO should be greater than 60% saturation.

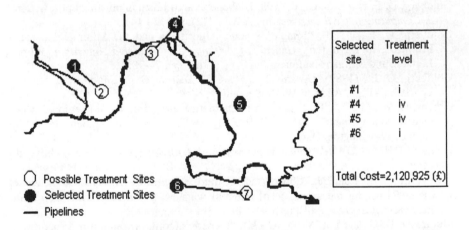

Figure 3. Optimal configuration considering individual discharge standards

Figure 3. shows that site #1 has to be chosen as a treatment site in order to meet the required water-quality objective when compared with fixed emission standards. In addition, sites #4, #5 should produce high-quality effluent which requires advance treatment. As a result, the eventual treatment cost is £2,120,925.

4   CONCLUSION

This paper advocates the use of a genetic algorithm for regional wastewater-treatment planning. In comparison with the traditional methods, the GA has the capability to overcome the complexity associated with mathematical programming and at the same time, reduce the computational burden and effort required by the user. In order to give some credence to the proposed methodology, the model has been applied to the upper Thames basin in southern England. The results show the advantages of this approach in deriving the optimal location of wastewater-treatment plants and when combined with an ANN, the waste load allocation and individual discharge standards to achieve whatever river water-quality is prescribed. Although a process-based simulation model is required for the latter to determine the cause-and-effect relationship, replacing it by an ANN is orders of magnitude more efficient computationally, thereby enabling a GA to be used for optimisation.

REFERENCES

Brill, E. D. and Nakamura, M., (1978), A branch and bound method for use in planning regional wastewater treatment systems. *Water Resources Research*, 14(1), pp. 109-118.

Deininger, R. A. and Su, S. Y., (1973), "Modelling regional wastewater treatment systems", *Water Res.* 7(4), pp. 633-646.

Goldberg D. E, and Samtani M. P (1986), "Engineering optimization via genetic algorithm", *Electronic Computation, Proceedings of the Ninth Conference*, Birmingham, AL, USA, 23-26, Feb. 1986, pp.471-482.

Goldberg, D. E. and Kuo, C. H., (1987), "Genetic algorithms in pipeline optimisation", *J. Computing in Civil Engineering*, ASCE, 1(2), pp. 128-141.

Goldberg, D. E., (1988), "Simple Genetic Algorithms and the Minimal Deceptive Problem", in Davis (ed), *Genetic Algorithms and Simulated Annealing*, Pitman, London, and Morgan Kaufmann Publishers, Inc., pp. 74-78.

Goldberg, D. E., (1989), *Genetic Algorithms in Search, Optimisation, and Machine Learning*, 412pp. Addison-Wesley, Reading, Mass.

Graves, G. W., (1972), "Extensions of mathematical programming for regional water-quality management", *EPA 16110 EGQ 04/72, U.S.* Environment Protection Agency, Washington, D. C.

Holland, J. H. (1975), *Adaptation in Natural and Artificial System*, 183.pp. University Michigan Press, Ann Arbor.

Joeres, E. F., Dressler, J., Choand, C. C. and Falkner, C. H., (1974), "Planning methodology for the designing of regional wastewater treatment systems", *Water Resources Research*. 10(4), pp. 643-649.

Mckinney, D. C. and Lin, M. D., (1994), "Genetic algorithm solution of groundwater management", *Water Resour. Res.* 30(6), pp. 1897-1906.

McNamara, J. R., (1976), "An optimisation model for regional water-quality management", *Water Resources Research*, 12(2), pp. 125-140.

Nakamura, M., Bill, E. D. and Liebman, J. C., (1981), "Multiperiod design of regional wastewater treatment systems: generating and evaluating alternative plans", *Water Resources Research,* 17(5), pp. 1339-1348.

Nikolay, S. V. and Boulos, P. F., (1993), "Heuristic screening methodology for regional wastewater treatment planning", *J. Environmental Engineering, Div., ASCE*, 199 (4), pp. 603-613.

Rao, Z. F. and Jamieson, D. G., (1997), "The use of neural networks and genetic algorithms for design of groundwater remediation schemes", *Hydrology and Earth Science systems*, 1(2), pp. 345-351

Ritzel, B. J. and Eheart, W., (1994), "Using genetic algorithms to solve a multiple objective groundwater pollution containment problem", *Water Resources Research*, 30(5), pp. 1589-1603.

Savic, D. A. and Walters, G. A., (1997), "Genetic algorithms for least-cost design of water distribution networks", *Journal of Water Resources Planning and Management*, Vol. 123, No. 2, pp. 67-77.

Simpson, A. R., Dandy, G. C. and Murphy, L. J., (1994), Genetic algorithms compared to other techniques for pipe optimisation", *J. Water Resources Planning and Management*, ASCE, 120(4), pp. 423-443.

Wang, C. G. and Jamieson, D. G., (1999), Use of artificial neural network for river water quality simulation. In *Computing and Control for Water Industry*, Research Study Press, UK, 317-327.

Wang, Q. J., (1991), "The genetic algorithm and its application to calibrating conceptual rainfall-runoff models", *Water Resources Research*, Vol. 27, No. 9, pp. 2467-2471

Wanielista, M. P. and Bauer, C. S., (1972), "Centralization of waste treatment facilities", *J. Water Poll. Contr. Fed.* 44(12), 2229-2238.

Water Resources Centre, UK, (1985), *Cost Information for Water Supply and Sewage Disposal*, Technical Report, TR61, Medmenham.

Keyword Index